Vibration measurement

Gheorghe Buzdugan
Elena Mihăilescu
Mircea Radeș
Strength of Materials Chair
Polytechnic Institute of Bucharest
Romania

1986 **MARTINUS NIJHOFF PUBLISHERS**
a member of the KLUWER ACADEMIC PUBLISHERS GROUP
DORDRECHT / BOSTON / LANCASTER

Distributors

for the United States and Canada: Kluwer Academic Publishers, 190 Old Derby Street, Hingham, MA 02043, USA
for the UK and Ireland: Kluwer Academic Publishers, MTP Press Limited, Falcon House, Queen Square, Lancaster LA1 1RN, UK
for all other countries: Kluwer Academic Publishers Group, Distribution Center, P.O. Box 322, 3300 AH Dordrecht, The Netherlands
for socialist countries: Editura Academiei, Calea Victoriei 125, R 79717, Bucharest, Romania

Library of Congress Cataloging in Publication Data

Buzdugan, Gh.
   Vibration measurement.

   (Mechanics, dynamical systems; 8)
   Updated English version of: Măsurarea vibrațiilor.
   Includes bibliographies and index.
   1. Vibration-Measurement. I. Mihăilescu, Elena.
II. Radeş, Mircea. III. Title. IV. Series: Monographs and textbooks on mechanics of solids and fluids.
Mechanics, dynamical systems; 8.
TA355.B8313   1986   620.3'2   84-25523
ISBN 90-247-3111-9

Book information

The English version represents the revised and updated translation of the Romanian work Măsurarea vibrațiilor published in 1979 by Editura Academiei.

Copyright

© 1986 by Martinus Nijhoff Publishers, Dordrecht and Editura Academiei, Bucharest.

All rights reserved. No part of this publication may be reproduced, stored in a retrieval system, or transmitted in any form or by any means, mechanical, photocopying, recording, or otherwise, without the prior written permission of the publishers,
Martinus Nijhoff Publishers, P. O. Box 163, 3300 AD Dordrecht, The Netherlands and
Editura Academiei, Calea Victoriei 125, R−79717, Bucharest, Romania.

PRINTED IN ROMANIA

Contents

Preface xiii

1. **Introduction** 1
   1.1 Definitions 1
   1.2 Object of vibration measurements 3
   1.3 Components of the instrumentation system. . . . 5
   1.4 Measured mechanical quantities 8
       References for Chapter 1 10

2. **Elements of the theory of vibrations** 13
   2.1 Classification of vibrations 13
   2.2 Characteristic vibration parameters 14
       2.2.1 Quantities describing the signal waveform . 14
       2.2.2 Correlation functions 16
       2.2.3 Frequency domain description of time signals 17
           2.2.3.1 Frequency analysis of periodic signals 18
           2.2.3.2 Frequency analysis of nonperiodic signals 19
           2.2.3.3 Frequency analysis of random signals 21
           2.2.3.4 The Discrete Fourier Transform . . 23
       2.2.4 Probability density function 25
   2.3 Estimation errors at random data analysis . . . 26
       2.3.1 Confidence probability. Confidence interval . 27
       2.3.2 Standard error 28

Contents

    2.4 Response of vibrating systems to various excitations   30
        2.4.1 Harmonic excitation 31
        2.4.2 Periodic excitation 36
        2.4.3 Transient excitation and shocks 37
        2.4.4 Stationary random excitation 42
    2.5 Natural frequencies 44
    2.6 Spring constants 48
        References for Chapter 2 50

3. **Effects of vibrations** 51
    3.1 Effects of vibrations on man 51
    3.2 Effects of vibrations on buildings 55
    3.3 Effects of vibrations on machinery 56
        References for Chapter 3 70

4. **Transducers and pickups for vibration measurement** . . 71
    4.1 Transducers for electrical measurement of vibrations   71
        4.1.1 Passive transducers 73
            4.1.1.1. Variable resistance transducers . . 73
            4.1.1.2. Capacitive transducers 75
            4.1.1.3. Inductive transducers 77
        4.1.2 Self-generating transducers 80
            4.1.2.1. Piezoelectric transducers 80
            4.1.2.2. Electrodynamic transducers . . . 83
            4.1.2.3. Electromagnetic transducers . . 83
    4.2 Vibration pickups 84
        4.2.1 Theory of seismic instruments 84
            4.2.1.1. Principle of operation 84
            4.2.1.2. Amplitude distortions 88
            4.2.1.3. Phase distortions 89
        4.2.2 Displacement and acceleration pickups . . . 90
            4.2.2.1. Displacement-measuring pickups . . 91
            4.2.2.2. Accelerometers 99
        4.2.3 Velocity pickups 105
        4.2.4 Force and torque gauges 107
        4.2.5 Dynamic pressure transducers 110
        4.2.6 Limits of vibration pickup performance . . 113
        4.2.7 Mounting 113
            References for Chapter 4 115

Contents

5 **Instrumentation for vibration measurement** 117

  5.1 General properties of measuring instruments 117

  5.2 Mechanical instruments for vibration measurement 120
      5.2.1 Tastograph 120
      5.2.2 Stoppani vibrograph 120
      5.2.3 Geiger vibrograph 121

  5.3 Conversion instruments 122
      5.3.1 Measuring bridges 122
          5.3.1.1. Bridge circuits 122
          5.3.1.2. Detection circuits 125
      5.3.2 Frequency discriminator circuits 128
      5.3.3 Amplifiers. 130
          5.3.3.1 Voltage amplifiers 130
          5.3.3.2 Amplifiers with feedback 131
          5.3.3.3 Charge amplifiers 131
          5.3.3.4 Impedance-transforming amplifiers. 132
      5.3.4 Integrators. Analog low-pass and high-pass filters. 132
      5.3.5 Analog-to-digital converters 135
          5.3.5.1 Sampling rate. Aliasing 136
          5.3.5.2 Number of discrete samples . . . 139
      5.3.6 Averagers 141
          5.3.6.1 Time averaging/integration of analog signals 141
          5.3.6.2 Time averaging/integration of digital signals 143
      5.3.7 R.M.S. detectors. 144

  5.4 Instruments for signal analysis 145
      5.4.1 Correlators 145
          5.4.1.1 Stepped correlators 145
          5.4.1.2 On-line correlators 146
          5.4.1.3 Real time correlators 148
      5.4.2 Bandpass filters 149
          5.4.2.1 Constant percentage bandwidth filters 150
          5.4.2.2 Constant bandwidth filters . . . 151
          5.4.2.3 Filter response time 152
      5.4.3 Non-real time spectrum analyzers 152
          5.4.3.1 Noise analyzers 152
          5.4.3.2 Tunable bandpass filters 153

Contents

    5.4.3.3 Heterodyne analyzers 153
    5.4.3.4 Synchronous filters 154
  5.4.4 Real time spectrum analyzers 155
    5.4.4.1 Parallel filter analyzers. 156
    5.4.4.2 Time compression analyzers . . . 157
    5.4.4.3 Weighting. 160
    5.4.4.4 Digital analyzers 164
  5.4.5 Shock spectrum analyzers 171
  5.4.6 Amplitude distribution analyzers 171
  5.4.7 Resolved component indicators 172
 5.5 Display and recording instruments 174
  5.5.1 Stroboscopes 174
  5.5.2 Analog meters 174
  5.5.3 Digital meters and printers 174
  5.5.4 Analog strip-chart recorders 177
  5.5.5 X—Y recorders 177
  5.5.6 Graphic level recorders 179
  5.5.7 Magnetic oscillographs 179
  5.5.8 Cathode-ray oscilloscopes 180
  5.5.9 Magnetic tape recorders 181
  5.5.10 Digital recorders 182
 5.6 Computerized vibration analysis systems 183
 References for Chapter 5 184

6 Vibration exciters 187

 6.1 Mechanical vibration exciters 187
  6.1.1 Reciprocating vibration exciters 187
  6.1.2 Rotating unbalance vibration exciters 188
 6.2 Electromagnetic vibration exciters 192
  6.2.1 Force generated by an electromagnet 192
  6.2.2 Measurement of electromagnet force 195
  6.2.3 General performance characteristics 197
 6.3 Electrodynamic vibration exciters 198
  6.3.1 Principle of operation 198
  6.3.2 Interaction between vibrator and tested structure 200
  6.3.3 Frequency response 202

	6.3.4 Measurement of the force applied to the structure	203
	6.3.5 Features of a vibrator used in structural testing	205
	6.3.6 Application	206
6.4	Hydraulic vibration exciters	207
	6.4.1 Construction	207
	6.4.2 Frequency response	208
	6.4.3 General performance characteristics	208
	References for Chapter 6	209

7 Instrument set-ups and techniques for vibration measurement 211

7.1	Selection of equipment	211
7.2	Basic set-ups for signal waveform measurement	213
7.3	Procedures for analysing random vibration records	217
	7.3.1 Analysis of a single record	218
	7.3.2 Analysis of a collection of records	220
7.4	Frequency analysis	221
	7.4.1 Frequency analysis of stationary signals	222
	7.4.1.1 Selective filtering	222
	7.4.1.2 Time compression analysis	223
	7.4.1.3 Digital analysis	225
	7.4.1.4 Analysis of stationary random signals	226
	7.4.2 Frequency analysis of shocks	227
	7.4.2.1 Pulse transformation into a pulse train	227
	7.4.2.2 Response of a very narrow bandpass filter	230
7.5	Vibration testing	231
	7.5.1 Sinusoidal tests	232
	7.5.2 Broadband random vibration tests	237
	7.5.3 Sweep narrowband random vibration tests	239
	7.5.4 Shock tests	240
7.6	Frequency response measurement	240
	7.6.1 Frequency response functions	240
	7.6.2 Sinusoidal test techniques	242
	7.6.2.1 Single-point excitation	242
	7.6.2.2 Multi-point excitation	247

Contents

    7.6.3 Transient test techniques 250
        7.6.3.1 Impact test technique 250
        7.6.3.2 Step relaxation technique 253
        7.6.3.3 Rapid frequency sweep excitation technique 254
    7.6.4 Random excitation techniques 255
        7.6.4.1 Measurement of the frequency response function 257
        7.6.4.2 Measurement of the impulse response function 259
    7.6.5 Experimental modal analysis 260
        7.6.5.1 Single-degree-of-freedom techniques 260
        7.6.5.2 Multi-degree-of-freedom techniques 263

    References for Chapter 7 267

8 **Calibration of transducers and instrumentation systems** 269

  8.1 Calibration of vibration pickups 269
    8.1.1 Static calibration 270
    8.1.2 "Direct" dynamic calibration 270
        8.1.2.1 Calibration of a piezoelectric accelerometer 271
        8.1.2.2 Calibration of a piezoelectric force transducer 273
    8.1.3 Reciprocity calibration 275
        8.1.3.1 Reciprocity procedure 276
        8.1.3.2 Theoretical background. 277
    8.1.4 Optical interferometry calibration 279
    8.1.5 Comparison calibration 281

  8.2 Simulation calibration of auxiliary circuits 282
    8.2.1 Substitution calibration 282
    8.2.2 Insert calibration 283
    8.2.3 Shunt-resistor calibration 285

  References for Chapter 8 286

9 **Examples of vibration measurements** 287

  9.1 Identification of vibration sources 287

  9.2 Measurements on prototypes 297
    9.2.1 Machine tools 297
    9.2.2 Rolling hoisting cranes 300

	9.2.3 Suspended pipelines	302
	9.2.4 Rotating machinery	303
	9.2.5 Cargo ships	310
9.3	Measurements for production control and acceptance	313
	9.3.1 Machine tools	313
	9.3.2 Reciprocating compressor piping	316
9.4	Measurements during machinery operation	318
	9.4.1 Forge hammers and machine foundations	318
	9.4.2 Machinery condition monitoring	320
9.5	Measurement of vibrations produced by blasting	324
9.6	Measurement of the dynamic characteristics of materials	326
9.7	Measurement of soil elastic characteristics by vibration methods	333
	9.7.1 Spring constants for rigid footings resting on soil	333
	9.7.2 Measurement of elastic constants for soils	334
	9.7.3 Resonance technique for determining the dynamic coefficients of subgrade reaction	336
	References for Chapter 9	339
	Subject index	345

Preface

Nowadays, the engineering practice raises far more vibration problems than can be theoretically explained or modelled. Because of this, measurements are used in almost all fields of industry, transportation and civil engineering in studies of mechanical and structural vibration. They are an invaluable tool for designing products and machines with high reliability and low noise level, vehicles and buildings with improved comfort and resistance to dynamic loads, as well as for obtaining increased safety of operation and optimum running parameters.

In order to cope with the increasing demand for experimental measurement of vibration characteristics, young engineers and designers need an introductory book with emphasis on "what has to be measured" and "by what means" before learning "how measurements are done". The expertise to perform vibration measurements must be gained in time, with every new investigation and studied problem.

A detailed presentation of instrumentation and measuring techniques is beyond the aim of this book. Such information can be found in product data sheets, application manuals and handbooks supplied by equipment manufacturers. Only general principles and widely used methods are presented herein, in order to provide the reader with an overview of the instrumentation and techniques encountered in vibration measurement.

The instrumentation had a rather spectacular development, from the purely mechanical vibrographs, used 40 years ago, to the analogic electrical equipment and, today, to the digital instruments, having high accuracy and increased operating speed. The advent of FFT analyzers, minicomputer-based data acquisition and data processing systems, and more recently of microprocessor-based measuring systems, had changed the conventional methods of vibration analysis.

The experimental techniques had a corresponding development, namely : 1) from the graphical analysis of time history

Preface

records and frequency response diagrams plotted point by point, to the modern real time on-line analysis and computer aided testing; 2) from the analysis of simple harmonic or periodic vibrations to the measurement and analysis of transient or random vibrations.

Herein, while considerable attention is paid to analogic instruments, detailed description is also given to the basic features of some digital instruments. As for the presented examples of application, these are somewhat limited to the authors' area of interest. Topics not treated include acoustic measurements, dynamic balancing, shock and vibration testing machines.

The book is organized in nine chapters covering the following topics: some elements from the theory of vibration, effects of vibration and allowable limits, transducers and pickups used in vibration measurement, instruments for signal conditioning, analysis, display and recording, exciters used in vibration testing systems, basic measurement set-ups and techniques, transducer and system calibration methods, examples of application of vibration measurement.

Although the book is written primarily for mechanical engineers, it is hoped that it will prove useful to students and researchers, as well as to specialists from all fields of engineering confronted with vibration problems.

The book is an updated English version of the original edition in Romanian published by Editura Academiei, Bucharest, in 1979. It is the result of authors' experience gained during the long term research activity in the benefit of industry.

The authors wish to thank all those who gave the permission to reproduce material from their publications or about their products. Thanks are also due to the publishers for their care and attention during the editing of this book.

THE AUTHORS

1

Introduction

1.1 Definitions

Speaking about mechanical vibrations, the following question arises: What is vibration? The answer differs according to the extent of the notion.

In the most restrictive meaning, *harmonic vibration* of a particle or rigid body is the motion in which the displacement varies continuously with time, according to the harmonic law

$$x = x_0 \sin pt, \qquad (1.1)$$

where x is the particle or body position (at time t) measured from the reference position, x_0 — displacement amplitude, p — angular frequency.

In a broader meaning, vibration is a motion to and fro (about an equilibrium position), which repeats itself at equal intervals of time, according to a given periodic law. This is a *periodic vibration*.

The content of the notion can be further extended, calling vibrations the damped motions, with decaying amplitudes, described by time functions of the form

$$x = x_0 \, e^{-nt} \sin pt, \qquad (1.2)$$

where n is a measure of the amount of damping. Unstable motions with increasing amplitude and transitory motions between steady-state regimes are also termed vibrations.

All afore mentioned motions, that can be described by an explicit mathematical relationship, are called *deterministic vibrations*.

Unlike these, motions whose instantaneous value is not specified by an explicit mathematical relationship, i.e. are not predictable as a time function, are termed *random vibrations*. These can be studied based only on finite length time history

1 Introduction

records, called *samples*, and by statistically processing sets of such records, called *ensembles*. Statistical data obtained from vibration records of finite length are only *estimates* of the vibration properties, so that *accuracy of estimation* must be considered along with the *accuracy of measurement* in any analysis of random vibrations.

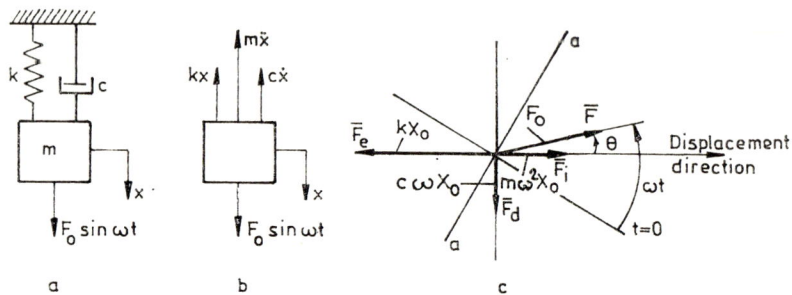

Figure 1.1

Generally, variations of a state parameter about the value corresponding to a stable equilibrium position (or trajectory) are called oscillations. Vibrations are oscillations due to an elastic restoring force. A flexible beam or string vibrates while a pendulum oscillates.

For lumped parameter linear systems, Stein [1] defines the vibration as a "linear weighted combination, specific to the system and to its components, of n-th order time derivatives and/or time integrals of displacements, observed at a specific point on the structure".

Let consider the viscously damped single degree of freedom linear system from Figure 1.1 a, acted upon by a harmonic force. In the free-body diagram from Figure 1.1 b, the forces acting on the mass are depicted: the excitation force $\bar{F} = F_0 \sin \omega t$, the inertia force $\bar{F}_i = -m\ddot{x}$, the damping force $\bar{F}_d = -c\dot{x}$ and the elastic force $\bar{F}_e = -k\bar{x}$. Their vector representation is shown in Figure 1.1 c.

The vibration, a continuous time variation of the displacement x and its derivatives $\dot{x}$, $\ddot{x}$, consists of a continuous variation in time of these forces, which are always in dynamic equilibrium. The phase angle θ, between the displacement vector and the excitation force vector, changes as a function of the dimensionless ratio $\dfrac{\omega}{p}$ between the excitation frequency and the natural frequency. For the three possible cases — excitation below, at and above resonance — the vector force diagrams have the form shown in Figure 1.2.

1.2 Object of vibration measurements

The basic problem in the study of vibrations is to determine the motion of a system subjected to a given excitation, i.e. to determine the corresponding input-output relationship. Both the excitation and the response can be expressed by either kinematic quantities — displacements, velocities, accelerations — or forces and torques.

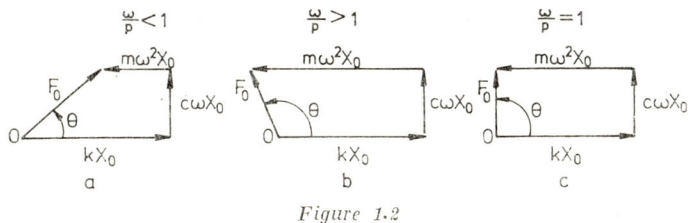

Figure 1.2

1.2 Object of vibration measurements

Nowadays, vibration studies are more often required in the engineering practice, being decisive for the design, production and operation of numerous machines, vehicles and buildings. Miniaturisation, the quest for greater economy in the use of engineering materials (leading towards light weight construction) and fabrication by welding (instead of bolting or riveting) give rise to *higher* resonance frequencies of individual components (and subsystems) and to less damped response amplitudes. On the other hand, speeds of machines and vehicles are increasing, with corresponding *increase* in the vibration frequencies of the excitation forces. These convergent trends of modern technological development are leading more frequently to resonance vibrations during transport or under service conditions, with detrimental effects on systems' reliability and man's comfort.

Apart from undesired vibrations, there are installations whose operation is based on vibratory motions, namely: concrete tampers, soil compaction machines, road drills, conveyers, vibrating screens, fatigue testing machines, etc.

It is well known that only a few oscillating systems, such as massive bodies exhibiting some symmetry, lumped parameter systems, uniform straight beams, flat plates, etc. can be treated completely theoretically. Machinery and complex structures raise more complicated problems and a purely theoretical model can differ substantially from the reality. In this case, experimental measurement of vibrations is the only way to solve the problem.

1 Introduction

There are three main types of vibration measurement [2], namely:

a. Measurement of *vibration levels*, i.e. of the system output, in order to compare them with standards or base-line data. For example, when vibrations of an elastic member are measured, one can determine the induced stresses, assessing any likely fatigue failure. Vibration measurement of machinery and installations — on foundation, casing, bearings or pipelines — may give quantitative elements which can be compared to allowable levels from the point of view of man comfort, machinery normal operation and buildings' safety. In the same group can be classified ground vibration measurements taken inside the ground area of a proposed building situated near quarry-blasting or on the floor of a room where sensible instrumentation has to be installed.

When the system description is given and the response is measured, it is possible to find the input which causes the response. Examples include the computation of unbalance forces of rotating shafts, based on displacement measurements, and the identification of earthquake acceleration records from a shock spectrum.

b. When the system response to a given excitation can be analytically determined, it is required to *measure the input*, which most often is an excitation force or torque. Measuring the forces generated by a machine, an *active isolation* system can be properly designed; measuring the vibration of a laboratory or workshop floor, one can design the *passive isolation* mounting of an instrument. Similarly, recording the road surface roughness one obtains useful data for vehicle suspension design.

c. The third class of measurements use known excitation and resulting response data. The problem is either to find a physically accomplishable system, which fits the input-output relationship as closely as possible, or to find a mathematical description (or model) of the system. Usually, a single force is used and measured, in addition to the response, so that it is possible to derive the *response characteristics* of the system or component under test. Multi-input multi-output techniques are used as well.

The object of this type of measurement can differ, aiming to perform [3]:

— identification of natural frequencies and mode shapes;
— measurement of specific dynamic properties of the system such as damping factors, equivalent masses and stiffnesses;
— check of theoretically derived response functions against measured data as a way of verifying an analytical method;
— formulation of a mathematical model of the analysed structure for the purpose of further analysis by simulation.

1.3 Instrumentation system

As for the excitation, though harmonic forces are most often used, broad band excitation is routinely used to speed up and simplify the testing, including pure random, periodic random, pseudorandom, impact, periodic chirps and swept sine excitation.

It is worth underlining that for the determination of vibration *levels* experienced by structures it is necessary to measure *absolute vibrations*, so that carefully calibrated instruments and measuring systems are needed. In other situations, only *relative vibration* measurements are required, such as in experiments performed to determine the *characteristics* of vibrating structures, where there is no need for calibration if modal stiffnesses are not required.

A first problem, preceding the measurement, is the clear definition of test objectives, selection of the most meaningful response characteristics (what is to be measured) and the measurement points.

After estimating the amplitude and frequency range of these quantities, as well as the possibilities of locating the equipment, the next step is selection of measuring instruments : transducers and instruments for signal conditioning, analysis, display and recording.

Depending on the purpose in view, other factors are then considered such as methods for data processing and analysis, data reduction and instrument calibration, as well as schedule and financial considerations.

1.3 Components of the instrumentation system

The functional organization of an instrumentation system used for vibration measurement, based on analog equipment, is shown

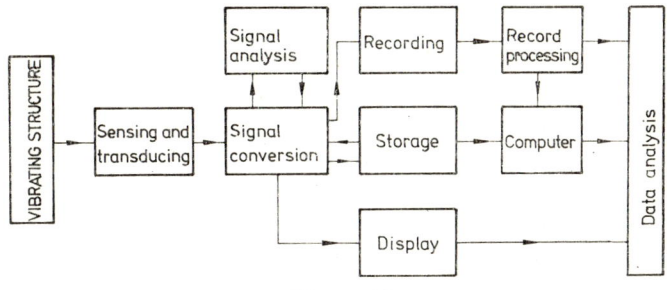

Figure 1.3

in Figure 1.3. From the *vibrating structure*, the motion (or the dynamic force) is sensed by a *vibration pickup* or *transducer* that converts motion into an electrical signal. Instruments for signal

1 Introduction

conversion and *analysis* are then used to amplify it to a value required by recorders and to extract vibration descriptors of interest to the experimenter. The output of these instruments is either displayed on a direct-reading instrument (or on the screen of a scope), or recorded as a hard copy, or stored and then introduced in a computer to get the final data in a format useful to visual interpretation and analysis.

This classification of instruments according to their function within the measuring system enables a large variety of combinations to be made, which increases their efficiency [4].

In the past, the whole instrumentation was concentrated into a single self-contained vibration-measuring instrument. This could serve a single purpose: connected to the vibrating structure, it supplied the result as an indication on a scale or as a vibrogram (time history record).

The advancement of the methods for the electrical measurement of mechanical quantities, accomplishment of the latest decades, allowed a wide development of the vibration measurement techniques.

Regarding the *vibration pickup*, a basic component of the instrumentation system, one should make distinction between:
— the *transducing element* — that part of the pickup which accomplishes the conversion of the change of the mechanical quantity (displacement, velocity, acceleration, force) into the change of an electrical quantity;
— the *pickup* — a device which besides the transducing element contains other components enabling signal processing and transmission to be carried out (see Chapter 4).

Generally, devices consisting basically of the transducing element are termed *measuring transducers*.

Considering either a self-contained instrument or a complete system, consisting of transducers and adequate electrical instruments, the vibration measuring equipment can be classified in different ways:

a. According to the mechanical connection between instrument and vibrating body, one can have *proximity* (non-contacting) instruments and *attached* instruments.

b. According to the type of data presentation, one can distinguish *direct-reading* (visual display) instruments and *recorders*.

c. According to the physical principle used in measurement, one may have *mechanical, optical, acoustical* and *electrical* instruments.

d. According to the measured quantity, instruments may be called: frequency meters, vibrometers, velocity meters, acce-

1.3 Instrumentation system

lerometers, torsiometers, tachometers, phasemeters, dynamometers, torquemeters, pressure gauges etc. When the instrument is provided with recording capabilities, the suffix *meter* is replaced by *graph*, for example vibrometer becomes vibrograph, etc.

e. According to the type of data, there are *analog, digital* and *hybrid* instruments.

f. According to the constructive principle, instruments may be *quasi-static, seismic* and *ballistic*. Digital analyzers can be stand alone instruments, like the hardwired FFT analyzers, and mini-computer-based systems, though hybrid type systems exist as well.

A basic function of *signal conversion instruments* is *amplification*, used especially when the output voltage of transducers is too weak to drive the recording equipment with satisfactory accuracy. *Multiplexing* is used when the number of available recorder channels is less than the number of signals to be simultaneously recorded.

When the transducer output is a direct current proportional to the physical variable being measured and only a magnetic recorder is available which is incapable of storing accurately d.c. signals, then a signal *modulation* is required. In order to extend the linear operating range of the measurement system, especially at low frequencies, capacitive and piezoelectric transducers are followed by preamplifiers, which also perform an *impedance transformation*, converting the transducer high output impedance into a smaller one, more adequate for measurement and analysis.

Besides these functions, conversion instruments also perform integration, multiplication, filtering, sampling, coding, averaging etc. (see Chapter 5).

After conversion and *analysis*, data can be either *displayed*, using oscilloscopes, readout meters and graphics terminals, or *recorded* on paper, moving film, magnetic tape, or by line printers, eventually *stored* on punched tape, cartridge or floppy disks, for further use in a computer (see Chapter 5).

When digital spectrum analyzers or mini-computers are used within the instrumentation system, *analog-to-digital* and *digital--to-analog converters* are introduced, as well as *anti-aliasing filters*, whose cutoff frequency is correlated to the sampling frequency, cyclic *memories* for frequency-translation of data,*weighting* circuits like the Hanning window, *averaging circuits*, etc.

Besides the instrumentation for vibration measurement, equipment for *vibration generation* is currently used, consisting of signal generators (oscillators), power amplifiers and electrodynamic or electrohydraulic vibrators (see Chapter 6).

1 Introduction

Instrumentation systems may also contain supplying sources, measuring transformers, voltage stabilizers, tachometers, frequency meters, time bases, chronometers, etc.

By today's standards, analog equipment is bulky, rather expensive and limited in performance. The development of the FFT algorithm and low cost mini-computers and associated peripherals led first to the appearance of general purpose digital analysis instruments, with improved accuracy and flexibility of data processing. During the past years, microprocessor based instruments have found increasing application in vibration measurement and analysis. This trend is accelerating rapidly as digital technology becomes less expensive. Though MOS microprocessors operate in a similar way to mini-computers, they are still slower and less powerful, but differences are quickly diminishing.

1.4 Measured mechanical quantities

The vibratory motion (translational or angular) of a point or a component of a structure (machine, building, foundation) can be defined by displacement, velocity, acceleration, jerk or dynamic strain. Besides these, measurements of forces, torques and dynamic pressures are also carried out for determining the dynamic structural response characteristics.

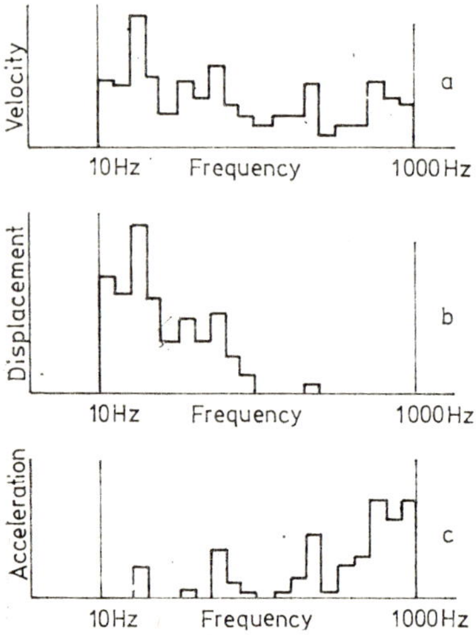

For simple harmonic motions, between the amplitudes of displacement x_{max}, velocity $\dot{x}_{max}$ and acceleration $\ddot{x}_{max}$ the following relationship holds

$$\ddot{x}_{max} = \omega \dot{x}_{max} = \omega^2 x_{max},$$

where ω is the angular frequency of vibration.

Basically, each of the three parameters contains the same amount of information, so that it seems that it does not matter which parameter is actually measured, they being deduced from one another by time integration or differentiation. However, it has been recently noticed that vibrations generated by most machinery have an approximately flat velocity spectrum over a broad frequency range between 10 and 1000 Hz (Fig. 1.4 a).

Figure 1.4

1.4 Measured quantities

Displacement measurements (Fig. 1.4 b) will tend to accentuate the low frequency components, generally corresponding to the machine speed. High frequency components, playing an important role in the safe operation, therefore in the preventive maintenance of machinery, being correlated with the machine noise and wear, are neglected, which could lead to erroneous conclusions on the machine quality.

Inversely, acceleration measurements (Fig. 1.4 c) will tend to accentuate the high frequency components, underestimating the others [5].

Figure 1.5 illustrates typical frequency regions of operation for different transducers [6].

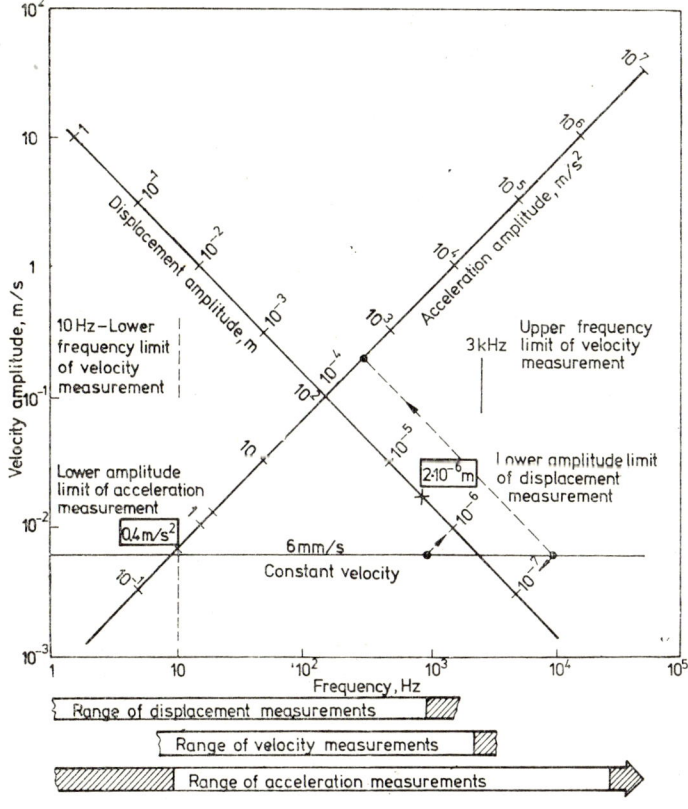

Figure 1.5

For a signal with a velocity level of 6 mm/s, the displacement amplitude is 1 μm at about 1000 Hz and disappears into the background noise of most commercially available measuring

1 Introduction

systems. In an extreme case, the 6 mm/s velocity at 10 kHz corresponds to an acceleration level of 400 m/s^2, i.e. approximately 40 g with a displacement of only 0.1 μm. Displacement is not an effective means of measuring high-frequency vibration because large forces are required at these frequencies to produce a measurable displacement and not because the instrumentation system is limited to a maximum frequency. Below about 20 Hz, the displacement amplitude necessary to produce an easily identifiable acceleration signal is so large that endangers the transducer mechanical integrity.

Thus, it is recommended to use displacement pickups from 0 to about 1000 Hz, velocity pickups from 10 to 3000 Hz and accelerometers from 20 Hz to well above 20 kHz, setting the lower limit for acceleration measurement at 0.4 m/s^2 and that for displacements at 2 μm [6]. The trend in the recent development of the measuring equipment is the extension of acceleration measurement at low frequencies. This is facilitated by the use of charge amplifiers and existence of new high-sensitivity accelerometers.

In rotating machinery applications, relative-motion *shaft displacement measurements* are recommended at rigid machines, having casing-to-rotor weight ratios of 20 : 1 or more and equipped with hydrodynamic bearings. *Casing* seismic-vibration *velocity* measurements are recommended for machinery having casing-to-rotor weight ratios of about 5 : 1 and equipped with heavily loaded or stiff bearings, mounted in a supporting structure, which is more flexible than the bearing itself [6].

In some cases, selection of the measured parameter is dictated by the available space for transducer mounting. For example, at the measurement of the torsional vibrations of a shaft, one can choose between using strain gauges and slip rings, i.e. measuring the dynamic torque, and attaching an angular vibration pickup at the shaft end, when it is accessible. Other remarks regarding the selection of transducers and, implicitly, of the measured parameter, can be found in Chapter 4 and Reference [7].

References for Chapter 1

1. Stein, P. K., *Measurement Engineering*, Imperial Litho Phoenix, Arizona, 1965.
2. Holzweissig, F., Meltzer, G., *Messtechnik der Maschinendynamik*, VEB Fachbuchverlag, Leipzig, 1973.

1.4 Measured quantities

3. Ewins D. J., *The whys and wherefores of mechanical impedance measurement*, Solartron.
4. Harris, C. M. and Crede, C. E. (Eds), *Shock and Vibration Handbook*, 2nd Ed., McGraw-Hill Book Comp., 1976.
5. * * * *Introduction aux vibrations*, Brüel & Kjaer Conference No. 80 (Nov. 1974).
6. Mitchell, J. S., *An Introduction to Machinery Analysis and Monitoring*, Penn Well Books, Tulsa, 1981.
7. Klyuev, V. V. (Red.), *Instruments and Systems for Measurement of Vibration, Noise and Shock*, Mashinostroyenye, Moscow, 1978 (In Russian).

2

Elements of the theory of vibrations

2.1 Classification of vibrations

Mechanical vibrations can be classified using various criteria.

a. As excitations as well as the motion kinematic parameters (displacement, velocity, acceleration) can or cannot be expressed by an explicit mathematical relationship, describing their time history, vibrations can be classified as being either *deterministic* or *random*.

b. According to the form of the *time history* record of the *kinematic parameters*, deterministic vibrations may be classified as being either *periodic* or *nonperiodic*. The harmonic motion is the simplest form of oscillation at which displacement, velocity and acceleration can be expressed in terms of a single trigonometric function "sin" or "cos".

c. According to the *mathematical form of the differential equations of motion*, vibrations can be *linear* or *nonlinear*.

d. According to the *number of independent coordinates* necessary to define the vibratory motion, one can have *single-*, *multi-degree-of-freedom* (discrete) and *continuous* systems. A large number of vibration problems can be treated with sufficient accuracy by reducing the system to one having a single degree of freedom.

e. According to the *nature of displacements* of the points of the oscillating system, vibrations may be *linear* or *angular*.

f. According to the *cause* producing or sustaining the vibratory motion, one can distinguish: *free vibrations*, produced by a shock or an initial displacement; *forced vibrations*, produced by external excitation forces; *parametric vibrations*, due to the change,

2 *Elements of theory*

produced by an external cause, of a system parameter; *selfexcited vibrations*, produced by a mechanism inherent in the system, by conversion of an energy obtained from a uniform source associated with the system to oscillatory excitation.

g. Oscillating systems can be classified as either *conservative* or *non-conservative*. Among the latter, *damped* systems are those in which energy is dissipated by friction and other resistances.

2.2 Characteristic vibration parameters

A large number of quantities occur in the study of vibrations, each being able to be measured directly or indirectly. From system physical parameters, forces and couples, and the simple motion kinematic parameters, to time-average properties, time domain descriptors and spectral properties of time signals, several quantities may be considered to characterize the vibration.

2.2.1 Quantities describing the signal waveform

Early experimental techniques for vibration measurement have been used for displaying and recording the time histories of observed phenomena. For harmonic or periodic (consisting of maximum 4 to 5 harmonic components) vibrations, analysis of time records has been used to get complete information on the vibratory motion.

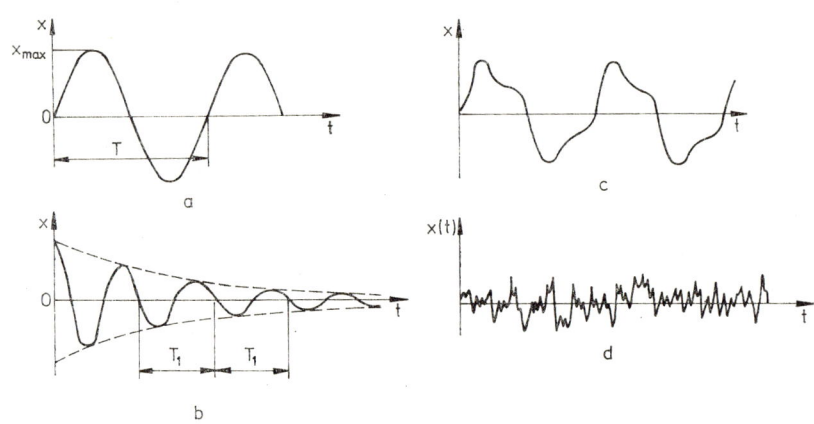

Figure 2.1

Figure 2.1 shows time records for several types of vibrations: steady-state harmonic motion (Fig. 2.1 a), decaying harmonic motion (Fig. 2.1 b), periodic motion (Fig. 2.1 c) and random motion (Fig. 2.1 d).

2.2 Characteristics of vibrations

There is evident that for the first three types of motion it is sufficient to specify the quantities defining the signal waveform: the amplitude x_{max}, the period T and the frequency f. As for the amount of *damping*, several quantities are currently used for its evaluation.

The *damping ratio* (fraction of critical damping) $\zeta = c/c_c$ is the ratio of the effective viscous damping coefficient c to the critical damping coefficient c_c (for which the motion becomes aperiodic).

For a single degree-of-freedom linear system, this coefficient is

$$c_c = \sqrt{4km} = 2\,pm,$$

where m is the mass, k — spring constant and p — natural angular frequency, so that

$$\zeta = \frac{c}{c_c} = \frac{c}{2pm}. \tag{2.1}$$

A convenient way to determine the amount of damping existing in a system is to measure the rate of decay of free vibrations, expressed in terms of the *logarithmic decrement*, defined as the natural logarithm of the ratio of any two successive amplitudes and given by

$$\Delta = \frac{2\pi\zeta}{\sqrt{1-\zeta^2}}. \tag{2.2}$$

For small values of ζ, the logarithmic decrement is proportional to ζ and is given by

$$\Delta \cong \frac{\pi c}{\sqrt{km}} = 2\pi\zeta. \tag{2.3}$$

An overall information on the vibration level can be obtained measuring the following time-average properties of waveforms:
— the *rectified-average value*, defined by

$$x_{|average|} = \frac{1}{T}\int_0^T |x|\,dt; \tag{2.4}$$

— the *effective value* or *root-mean-square* (rms) *value*

$$x_{eff} = x_{RMS} = \sqrt{\frac{1}{T}\int_0^T x^2\,dt}. \tag{2.5}$$

2   *Elements of theory*

For a pure harmonic motion, the following relationships exist between the various values:

$$x_{\text{RMS}} = \frac{\sqrt{2}}{2} x_{\max} = \frac{\pi}{2\sqrt{2}} x_{|\text{average}|} = 1.11\, x_{|\text{average}|},$$

$$x_{|\text{average}|} = \frac{2}{\pi} x_{\max} = 0.636\, x_{\max} = 0.9\, x_{\text{RMS}}.$$

2.2.2   Correlation functions

The waveform analysis using time records is difficult and inconclusive in the case of complex periodic and/or random vibrations. This has determined the improvement of experimental techniques and, implicitly, of the instrumentation [1].

Correlation techniques have been developed, i.e. measurement of the *autocorrelation function* of a waveform and/or the *crosscorrelation function* of two waveforms. The special time average functions are defined by the following formulae:

— the autocorrelation function of a signal $x(t)$

$$\bar{R}_{xx}(\tau) = \lim_{T \to \infty} \frac{1}{T} \int_{-T/2}^{+T/2} x(t)x(t-\tau)\, dt; \tag{2.6}$$

— the cross-correlation function of signals $x(t)$ and $y(t)$

$$\bar{R}_{xy}(\tau) = \lim_{T \to \infty} \frac{1}{T} \int_{-T/2}^{+T/2} x(t)y(t-\tau)\, dt. \tag{2.7}$$

The autocorrelation function is real, even, having a maximum at $\tau = 0$ equal to the mean square value of $x(t)$

$$\bar{R}_{xx}(0) = \lim_{T \to \infty} \frac{1}{T} \int_{-T/2}^{T/2} x^2(t)\, dt = \overline{x^2(t)}. \tag{2.8}$$

For large values of τ, the autocorrelation function tends to the square of the mean value of $x(t)$

$$\bar{R}_{xx}(\infty) = [\overline{x(t)}]^2.$$

The autocorrelation function of a "white noise" (having the auto power spectrum constant over the whole frequency range) has the form of the unit impulse, being zero for any nonzero value of τ. If the signal $x(t)$ is periodic, the autocorrelation function is also periodic, with the same period. Random noise will correlate only around $\tau = 0$.

The cross-correlation function of two independent signals is zero. If two signals $x(t)$ and $y(t)$ are periodic, with different periods, the crosscorrelation function is also periodic, with a period equal to that of the "beats" resulting by their addition.

At the experimental measurement of correlation functions, the time interval T (in eqs 2.6 and 2.7) has a finite length which, depending on the signal nature, must satisfy the following conditions :
— for periodic signals, T equals the signal period or an integer multiple of the period, in order not to introduce estimation errors :

$$\bar{R}_{xx}(\tau) = \frac{1}{T} \int_{-T/2}^{T/2} x(t)x(t - \tau)\, dt.$$

— for random signals, the record length is determined by the imposed estimation error (see Section 2.3).

The autocorrelation function of transient time signals

$$R_{xx}(\tau) = \int_{-\infty}^{+\infty} x(t)x(t - \tau)\, dt \qquad (2.9)$$

is calculated without division to the integration time, since the time average will tend to zero when the observation time increases.

By definition, the cross-correlation function of two transient signals $x(t)$ and $y(t)$ is

$$R_{xy}(\tau) = \int_{-\infty}^{+\infty} x(t)y(t - \tau)\, dt. \qquad (2.10)$$

For a detailed treatment of this subject see Reference [2].

2.2.3 Frequency domain description of time signals

The frequency of a pure harmonic signal (Fig. 2.1 a) can be determined from the period measured on the time record ($f = 1/T$). For complex signals this is practically impossible and other methods must be used for describing their frequency content. *Frequency analysis* is the experimental technique used for determining the spectral properties of time functions and spectrum analyzers are the instruments used to carry it out.

2 Elements of theory

2.2.3.1 Frequency analysis of periodic signals. Periodic motions, described by time functions satisfying the Dirichlet conditions, can be represented by a combination of a number of pure harmonic motions (with harmonically related frequencies), resulting from the Fourier series of the periodic time function.

The complex Fourier series of a periodic function $x(t)$ can be written

$$x(t) = \sum_{n=-\infty}^{+\infty} c_n e^{i\omega_n t}, \quad \omega_n = n\omega_0. \qquad (2.11)$$

The coefficients c_n are given by

$$c_n(i\omega_n) = \frac{1}{T_0} \int_{-T_0/2}^{T_0/2} x(t) e^{-i\omega_n t} dt, \qquad (2.12)$$

where $T_0 = \dfrac{2\pi}{\omega_0}$ is the fundamental period of $x(t)$.

The plot of the magnitude of the complex coefficients x_n versus frequency is called the *amplitude spectrum* of the periodic function $x(t)$; since the index n assumes only integers, this is a discrete frequency spectrum (line spectrum) (Fig. 2.2).

Equation

$$\int_{-T_0/2}^{T_0/2} x^2(t) dt = T_0 \sum_{n=-\infty}^{+\infty} |c_n|^2 \qquad (2.13)$$

is known as Parseval's theorem [1].

The integral from the above equation is a measure of the energy accumulated by the system during a period T_0.

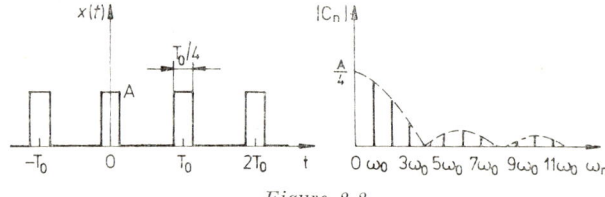

Figure 2.2

The power content of the function $x(t)$ in the period T_0 (the average power per period) is defined by the mean-square value

$$\overline{x^2(t)} = \frac{1}{T_0} \int_{-T_0/2}^{T_0/2} x^2(t) dt = \sum_{n=-\infty}^{+\infty} |c_n|^2. \qquad (2.14)$$

2.2 Characteristics of vibrations

The plot of $|c_n|^2$ against frequency is called the *power spectrum* of $x(t)$ (Fig. 2.3).

Equation (2.14) can also be written

$$\overline{x^2(t)} = \sum_{n=-\infty}^{+\infty} \left[\frac{|c_n|^2}{\Delta f}\right] \Delta f =$$

$$= \sum_{n=-\infty}^{+\infty} S(f_n) \Delta f, \qquad (2.14\,\text{a})$$

where $S(f_n)$ is the *power spectral density* of $x(t)$. The word *power* is generally used for the mean value of a squared quantity.

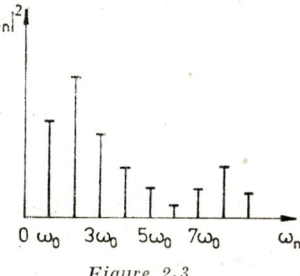

Figure 2.3

2.2.3.2 Frequency analysis of nonperiodic signals. Starting with a periodic function $x_{T_0}(t)$ of period T_0 and letting T_0 approach infinity, the resulting function $x(t) = \lim_{T_0 \to \infty} x_{T_0}(t)$ is no longer periodic. As T_0 increases, the distance $\omega_0 = \dfrac{2\pi}{T_0}$ between neighbouring harmonics decreases. In the limit, when $T_0 \to \infty$, $\omega_{n+1} - \omega_n = d\omega \to 0$, the frequency of any "harmonic" ω_n corresponds to the general frequency variable, the summation in equation (2.11) becomes an integration over ω, and equation (2.12) becomes

$$X(i\omega) = \int_{-\infty}^{+\infty} x(t)\, e^{-i\omega t}\, dt. \qquad (2.15)$$

or

$$X(if) = \int_{-\infty}^{+\infty} x(t)\, e^{-i2\pi ft}\, dt. \qquad (2.15\,\text{a})$$

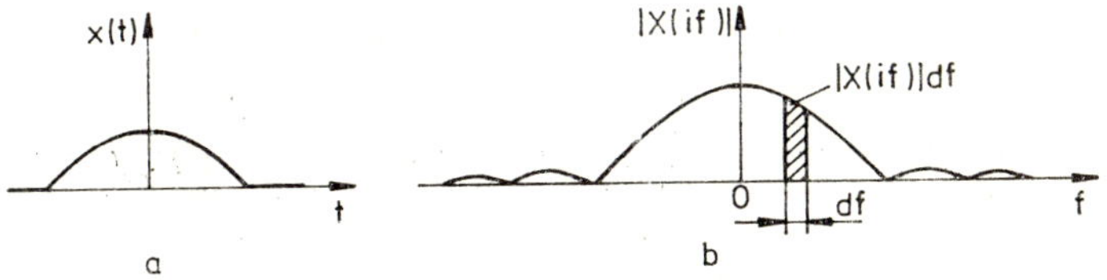

Figure 2.4

The function $X(if)$ defined by (2.15 a) is known as the *Fourier transform* of $x(t)$ and is often denoted by $\mathscr{F}[x(t)]$. The plot of the modulus $|X(if)|$ against frequency (Fig. 2.4) is called the *linear Fourier spectrum* or the *magnitude spectrum* of $x(t)$, being a continuous spectrum.

Elements of theory

For nonperiodic functions, Parseval's theorem becomes

$$E = \int_{-\infty}^{+\infty} x^2(t)\,dt = \frac{1}{2\pi}\int_{-\infty}^{+\infty}|X(i\omega)|^2\,d\omega = \int_{-\infty}^{+\infty}|X(if)|^2\,df =$$

$$= \int_{-\infty}^{+\infty} X(if)\cdot X^*(if)\,df. \tag{2.16}$$

As the total energy content E of a transient time signal is finite, the average power over an infinite period is infinitesimal. The integral square of the time function over an infinite interval is finite, but the mean square over this infinite interval is an infinitesimal quantity.

The autospectrum $S_{xx}(f)$ of the time signal $x(t)$ is defined as

$$S_{xx}(f) = X(if)\cdot X^*(if) \tag{2.17}$$

where * indicates complex conjugation. For a transient type signal it represents the *energy spectral density* and is measured in terms of some unit squared times seconds per Hertz.

The cross spectrum $S_{xy}(if)$ of two time signals $x(t)$ and $y(t)$ is defined as

$$S_{xy}(if) = X(if)\cdot Y^*(if). \tag{2.18}$$

The autocorrelation function given by equation (2.9) and the energy spectral density given by (2.17) constitute a Fourier transform pair, i.e.

$$S_{xx}(f) = \mathscr{F}[R_{xx}(\tau)] = \int_{-\infty}^{+\infty} R_{xx}(\tau)\,e^{-i2\pi f\tau}\,d\tau \tag{2.19}$$

and

$$R_{xx}(\tau) = \mathscr{F}[S_{xx}(f)] = \int_{-\infty}^{+\infty} S_{xx}(f)\,e^{i2\pi f\tau}\,df = \frac{1}{2\pi}\int_{-\infty}^{+\infty} S_{xx}(\omega)e^{i\omega\tau}\,d\omega =$$

$$= \int_{-\infty}^{+\infty} X(if)\cdot X^*(if)\,e^{i2\pi f\tau}\,df. \tag{2.20}$$

Equations (2.19) and (2.20) are commonly referred to as the Wiener-Khintchine relations.

2.2 Characteristics of vibrations

The cross power spectrum (cross spectral density) can be obtained as the Fourier transform of the cross-correlation function (2.10):

$$S_{xy}(if) = \mathscr{F}[R_{xy}(\tau)] = \int_{-\infty}^{+\infty} R_{xy}(\tau)\, e^{i2\pi f\tau}\, d\tau. \qquad (2.21)$$

2.2.3.3 Frequency analysis of random signals. Random time signals do not have finite energy content. The average power over an interval T approaches zero as $T \to \infty$ so that a *power spectral density* is considered instead of the energy spectral density.

The auto power spectrum $S_{xx}(f)$ and the autocorrelation function $\bar{R}_{xx}(\tau)$ are defined by equations (2.19) and (2.20) where $R_{xx}(\tau)$ is replaced by $\bar{R}_{xx}(\tau)$ given by equation (2.6).

For $\tau = 0$, equations (2.6) and (2.20) yield

$$\bar{R}_{xx}(0) = \lim_{T\to\infty} \frac{1}{T} \int_{-T/2}^{T/2} x^2(t)\, dt = \overline{x^2(t)} = \int_{-\infty}^{+\infty} S_{xx}(f)\, df =$$

$$= \frac{1}{2\pi} \int_{-\infty}^{+\infty} S_{xx}(\omega)\, d\omega. \qquad (2.22)$$

The physical signification of the quantities and functions entering the relationships (2.22) can be assessed by comparison to the corresponding quantities defined for harmonic functions:

— the integral $\int_{-T/2}^{T/2} x^2(t)\, dt$ is a measure of the energy over the interval $\left(-\dfrac{T}{2},\, +\dfrac{T}{2}\right)$; division by T gives the average power over the interval T;

— the mean square $\overline{x^2(t)}$ represents the total average power over the whole duration of the sample record $x(t)$;

— the function $S_{xx}(f)$ shows how is the mean-square (and therefore the average power) distributed per unit frequency; it is the power spectral density and is measured in terms of some unit squared per Hertz.

It can be shown that $S_{xx}(f)$ is real and even. The information at negative frequencies is the same as that at positive frequencies so that one-sided spectra are used. The one-sided autospectrum $W_{xx}(f)$ is related to the value given in (2.19) by the equation

$$W_{xx}(f) = 2S_{xx}(f) \qquad \text{for } f > 0.$$

2 Elements of theory

In this notation, the mean square value is given by

$$\overline{x^2(t)} = \lim_{T\to\infty} \frac{1}{T} \int_{-T/2}^{T/2} x^2(t)\,dt = 2\int_0^\infty S_{xx}(f)\,df = \int_0^\infty W_{xx}(f)\,df. \quad (2.23)$$

The cross power spectrum $S_{xy}(if)$ is defined as the Fourier transform of the cross-correlation function given by (2.7).

Because $\overline{R}_{xy}(\tau)$ is not in general an even function of τ, $S_{xy}(if)$ is a complex function which can be written

$$S_{xy}(if) = C_{xy}(f) - iQ_{xy}(f)$$

where $C_{xy}(f)$ is the *co-spectral density* function (coincident spectrum) and $Q_{xy}(f)$ is the *quad-spectral density* function (quadrature spectrum).

The real quantity

$$\gamma_{xy}^2(f) = \frac{|S_{xy}(if)|^2}{S_{xx}(f) \cdot S_{yy}(f)}, \quad 0 \leqslant \gamma_{xy}^2 \leqslant 1$$

is the *coherence function* of random processes $x(t)$ and $y(t)$. When $x(t)$ and $y(t)$ are statistically independent, then $\gamma_{xy}^2(f) = 0$; when the two processes are totally coherent, then $\gamma_{xy}^2(f) = 1$.

The power spectral density of random vibrations can be measured by analog selective filtering. The signal $x(t)$ is fed to a bandpass filter (see Section 5.4.2) of bandwidth B and centred on frequency f. At the output one obtains the signal $x_B(f, t)$, which has frequency components between $\left(f - \dfrac{B}{2}\right)$ and $\left(f + \dfrac{B}{2}\right)$.

Assuming that no attenuation takes place in the filter, equation (2.23) may be written

$$\int_0^\infty W_{xx}(f)\,df = \int_{f-B/2}^{f+B/2} W_{xx}(f)\,df = \lim_{T\to\infty} \frac{1}{T} \int_{-T/2}^{T/2} x_B^2(f, t)\,dt, \quad (2.24)$$

where $\int_0^\infty W_{xx}(f)\,df$ is the total power of the input signal $x(t)$, and $\int_{f-B/2}^{f+B/2} W_{xx}(f)\,df$ is the power corresponding to the frequencies within $\left(f - \dfrac{B}{2}\right)$ and $\left(f + \dfrac{B}{2}\right)$.

2.2 Characteristics of vibrations

If B is made so small that $W_{xx}(f)$ can be considered constant within this frequency range, then

$$\int_{f-B/2}^{f+B/2} W_{xx}(f)\, \mathrm{d}f = W_{xx}(f) \cdot B.$$

In the limiting case, when $B \to 0$,

$$W_{xx}(f) = \lim_{B \to 0} \lim_{T \to \infty} \frac{1}{BT} \int_{-T/2}^{T/2} x_B^2(f, t)\, \mathrm{d}t. \tag{2.25}$$

Similar equations are used for the measurement of the cross--spectral density of two signals [3].

Another way of determining the power spectral density of random signals is to use the Fourier transform applied to the signal divided in short duration segments.

The time record of length T is divided into n segments of duration T'. Let $x_k(t, T')$ be the signal corresponding to the k-th interval and denote its Fourier transform by $X_k(\mathrm{i}f, T')$. The auto-spectral density $S_{x_k x_k}(f, T')$ of the truncated signal $x_k(t, T')$ is given by

$$S_{x_k x_k}(f, T') = \frac{1}{T'} |X_k(\mathrm{i}f, T')|^2.$$

The auto-spectral density of the signal $x(t)$, over the whole duration T, can be obtained as an ensemble average of the n values

$$S_{xx}(f) = \frac{1}{n} \sum_{k=1}^{n} S_{x_k x_k}(f, T') = \frac{1}{n \cdot T'} \sum_{k=1}^{n} |X_k(\mathrm{i}f, T')|^2. \tag{2.26}$$

Similarly, the cross spectral density of the signals $x(t)$ and $y(t)$ is given by

$$S_{yx}(\mathrm{i}f) = \frac{1}{n \cdot T'} \sum_{k=1}^{n} X_k^*(\mathrm{i}f, T') \cdot Y_k(\mathrm{i}f, T'), \tag{2.27}$$

where $Y_k(\mathrm{i}f, T')$ is the Fourier transform of the signal $y_k(t, T')$ of duration T'. A similar method is used in FFT analyzers which work on time records of finite length.

2.2.3.4 *The Discrete Fourier Transform.* Modern Fourier analyzers utilize digital computers to calculate the Fourier transform digitally. Assume that $x(t)$ is of length T and is sampled at

2 Elements of theory

N points spaced Δt apart such that $T = N \cdot \Delta t$. The truncated version of equation (2.15, a) becomes

$$X(if) = \Delta t \sum_{n=0}^{N-1} x(n \cdot \Delta t) \, e^{-i2\pi f n \Delta t}, \qquad (2.28)$$

where $x(n \cdot \Delta t)$ are the measured values of the input function.

If the frequency function $X(if)$ is sampled N times at $\Delta f = 1/T$ intervals, assuming that $X(if)$ has non-zero values only at the frequencies $f = m \cdot \Delta f$ for N values of m, then the Discrete Fourier Transform (DFT) becomes

$$X(im \cdot \Delta f) = \Delta t \sum_{n=0}^{N-1} x(n \cdot \Delta t) \, e^{-i 2\pi mn/N}, \quad m = 0, \ldots, N-1 \qquad (2.29)$$

where $\Delta f \cdot \Delta t = 1/N$ and Δf is the frequency resolution.

Likewise, the Inverse Discrete Fourier Transform can be written

$$x(n \cdot \Delta t) = \Delta f \sum_{m=0}^{N-1} X(im \cdot \Delta f) \, e^{i 2\pi mn/N}, \quad n=0, \ldots, N-1. \qquad (2.30)$$

The independent variables are the pair of indices m and n. If $x(n \cdot \Delta t)$ is real, then $X(m \cdot \Delta f)$ will be Hermitian, so that only the positive half-period of $X(m \cdot \Delta f)$ need be calculated.

The DFT gives the *correct* Fourier transform for sampled periodic data. If the time function $x(t)$ is indeed periodic, such that the time window length T is an integer number of periods, and if the frequency spectrum is band-limited so as not to exceed the Nyquist frequency $\dfrac{1}{2\Delta t} = \dfrac{N}{2} \Delta f$ (see Chapter 5), then the DFT will produce the correct spectrum without errors (like leakage or aliasing).

The implementation of DFT became practical in 1965 when Cooley and Tukey developed the so-called Fast Fourier Transform (FFT) algorithm permitting the efficient computation of DFT with time savings of 100 : 1[4]. The power spectra are obtained by Fourier transforming the time signals. The autopower spectrum is calculated from

$$S_{xx}(m \cdot \Delta f) = X(m \cdot \Delta f) \cdot X^*(m \cdot \Delta f) \qquad (2.31)$$

and the cross-power spectrum from

$$S_{xy}(m \cdot \Delta f) = X(m \cdot \Delta f) \cdot Y^*(m \cdot \Delta f),$$
$$S_{yx}(m \cdot \Delta f) = Y(m \cdot \Delta f) \cdot X^*(m \cdot \Delta f),$$

2.2 Characteristics of vibrations

where $X(m \cdot \Delta f)$ and $Y(m \cdot \Delta f)$ are Fourier transforms of the signals $x(t)$ and $y(t)$. In practice these values are averaged over a finite number of records [5].

Instead of estimating spectra by first determining correlation functions and then calculating their Fourier transforms, it is now quicker and more accurate to calculate spectral estimates directly from the time records.

A more detailed discussion of this subject is found in References [1] and [6].

2.2.4 Probability density function

Considering a sample record $x(t)$ of duration T (Fig. 2.5), one can determine the proportion of time spent by the signal at all possible amplitudes during the finite period of time T. This is done totalising the time spent by the signal in a series of narrow (Δx) amplitude windows, and then dividing the total for each window by the averaging time T.

The ratio

$$P(x, x + \Delta x) = \frac{\sum_{i=1}^{n} \Delta t_i}{T} \qquad (2.32)$$

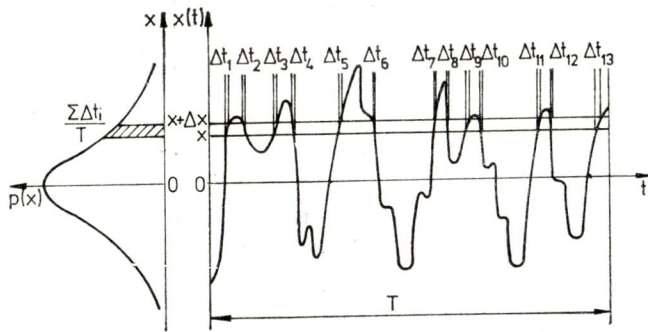

Figure 2.5

represents the probability that the signal will lie in a particular window Δx wide, being between x and $x + \Delta x$, a time interval equal to the sum $\Delta t_1 + \Delta t_2 + \ldots + \Delta t_n$. The longer the duration T, the closer the probability to that one corresponding to the development of the whole phenomenon. The amplitude probability

2 Elements of theory

is a function of x and Δx. The *probability density function* is defined by

$$p(x) = \lim_{\Delta x \to 0} \frac{P(x) - P(x + \Delta x)}{\Delta x} = \lim_{\Delta x \to 0} \lim_{T \to \infty} \frac{1}{T \Delta x} \sum_{i=1}^{n} \Delta t_i \quad (2.33)$$

and is plotted on the left of Figure 2.5. The bell-shaped curve corresponds to the Gaussian (normal) distribution, which is the most commonly encountered probability density function for naturally occurring signals.

Numerical characteristics of random variables can give indications on their distribution, expressing in compact form some basic properties. Table 2.1 gives the main numerical characteristics of the random variables and their theoretical formulae.

TABLE 2.1 *Characteristics of Random Variables*

Parameter	Notation	Definition using	
		the probability density function, $p(x)$	the time history of a sample record, $x(t)$
Mean value Expected value	$\bar{x}$	$\bar{x} = \int_{-\infty}^{+\infty} x p(x) \, dx$	$\bar{x} = \lim_{T \to \infty} \frac{1}{T} \int_{0}^{T} x(t) \, dt$
Mean square value	$\overline{x^2}$	$\overline{x^2} = \int_{-\infty}^{+\infty} x^2 p(x) \, dx$	$\overline{x^2} = \lim_{T \to \infty} \frac{1}{T} \int_{0}^{T} x^2(t) \, dt$
Variance Dispersion	σ^2	$\sigma^2 = \int_{-\infty}^{+\infty} (x - \bar{x})^2 p(x) \, dx$	$\sigma^2 = \lim_{T \to \infty} \frac{1}{T} \int_{0}^{T} [x(t) - \bar{x}]^2 \, dt$
Mean square error Standard deviation	σ	$\sigma = \sqrt{\sigma^2}$	$\sigma = \sqrt{\sigma^2}$

The relationship $\sigma^2 = \overline{x^2} - [\bar{x}]^2$ is evident.

2.3 Estimation errors at random data analysis

For different sample records of a random process, taken in identical measurement conditions, the measured parameter differs from one sample record to another, the randomness of the process being reflected by the stochastic distribution of measurement data.

2.3 Estimation errors

Theoretically, the statistical parameters and functions have been defined in Section 2.2, for an analysis time tending to infinity. Practical use of finite length sample records gives rise to *estimation errors*.

Under these circumstances, it is not sufficient to find experimentally numerical values of the unknown parameter or a series of numerical values for the statistical function, but one has to estimate the measurement accuracy.

2.3.1 Confidence probability. Confidence interval

Let $\{x(t)\}$ be a stationary random process. Suppose that Φ is the true value of an unknown parameter of the random process $\{x(t)\}$ and $\hat{\Phi}$ is an estimate of Φ obtained from a measurement made on a particular sample record $x(t)$ extending over a finite time interval T. The values of $\hat{\Phi}$ computed for different sample records from $\{x(t)\}$ will vary randomly.

The probability that the modulus of the difference between the true value Φ and the estimated value $\hat{\Phi}$ be smaller than a given quantity ε_β is called *confidence probability* and is denoted by β.

Hence, the true unknown value of the parameter Φ will lie, with a probability β, within

$$I_\beta = (\hat{\Phi} - \varepsilon_\beta, \hat{\Phi} + \varepsilon_\beta)$$

called *confidence interval*.

Supposing that measurement data have a Gaussian (normal) distribution, the error ε_β is expressed as a product of the mean square error of the estimated value $\sigma_{\hat{\Phi}}$ and a function γ, given in tabled form for different values of the confidence probability β (Table 2.2)

$$\varepsilon_\beta = \sigma_{\hat{\Phi}} \gamma. \tag{2.34}$$

TABLE 2.2 *Values of γ as a Function of the Confidence Probability β*

β	γ	β	γ	β	γ	β	γ
0.80	1.282	0.86	1.475	0.92	1.750	0.97	2.169
0.81	1.310	0.87	1.513	0.93	1.810	0.98	2.325
0.82	1.340	0.88	1.554	0.94	1.880	0.99	2.576
0.83	1.371	0.89	1.597	0.95	1.960	0.9973	3.000
0.84	1.404	0.90	1.643	0.96	2.053	0.999	3.29
0.85	1.439	0.91	1.694				

2 Elements of theory

A complete result is given under the form

$$\Phi = \hat{\Phi} \pm \varepsilon_\beta. \tag{2.35}$$

The above relationship can be expressed in words as follows: the real value of the parameter Φ is given by the estimated value $\hat{\Phi}$ with an error of $\pm \varepsilon_\beta$, within a probability β.

2.3.2 Standard error

Usually, the *standard error* of the estimate is used, denoted by ε and defined by

$$\varepsilon = \frac{\sigma_{\hat{\Phi}}}{\Phi} \cdot 100 \ [\%]. \tag{2.36}$$

Equations (2.34) to (2.36) yield

$$\frac{\Phi - \hat{\Phi}}{\Phi} = \pm \frac{\sigma_{\hat{\Phi}}}{\Phi} \gamma = \pm \varepsilon \gamma. \tag{2.37}$$

The product $\varepsilon\gamma$ represents the maximum percentage deviation of the parameter estimate with respect to the real (unknown) value. Imposing a certain measurement accuracy (confidence probability, maximum deviation of the measured value from the true value), the standard error can be calculated.

Table 2.3 gives the expressions for the standard errors of some statistical parameter estimates [1]. These formulae have been derived for bandwidth limited Gaussian white noise data and for a record of finite time T.

The standard error of the power spectral density function estimates is calculated in a narrow bandwidth B_e about the frequency f of interest, corresponding to the filter bandwidth.

The standard error of the probability density function estimates tends to a minimum for the most probable value of the random variable, which is obtained for the maximum value of $p(x)$. The amplitude window Δx from equation (2.33) appears as a finite quantity at the denominator of the standard error. This error increases with decreasing Δx, i.e. in the case of a smooth analysis of the signal amplitude distribution.

Examining the standard error formulae from Table 2.3, it is noticed that ε decreases as T increases. These formulae may be used in two ways: either to evaluate the errors in estimates for a given sample length T or to determine the duration T so that a desired standard error will be achieved.

2.3 *Estimation errors*

TABLE 2.3 *Standard Errors for Statistical Parameters and Functions*

Statistical parameter or function	Mean square value	Root mean square value	Probability density	Autocorrelation function	Cross-correlation function	Power spectral density
	$\overline{x^2}$	$\sqrt{\overline{x^2}}$	$p(x)$	$\overline{R}_{xx}(\tau)$	$\overline{R}_{xy}(\tau)$	$W_{11}(f)$
Standard error ε	$\dfrac{1}{\sqrt{BT}}$	$\dfrac{1}{2\sqrt{BT}}$	$\dfrac{1}{\sqrt{BT}\,p(x)\,\Delta x}$	$\sqrt{\dfrac{1+\dfrac{\overline{R}_{xx}^2(0)}{\overline{R}_{xx}^2(\tau)}}{2BT}}$	$\sqrt{1+\dfrac{\overline{R}_{xx}(0)\overline{R}_{yy}(0)}{\overline{R}_{xy}^2(\tau)}}{2BT}$	$\dfrac{1}{\sqrt{B_e T}}$

T — signal analysis time; B — the total "equivalent white noise" bandwidth occupied by all the data in the record; Δx — amplitude window used at probability density measurements; B_e — filter bandwidth used at signal frequency analysis.

2   *Elements of theory*

   Example. The minimum sample length T is required for the measurement of the mean square of a signal so that the true value of $\overline{x^2}$ should be within $\pm 8\%$ of the measured value, with 90% confidence. The signal is supposed to have components between 30 and 80 Hz.
   Solution. From Table 2.2, for $\beta = 90\%$ one obtains $\gamma = 1.6$. For $\varepsilon\gamma = 8\%$ the standard error will be $\varepsilon = 5\%$. From Table 2.3, for mean square estimates one obtains $\varepsilon = \dfrac{1}{\sqrt{BT}}$. The signal bandwidth is $B = 80 - 30 = 50$ Hz so that $T = \dfrac{1}{\varepsilon^2 B} = \dfrac{1}{0.05^2 \cdot 50} = 8\text{s}.$

2.4 Response of vibrating systems to various excitations

As previously shown, a vibrating system can be subjected to harmonic or periodic forces, shocks, transients, or random excitations. In each case the system motion can be described by certain characteristics of the response (to the impressed excitation). The response depends on both the excitation characteristics and the system dynamic parameters.

   Let consider a constant parameter linear system (Fig. 2.6), with an *input* $x(t)$, which constitutes the *excitation*, and an *output* $y(t)$, which constitutes the *response*. Both $x(t)$ and $y(t)$ may be forces, displacements, velocities, accelerations, or a mixture of these.

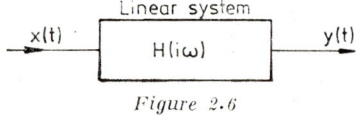

Figure 2.6

The Fourier transforms of the input and of the output

$$X(i\omega) = \int_{-\infty}^{+\infty} x(t)\, e^{-i\omega t}\, dt,$$

$$Y(i\omega) = \int_{-\infty}^{+\infty} y(t)\, e^{-i\omega t}\, dt, \qquad (2.38)$$

2.4 Response of vibrating systems

are related by
$$Y(i\omega) = H(i\omega) \cdot X(i\omega), \tag{2.39}$$

where $H(i\omega)$ is the *frequency response function* of the vibrating system. Its dimensions are dependent on the physical nature of the quantities $x(t)$ and $y(t)$.

The equation of motion of a linear viscously damped system, acted upon by the force $F(t)$, can be written

$$m\ddot{y} + c\dot{y} + ky = F(t). \tag{2.40}$$

Denoting

$$p^2 = \frac{k}{m}, \quad c = \zeta c_c, \quad c_c = 2\,mp,$$

equation (2.40) becomes

$$\ddot{y} + 2p\zeta\dot{y} + p^2 y = \frac{F(t)}{m} = x(t). \tag{2.41}$$

In this case, the frequency response function is

$$H(i\omega) = \frac{1}{p^2 - \omega^2 + i2\zeta\omega p} \tag{2.42}$$

or

$$H(i\omega) = \frac{1}{p^2} A$$

where the dimensionless quantity

$$A = \frac{1}{1 - \dfrac{\omega^2}{p^2} + i2\zeta\dfrac{\omega}{p}} \tag{2.43}$$

is called a *complex magnification factor*.

2.4.1 Harmonic excitation

Denoting by x the displacement, the equation of motion for a single-degree-of-freedom linear system (Fig. 1.1), when a *harmonic force* is applied to the mass m, can be written as

$$m\ddot{x} + c\dot{x} + kx = F_0 \sin \omega t, \tag{2.44}$$

respectively

$$\ddot{x} + 2p\zeta\dot{x} + p^2 x = \frac{F_0}{m} \sin \omega t. \tag{2.45}$$

2 Elements of theory

The steady-state solution of equation (2.45) is the harmonic displacement

$$x(t) = X_0 \sin(\omega t - \theta) \tag{2.46}$$

of amplitude

$$X_0 = x_s A_1 = \frac{F_0}{k} A_1 \tag{2.47}$$

where $\left(\text{denoting } \frac{\omega}{p} \text{ by } \eta\right)$ the *magnification factor* A_1 and the phase angle are given respectively by

$$A_1 = \frac{1}{\sqrt{(1 - \eta^2)^2 + (2\zeta\eta)^2}}, \tag{2.48}$$

$$\theta = \tan^{-1} \frac{2\zeta\eta}{1 - \eta^2}. \tag{2.49}$$

The frequency response is given by the plots of the functions $A_1(\omega)$ and $\theta(\omega)$, respectively $A_1(\eta)$ and $\theta(\eta)$ (Figs 2.7 and 2.8).

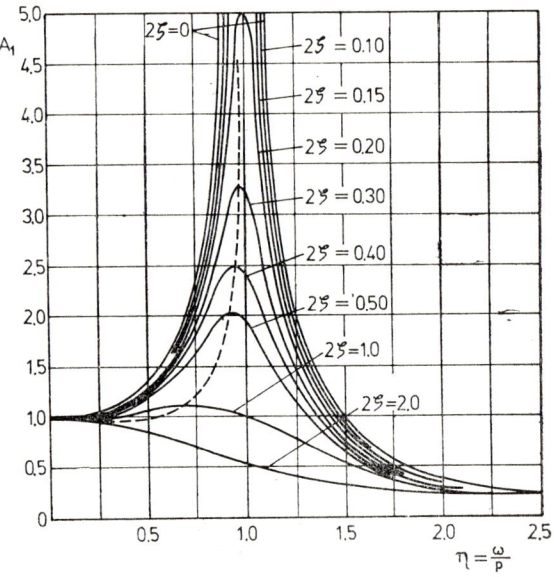

Figure 2.7

If the exciting force is supplied by the rotation of an eccentric mass m_0, with eccentricity r_0, the inertia force is

$$F(t) = m_0 r_0 \omega^2 \sin \omega t, \tag{2.50}$$

2.4 *Response of vibrating systems*

and the motion of the mass m is described by equation (2.46), the amplitude X_0 being given by

$$X_0 = A_2 \frac{m_0 r_0}{m + m_0} \qquad (2.51)$$

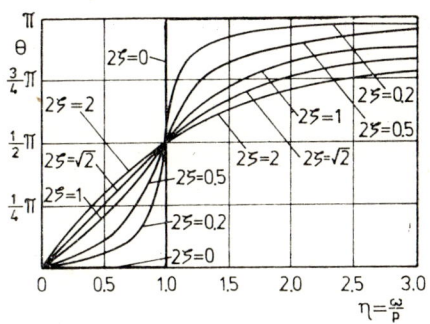

Figure 2.8

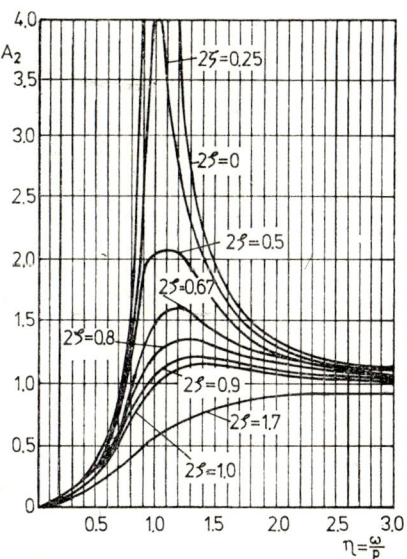

Figure 2.9

where the magnification factor A_2, plotted versus η in Figure 2.9, is

$$A_2 = \frac{\eta^2}{\sqrt{(1-\eta^2)^2 + (2\zeta\eta)^2}}. \qquad (2.52)$$

2 Elements of theory

The magnification factor A_3, defined for systems excited through a dashpot, is given by

$$A_3 = \frac{2\zeta\eta}{\sqrt{(1-\eta^2)^2 + (2\zeta\eta)^2}} \qquad (2.53)$$

and is shown graphically in Figure 2.10.

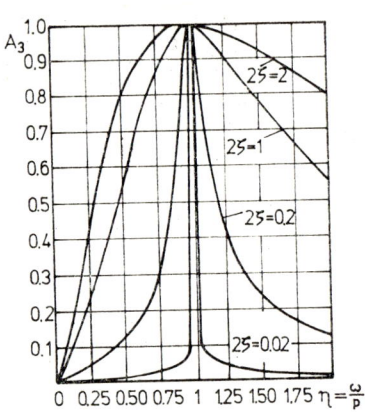

Figure 2.10

At systems with hysteretic (structural) damping, one can introduce the complex moduli of elasticity

$$E^* = E_1 + iE_2, \quad G^* = G_1 + iG_2,$$

where E_1 and G_1 are *dynamic moduli of elasticity* (storage moduli), E_2 and G_2 are *loss moduli*, and the *hysteretic damping factor* is defined as

$$g = \frac{E_2}{E_1} = \frac{G_2}{G_1}.$$

Likewise, a complex stiffness is defined by

$$k^* = k + ih = k\left(1 + i\frac{h}{k}\right) = k(1 + ig), \qquad (2.54)$$

where k is the *dynamic spring constant* and h is the *hysteretic damping coefficient*.

Substituting an equivalent viscous damping coefficient

$$c_i = \frac{k}{\omega}g \qquad (2.55)$$

2.4 Response of vibrating systems

into equation (2.44), the magnification factor — for constant amplitude force excitation — is given by

$$A_1 = \frac{1}{\sqrt{(1-\eta^2)^2 + g^2}} \tag{2.56}$$

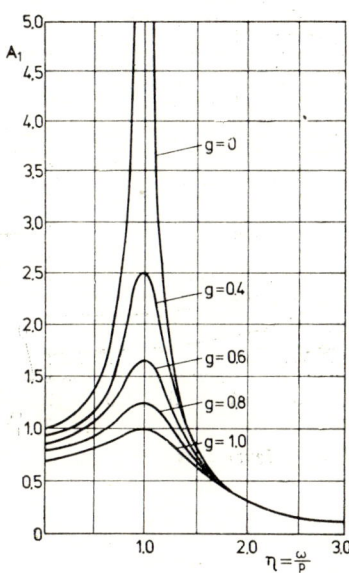

Figure 2.11

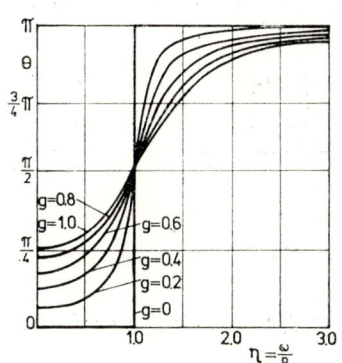

Figure 2.12

and the phase angle by

$$\theta = \tan^{-1}\frac{g}{1-\eta^2}. \tag{2.57}$$

Equations (2.56) and (2.57) are shown graphically in Figures 2.11 and 2.12.

The system response can be also expressed in terms of the ratio between the *output force* and the *input force*, called *transmissibility*. For the system from Figure 2.13, the force transmissibility is

$$T = \frac{F_{T_0}}{F_0} = \sqrt{\frac{1+(2\zeta\eta)^2}{(1-\eta^2)^2 + (2\zeta\eta)^2}} \tag{2.58}$$

and is shown graphically in Figure 2.14.

The motion transmissibility, equal to the amplitude ratio for the base excited system, has the same expression (2.58).

2 Elements of theory

The amplitude-frequency plot (like that from Fig. 2.7) for a system with several degrees of freedom has more peaks, corresponding to the resonances of the system.

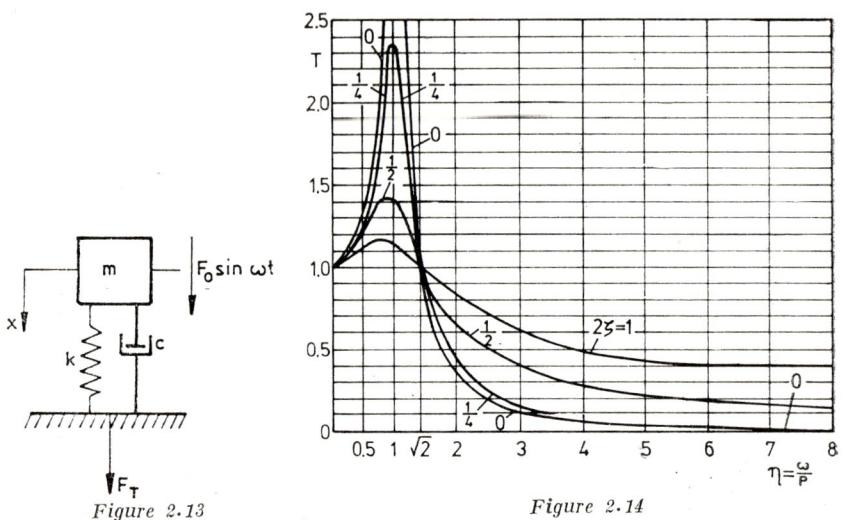

Figure 2.13   Figure 2.14

2.4.2 Periodic excitation

Periodic vibrations are analysed using the Fourier series representation.

Examining the *system response* to a periodic excitation, the *magnification factor* from equation (2.47) can be written

$$A_{1_N} = \frac{1}{\sqrt{\left[1 - \left(\frac{N\omega_0}{p}\right)^2\right]^2 + \left[2\zeta\frac{N\omega_0}{p}\right]^2}}, \qquad (2.59)$$

where p is the natural angular frequency of the (single-degree-of-freedom) system and $N\omega_0$ is the angular frequency of the N-th harmonic component of the excitation.

If $f(t)$ is the excitation time function, its linear frequency spectrum $|F(i\omega)|$ can be plotted as in Figure 2.15a. Equation (2.59) gives the magnification factor for each harmonic of angular frequency $N\omega_0$. This enables plotting of the response frequency spectrum $|X(i\omega)|$ to be done as in Figure 2.15b. Generally, the two spectra do not resemble each other due to the frequency dependence of the magnification factor.

2.4 Response of vibrating systems

If the vibratory system is a vibrograph, then the input is the measurand and the output is the time history record. The difference between the two spectra is due to the instrument induced distortions.

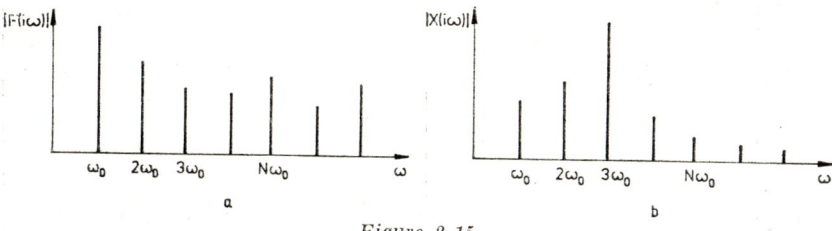

Figure 2.15

2.4.3 Transient excitation and shocks

Shocks are suddenly applied nonperiodic excitations transmitting kinetic energy from a source to a system in a relatively short time compared with the natural period of vibration of the system. *Transients* may last for several periods of vibration of the system.

Shocks and transients may be described in terms of force, displacement, velocity or acceleration, expressed by a nonperiodic time function $f(t)$, defining their waveform.

For three usual shock time functions, i.e. the rectangular, the half-sine and the final peak sawtooth shock pulses, the amplitude spectra $|F(i\omega)|$ are given in Table 2.4, where A is the pulse magnitude and T is the overall pulse duration.

Let consider a single degree of freedom system with a *step excitation*, i.e. a suddenly applied force $F(t) = F_0$ (Fig. 2.16a), which remains constant indefinitely. For zero initial conditions, the system response is the displacement

$$x(t) = \frac{F_0}{k}\left[1 - e^{-p\zeta t}\left(\cos p\sqrt{1-\zeta^2}\,t + \frac{\zeta}{\sqrt{1-\zeta^2}}\sin p\sqrt{1-\zeta^2}\,t\right)\right] \quad (2.60)$$

shown graphically in Figure 2.16b. In equation (2.60) $p = \sqrt{k/m}$ is the natural angular frequency of the undamped system.

The maximum displacement value takes place at $t_0 = \dfrac{\pi}{p\sqrt{1-\zeta^2}}$ and is

$$x(t_0) = x_0\left(1 + e^{-\frac{\zeta}{\sqrt{1-\zeta^2}}\pi}\right) \quad (2.61)$$

2 *Elements of theory*

TABLE 2.4 *Time Functions and Amplitude Spectra for Three Pulse Shapes*

Time function	Amplitude spectrum	Equation of amplitude spectrum
rectangular pulse	sinc-shape spectrum	$\dfrac{|F(i\omega)|}{AT} = \left|\dfrac{\sin\dfrac{\omega T}{2}}{\dfrac{\omega T}{2}}\right|$
triangular (ramp) pulse	spectrum	$\dfrac{|F(i\omega)|}{AT} = \dfrac{1}{\omega T}\sqrt{1 - \dfrac{2\sin\omega T}{\omega T} + \left(\dfrac{\sin\dfrac{\omega T}{2}}{\dfrac{\omega T}{2}}\right)^{2}}$
cosine (half-sine) pulse	spectrum	$\dfrac{|F(i\omega)|}{AT} = \dfrac{2}{\pi}\left|\dfrac{\cos\dfrac{\omega T}{2}}{1 - \left(\dfrac{\omega T}{\pi}\right)^{2}}\right|$

38

2.4 Response of vibrating systems

where

$$x_0 = \frac{F_0}{k} = \lim_{t \to \infty} x(t). \tag{2.62}$$

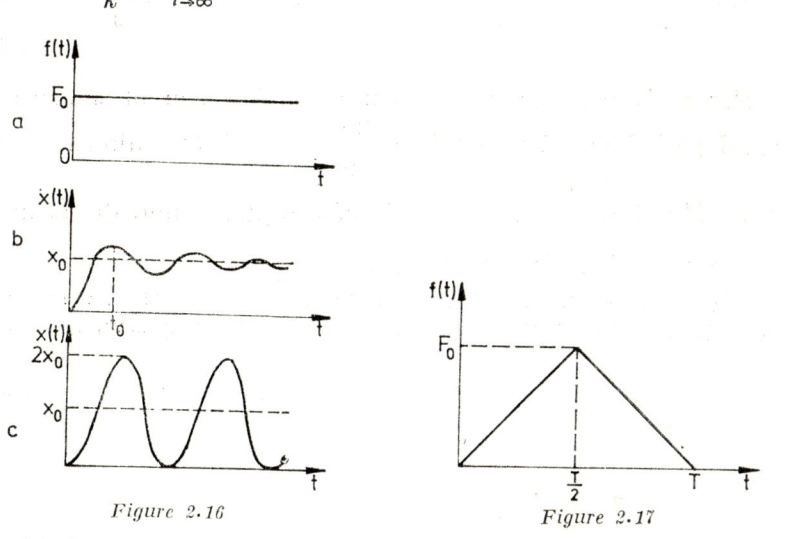

Figure 2.16   Figure 2.17

If the system is undamped, equation (2.60) becomes

$$x(t) = x_0(1 - \cos pt) \tag{2.63}$$

and is graphically shown in Figure 2.16c. The effect of a suddenly applied constant force is to double the static deflection x_0 of the undamped system.

Another case is the double slope triangular impulsive excitation (Fig. 2.17). Two stages of the motion can be distinguished: the first, which takes place while the shock pulse is still acting, called *initial response* or *initial shock* and the second, after the pulse has occurred, called *residual response* or *residual shock*. Using notation from (2.62), the motion resulting during the two stages is:

— initial response

$$x(t) = \frac{2x_0}{pT}(pt - \sin pt), \quad \text{for } 0 \leqslant t \leqslant \frac{T}{2}, \tag{2.64}$$

$$x(t) = \frac{2x_0}{pT}\left[pT - \sin pt - pt + 2\sin p\left(t - \frac{T}{2}\right)\right],$$

for $\dfrac{T}{2} \leqslant t \leqslant T$; \hfill (2.65)

2 Elements of theory

— residual response

$$x(t) = \frac{2x_0}{pT}\left[2\sin p\left(t - \frac{T}{2}\right) - \sin p(t-T) - \sin pt\right], \quad t \geq T. \tag{2.66}$$

For a given shock, the response is a function of the system natural period of vibration $T_0 = \frac{2\pi}{p}$ and of the dimensionless factor $pT = 2\pi\frac{T}{T_0}$, i.e. the ratio of the pulse time duration T and the natural period of the system T_0.

Figure 2.18 illustrates (in dimensionless form) the response to a triangular pulse for $T/T_0 = 1, 2,$ and 5. The system does not

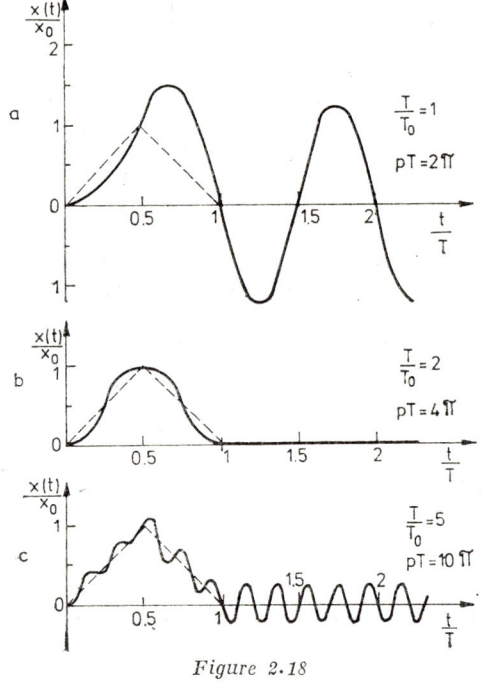

Figure 2.18

oscillate when the pulse duration is an even multiple of the natural period (Fig. 2.18b). It can be seen that the maximum amplitude of both the initial shock (for $0 \leq t \leq T$) and the residual shock is a function of the ratio T/T_0.

2.4 Response of vibrating systems

Figure 2.19 shows the *initial shock spectrum* and the *residual shock spectrum*, i.e. the variation of response maximum amplitude against T/T_0, for the pulse shape from Figure 2.17. Residual

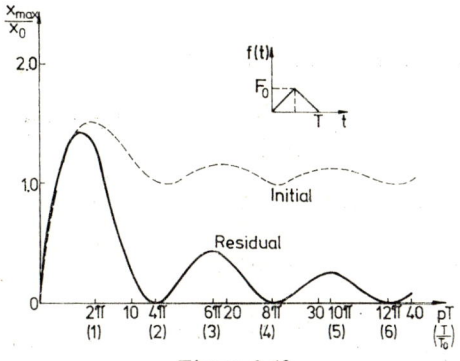

Figure 2.19

shock spectra are shown in Figure 2.20 for four pulse shapes which are equivalent from the point of view of the momentum applied to the vibrating system

$$\int_0^T f(t)\,dt = \text{const.}$$

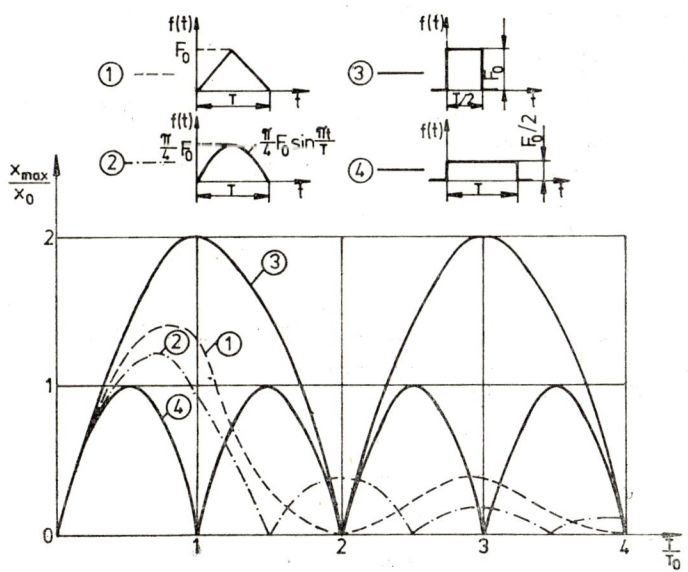

Figure 2.20

41

2   Elements of theory

Shock spectra are used to investigate either the response of a given system to various types of pulses, or the responses of various systems to a given type of pulse.

2.4.4   Stationary random excitation

Let consider a single degree of freedom linear system, defined by the frequency response function

$$H(i\omega) = \frac{1}{p^2 - \omega^2 + i2\zeta\omega p} \qquad (2.67)$$

and the impulse response function

$$h(t) = \frac{1}{mp\sqrt{1-\zeta^2}} e^{-p\zeta t} \sin p\sqrt{1-\zeta^2}\, t. \qquad (2.68)$$

The system is acted upon by an ergodic stationary random force $f(t)$. The response can be expressed as

$$x(t) = \int_{-\infty}^{t} f(\tau) h(t-\tau)\, d\tau \qquad (2.69)$$

where τ is a dummy time variable, $0 \leq \tau \leq t$.

The *mean value* of the response is

$$\overline{x(t)} = H(0) \cdot \overline{f(t)} \qquad (2.70)$$

where $\overline{f(t)}$ is the mean value of the excitation and $H(0)$ is the value of the frequency response function at $\omega = 0$.

The *power spectral density* of the response is

$$S_{xx}(\omega) = \int_{-\infty}^{+\infty} \bar{R}_{xx}(\tau) e^{-i\omega\tau}\, d\tau = H^*(i\omega) \cdot H(i\omega) \cdot S_{ff}(\omega) \qquad (2.71)$$

where $H^*(i\omega)$ is the complex conjugate of $H(i\omega)$ and $S_{ff}(\omega)$ is the power spectral density of the input $f(t)$.

Equation (2.71) may also be written

$$S_{xx}(\omega) = |H(i\omega)|^2 \cdot S_{ff}(\omega). \qquad (2.72)$$

The *mean square* of the response is

$$\overline{x^2(t)} = \bar{R}_{xx}(0) = \frac{1}{2\pi} \int_{-\infty}^{+\infty} S_{xx}(\omega)\, d\omega.$$

2.4 Response of vibrating systems

If plots of $S_{ff}(\omega)$ and $|H(i\omega)|^2$ against frequency are given (Fig. 2.21), then the plot of $S_{xx}(\omega)$ versus ω can be obtained multiplying ordinate by ordinate the given frequency domain functions. Measuring $S_{ff}(\omega)$ and $S_{xx}(\omega)$, the absolute value of the frequency response function can be obtained from equation (2.72).

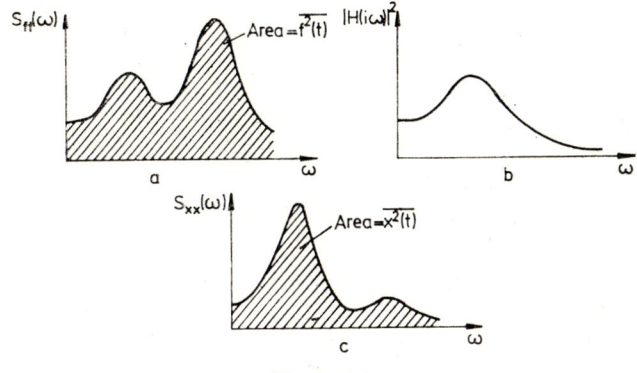

Figure 2.21

The cross-power density $S_{xf}(i\omega)$ between $x(t)$ and $f(t)$ is given by

$$S_{xf}(i\omega) = H(i\omega) \cdot S_{ff}(\omega) \tag{2.73}$$

where $S_{ff}(\omega)$ is the power spectral density of the input. Similarly $S_{xx}(\omega) = H(i\omega)S_{fx}(i\omega)$.

The *coherence function* between $x(t)$ and $f(t)$ is defined by equation

$$\gamma_{fx}^2(\omega) = \frac{|S_{fx}(i\omega)|^2}{S_{xx}(\omega) \cdot S_{ff}(\omega)} \tag{2.74}$$

being a measure of the degree of causality between system input and output. It is bounded between zero and unity, that is

$$0 \leqslant \gamma_{fx}^2(\omega) \leqslant 1.$$

A value $\gamma_{fx}^2(\omega) = 1$ indicates that all of the output signal at the frequency ω is due to the measured input. If $\gamma_{fx}^2(\omega) < 1$, there are extraneous input signals not being measured, noise present in the system, time delays or even nonlinearities.

Multiplying $S_{xx}(\omega)$ by $\gamma_{fx}^2(\omega)$ gives the *coherent output power* $S_{xx}(\omega) \cdot \gamma_{fx}^2(\omega)$, which is the power at the output due to the input, being useful in isolating the contributions of multiple sources.

43

2   Elements of theory

At the measurement of the frequency response, a small value of $\gamma_{fx}^2(\omega)$ means that a lot of averaging must be done to improve the signal-to-noise ratio.

2.5   Natural frequencies

For systems which can be analytically modelled and whose equation of motion can be written, the natural frequencies are well known [7].

For the single degree of freedom system (Fig. 2.22) in translational motion, the natural angular frequency is

$$p = \sqrt{\frac{k}{m}} = \sqrt{\frac{g}{\delta_s}}, \qquad (2.75)$$

where k is the spring constant (stiffness), m — oscillating mass, g — acceleration of gravity, δ_s — spring statical deflection as a result of the gravity force of the mass. This relation also applies when the elastic element is a beam in bending or a bar in tension.

For the *torsional pendulum* (Fig. 2.23), the natural frequency is

$$p = \sqrt{\frac{k}{J}} \qquad (2.76)$$

where k is the rod torsional stiffness and J is the disc mass moment of inertia with respect to the rotation axis.

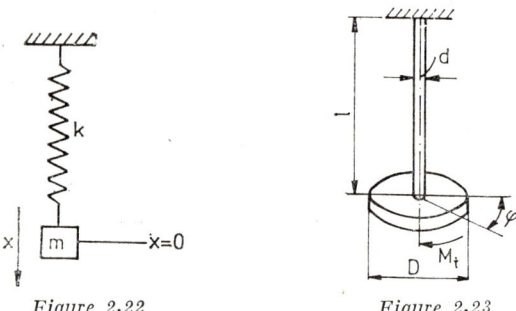

Figure 2.22   Figure 2.23

The natural frequencies of the two-degree-of-freedom system (Fig. 2.24a) are the roots of the equation

$$p^4 - \left(\frac{k_1 + k_{12}}{m_1} + \frac{k_2 + k_{12}}{m_2}\right)p^2 + \frac{k_1 k_2 + k_1 k_{12} + k_2 k_{12}}{m_1 m_2} = 0. \qquad (2.77)$$

2.5 Natural frequencies

For the *system with n lumped masses* (Fig. 2.24b), these are the roots of the determinantal equation

$$\Delta = \begin{vmatrix} k_{12}-m_1 p^2 & -k_{12} & \cdots & 0 & 0 \\ -k_{12} & k_{12}+k_{23}-m_2 p^2 & \cdots & 0 & 0 \\ \vdots & \vdots & & \vdots & \vdots \\ & & & & -k_{n-1,n} \\ 0 & 0 & \cdots & -k_{n-1,n} & k_{n-1,n}-m_n p^2 \end{vmatrix} = 0. \tag{2.78}$$

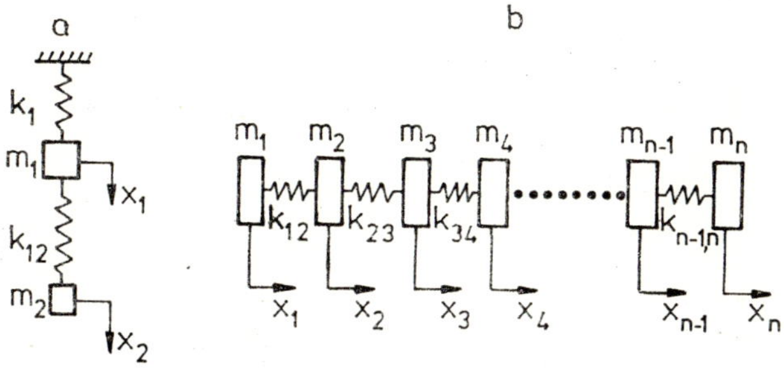

Figure 2.24

Generally, the n-mass system has n degrees of freedom, therefore n natural frequencies; excepted are the "free-free" systems, like that depicted in Figure 2.24b, which generally have $(n-1)$ non-zero natural frequencies. Thus, for the "free-free" torsional systems having two or three discs, the expressions of the natural frequencies are given in Table 2.5, where k, k_{12}, k_{23} are torsional spring constants [8].

TABLE 2.5 *Natural Frequencies of Torsional Systems With Two and Three Discs*

Oscillating system	Natural angular frequencies
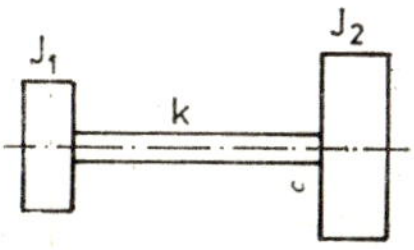 J_1, k, J_2	$p = \sqrt{\dfrac{k(J_1 + J_2)}{J_1 J_2}}$
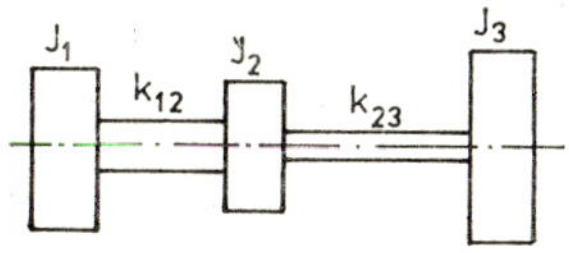 J_1, k_{12}, J_2, k_{23}, J_3	$p_{1,2} = \dfrac{1}{2}\left[A \pm \sqrt{A^2 - \dfrac{4 k_{12} k_{23}}{J_1 J_2 J_3}(J_1 + J_2 + J_3)} \right]$ where $A = \dfrac{k_{12}}{J_1} + \dfrac{k_{23}}{J_3} + \dfrac{k_{12} + k_{23}}{J_2}$

Beams of uniform section, undergoing lateral vibrations, have an infinite number of natural frequencies. Table 2.6 gives the first five natural frequencies and mode shapes for uniform beams having various end conditions.

45

2 Elements of theory

TABLE 2.6 *Natural Frequencies and Mode Shapes for Lateral Vibrations of Uniform Beams* [9]

Boundary	1st mode	2nd mode	3rd mode	4th mode	5th mode
Clamped-free (cantilever)	A=3.52	0.774; A=22.4	0.500 0.868; A=61.7	0.356 0.644 0.906; A=121.0	0.279 0.500 0.723 0.926; A=200.0
Hinged-hinged	A=9.87	0.500; A=39.5	0.333 0.667; A=88.9	0.25 0.50 0.75; A=158	0.20 0.40 0.60 0.80; A=247
Clamped-clamped	A=22.4	0.500; A=61.7	0.359 0.641; A=121	0.278 0.500 0.722; A=200	0.227 0.409 0.591 0.773; A=298
Free-free	0.224 0.776; A=22.4	0.132 0.500 0.868; A=61.7	0.094 0.356 0.644 0.906; A=121	0.073 0.277 0.500 0.723 0.927; A=200	0.060 0.227 0.409 0.591 0.773 0.940; A=298
Clamped-hinged	A=15.4	0.560; A=50.0	0.384 0.692; A=104	0.294 0.529 0.765; A=178	0.238 0.429 0.619 0.810; A=272
Hinged-free	0.736; A=15.4	0.446 0.853; A=50.0	0.308 0.616 0.898; A=104	0.235 0.471 0.707 0.922; A=178	0.190 0.381 0.581 0.763 0.937; A=272

Angular natural frequencies are given by $p = A\sqrt{\dfrac{EI}{\rho Sl^4}}$ [rad/s], where A is a coefficient given in the table; E — Young's modulus, N/m^2; I — moment of inertia of beam cross section, m^4; ρ — mass density, kg/m^3; S — cross section area, m^2; l — beam length, m.

TABLE 2.7 *Natural Frequencies and Nodal Patterns for Square Plates (After D. Young)*

		First mode	Second mode	Third mode	Fourth mode	Fifth mode	Sixth mode
One edge clamped, three edges free	$\omega_n\sqrt{Dg/\gamma h a^4}$	3.494	8.547	21.44	27.46	31.17	
	Nodal lines						
All edges clamped	$\omega_n\sqrt{Dg/\gamma h a^4}$	35.99	73.41	108.27	131.64	132.25	165.15
	Nodal lines						
Two edges clamped, two edges free	$\omega_n\sqrt{Dg/\gamma h a^4}$	6.958	24.08	26.80	48.05	63.14	
	Nodal lines						

$p_n = A\sqrt{\dfrac{Dg}{\gamma h a^4}}$ where: A is the coefficient given in the table;

$D = \dfrac{Eh^3}{12(1-\nu^2)}$ is the plate bending stiffness; g — acceleration of gravity; γ — weight density; h — plate thickness; a — plate length; ν — Poisson's ratio; E — modulus of elasticity.

2.5 Natural frequencies

Natural frequencies of thin flat *square plates* are given in Table 2.7 [9].

Table 2.8 may be used for calculating natural frequencies of *cantilevered rectangular plates*. The same notations are used as

TABLE 2.8 *Natural Frequencies and Nodal Patterns of Cantilevered Rectangular Plates* ($\nu=0.3$) (*After M. V. Barton*)

Mode a/b	1/2	1	2	3
First	3.508	3.494	3.472	3.450
Second	5.372	8.547	14.93	34.73
Third	21.96	21.44	21.61	21.52
Fourth	10.26	27.46	94.49	563.9
Fifth	24.85	31.17	48.71	105.9

for Table 2.7 and *a* is the length of the edge perpendicular to the clamped edge. For *rectangular plates* with various edge conditions, natural frequencies can be determined using Table 2.9.

2 Elements of theory

TABLE 2.9 *Natural Frequencies of Rectangular Plates (After R. F. S. Hearmon)*

Shape		b/a or a/b	1.0	1.5	2.0	2.5	3.0	∞
s s s / s		b/a	1.0	1.5	2.0	2.5	3.0	∞
		A	19.74	14.26	12.34	11.45	10.97	9.87
c s s / s		b/a	1.0	1.5	2.0	2.5	3.0	∞
		A	23.65	18.90	17.33	16.63	16.26	15.43
		a/b	1.0	1.5	2.0	2.5	3.0	∞
		A	23.65	15.57	12.92	11.75	11.14	9.87
c s c / s		b/a	1.0	1.5	2.0	2.5	3.0	∞
		A	28.95	25.05	23.82	23.27	22.99	22.37
		a/b	1.0	1.5	2.0	2.5	3.0	∞
		A	28.95	17.37	13.69	12.13	11.36	9.87
c c c / c		b/a	1.0	1.5	2.0	2.5	3.0	∞
		A	35.98	27.00	24.57	23.77	23.19	22.37

s denotes simply supported edge; c denotes built-in or clamped edge. Natural frequencies are given by equation from Table 2.7.

2.6 Spring constants

In the equation of motion of a single degree of freedom linear system

$$m\ddot{x} + c\dot{x} + kx = F(t),$$

the three basic parameters are the mass m, the viscous damping coefficient c and the spring constant k. In various forms, these parameters are found in all equations of vibratory motions.

For elements with simple geometrical shape, the spring constants can be determined analytically:
— for an elastic wire or a slender rod in tension (or in compression)

$$k = \frac{EA}{l} \quad \left[\frac{N}{m}\right] \tag{2.79}$$

where E is the dynamic modulus of elasticity, A — cross section area, l — length;

2.6 Spring constants

— for a cylindrical close-coiled helical spring, loaded along its axis,

$$k = \frac{Gd^4}{64\,R^3 n} \quad \left[\frac{\text{N}}{\text{m}}\right] \qquad (2.80)$$

where d is the wire diameter, R — mean coil radius, n — number of coils (turns);

— for a uniform cantilever massless beam, with a concentrated mass at the free end,

$$k = \frac{3EI}{l^3} \quad \left[\frac{\text{N}}{\text{m}}\right] \qquad (2.81)$$

where I is the area moment of inertia of beam cross section and l — length.

Resultant spring constants for systems of springs connected in series or/and in parallel are given by formulae similar to those established for electrical condensers.

For a cylindrical rod, the torsional spring constant is

$$k = \frac{GI_p}{l} \quad [\text{N·m}] \qquad (2.82)$$

where G is the shear modulus, I_p — the polar moment of inertia of the cross section, l — rod length.

As for the damping, various types of mathematical models are used in the theory of vibrations [10]:

— *viscous damping*, for which the damping force is proportional to the relative velocity across the damper, so that the equation of motion for forced vibrations of the single degree of freedom system is

$$m\ddot{x} + c\dot{x} + kx = F(t).$$

— *Coulomb damping*, for which the damping force R is of constant magnitude, independent of displacement, and changes the sign when the relative velocity across the damper does it. The equation of motion becomes

$$m\ddot{x} + R\,\text{sgn}(\dot{x}) + kx = F(t).$$

— *hysteretic damping*, for which the damping force is in phase with the relative velocity but is proportional to the relative displacement across the damper. The equation of motion can be written

$$m\ddot{x} + \frac{k}{\omega} g\dot{x} + kx = F(t)$$

and can be used for harmonic excitation only.

2 Elements of theory

References for Chapter 2

1. Bendat J. S. and Piersol A. G., *Engineering Applications of Correlation and Spectral Analysis*, John Wiley and Sons, New York, 1980.
2. Wehrmann W. et al., *Korrelationstechnik*, Kontakt und Studium, Band 14, Lexica-Verlag, Grafenau, 1977.
3. Broch J. T., *Mechanical Vibration and Shock Measurements*, Brüel & Kjaer, 1972.
4. Brigham E. O., *The Fast Fourier Transform*, Prentice Hall Inc., New Jersey, 1974.
5. Thrane N., The Discrete Fourier Transform and FFT Analyzers, *Brüel & Kjaer Technical Review*, 1 (1979).
6. Otnes R. K. and Enochson L., *Applied Time Series Analysis*, John Wiley and Sons, New York, 1978.
7. Blevins R. D., *Formulas for Natural Frequencies and Mode Shapes*, Van Nostrand–Reinhold, New York, 1979.
8. Ker Wilson W., *Practical Solution of Torsional Vibration Problems*, Chapman & Hall Ltd., London, 1956.
9. Harris C. and Crede Ch. (Reds), *Shock and Vibration Handbook*, McGraw-Hill Book Company Inc., New York, 1961.
10. Buzdugan Gh., Fetcu L. and Radeş M., *Vibraţiile sistemelor mecanice*, Editura Academiei, Bucureşti, 1975.
11. Buzdugan Gh., *Rezistenţa materialelor*, Editura tehnică, Bucureşti, 1980.

3

Effects of vibrations

Vibration measurement data are interpreted according to their nocivity. Beyond certain limits, vibrations are harmful to people, can produce damages in buildings or may disturb the normal operation of machinery. Studies in this field refer to persons, buildings and machinery. Persons are the most sensitive to vibrations. Safe limits and vibration severity criteria are presented in the following.

3.1 Effects of vibrations on man

Based on Dieckmann's work [1], a *coefficient of vibration perception* K has been established, used as an indication of the effects of vibrations on man. The vibration perception ranges, according to VDI-2057 Richtlinien, are listed in Table 3.1. The dimensionless coefficient K can be taken from Figure 3.1 — as a function of vibration acceleration and frequency, from Figure 3.2 — as a function of vibration velocity and frequency, and from Figure 3.3 — as a function of vibration displacement and frequency. Both peak values and r.m.s. values are given in the three diagrams.

According to Koch, vibrations can be classified depending on the *level of perception*, measured in Pal, as in Table 3.2.

According to Zeller [2], defining the *vibration intensity* by

$$Z = \frac{a_0^2}{f} = 16\pi^4 x_0^2 f^3 \quad [\text{cm}^2/\text{s}^3] \tag{3.1}$$

where a_0 is the acceleration amplitude, x_0 — displacement amplitude for harmonic motion, f — vibration frequency, the *vibration*

3 Effects of vibrations

TABLE 3.1 *Levels of Vibration Perception by Man* (*After VDI 2057*)

Coefficient of perception K	Range	Vibration category
—0.1—	A	Imperceptible Lower limit of perception
—0.25—	B	Just perceptible
—0.63—	C	Perceptible
—1.4—	D	Clearly perceptible
—4.0—	E	Annoying
—10.0—	F	
—25—	G	Unpleasant
—63—	H	Painful
	I	

TABLE 3.2 *Vibration Perception Levels* (*After Koch*)

Vibration level, Pal	Effect on man Vibration source
0...10	Threshold of perception, function of body position
10...20	General perception
20...30	Vibrations from traffic, inadmissible in buildings
30...40	Vibrations in smoothly running vehicles
40...50	Vibrations in vehicles and lifts
50...60	Severe vibrations in vehicles, allowable for a short time
60...80	Harmful vibrations, producing seasickness and painful sensations

3.1 Effects on man

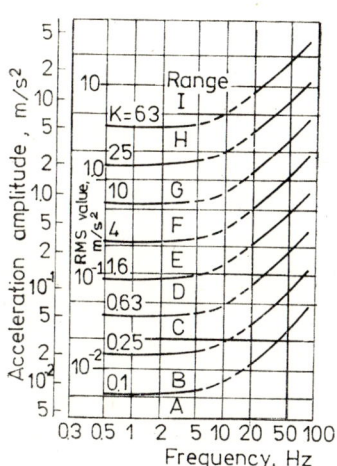

Figure 3.1

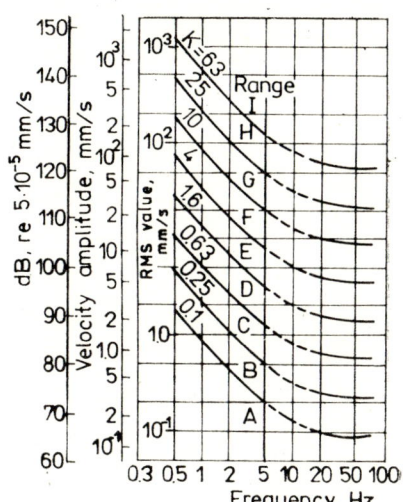

Figure 3.2

level is given by

$$P = 10 \log \frac{Z}{Z_1} \qquad (3.2)$$

where $Z_1 = 0.5$ cm²/s³. Substituting Z_1 in equation (3.2), one obtains

$$P = 10 \log 2Z. \quad [\text{Pal}] \qquad (3.3)$$

The vibration level can be expressed as a function of velocity or displacement

$$\begin{aligned} P &= 20 \log 22.4 \, v_0, \\ P &= 20 \log 140 \, x_0 f. \end{aligned} \qquad (3.4)$$

The effect of vibrations on man depends on the exposure time. Vertical vibration exposure criteria curves are given in Figure 3.4 (after ISO 2631 — 1978), defining equal fatigue-decreased proficiency boundaries for sitting or standing persons. From Figure 3.4 b it can be seen that tolerable levels increase with decreasing exposure time.

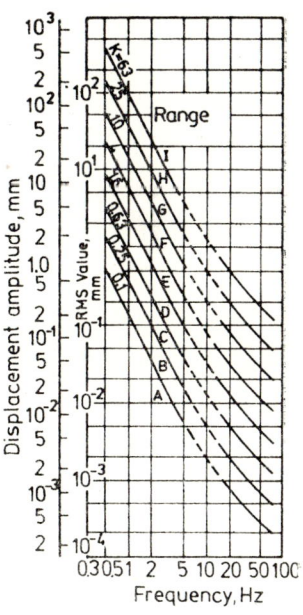

Figure 3.3

3 Effects of vibrations

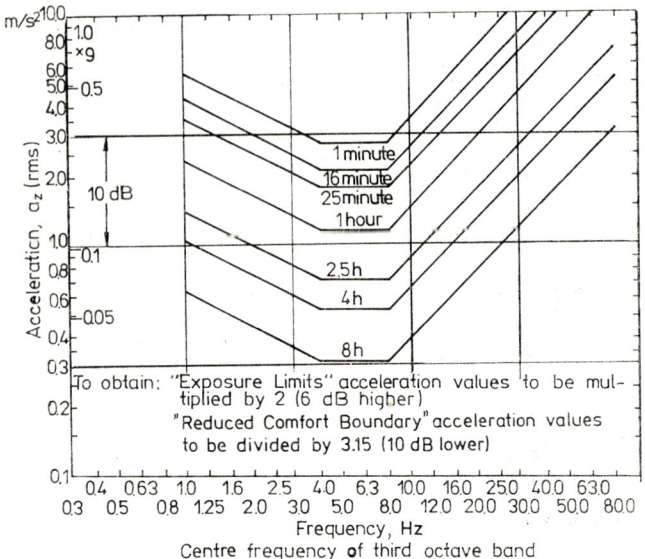

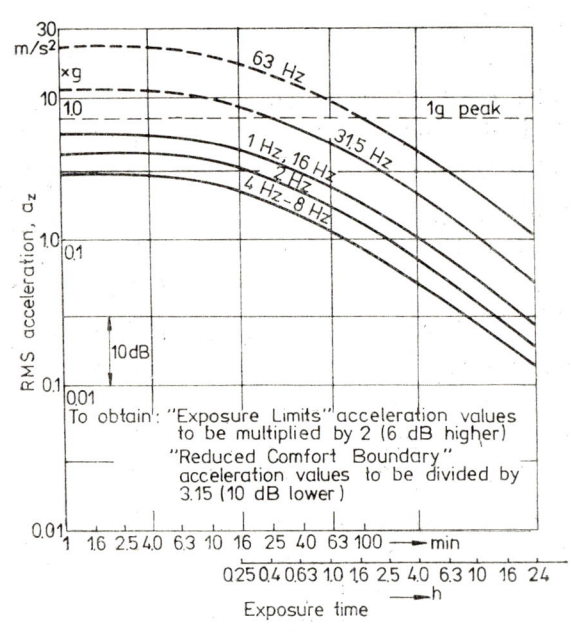

Figure 3.4

3.2 Effects of vibrations on buildings

Table 3.3 [3] lists the probabilities of damaging buildings as a function of vibration velocity amplitude.

Diagrams from Figures 3.5 and 3.6 illustrate the effect on buildings of vibrations having different velocity and displacement amplitudes, and frequencies up to 100 Hz.

TABLE 3.3 *Damage to Buildings as a Function of Vibration Amplitude*

Vibration velocity cm/s	Building damage
0.5	scarcely probable
1.0	improbable
5.0	probable
10.0	very probable

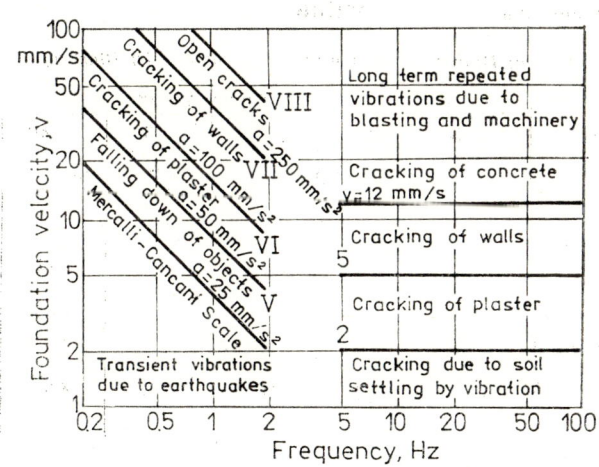

Figure 3.5

Often, the problem of vibrations produced by blasting in neighbouring structures is of considerable importance. Crandell [4] established as criterion the "energy ratio" and the allowable limit $a^2/f^2 = 0.28$ (m/s)^2, where a is the acceleration, in m/s^2,

3 Effects of vibrations

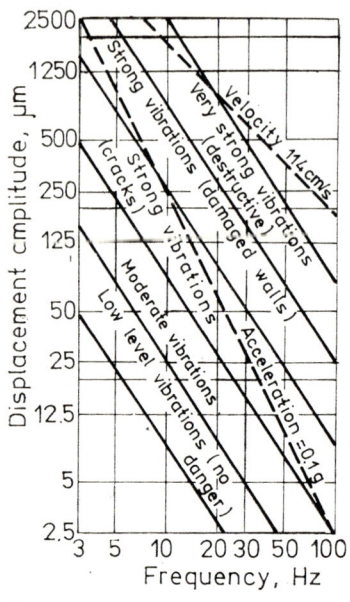

Figure 3.6

and f — frequency, in Hz. According to Edwards and Northwood, the safe limit corresponds to a velocity of 11.5 cm/s. Allowable displacement amplitudes obtained using the two criteria are given in Table 3.4 for several frequencies.

The effect of vibrations on buildings can be also assessed calculating their *intensity* from measurement data and using Table 3.5.

Vibration intensity, expressed in Vibrar, is defined by

$$S = 10 \log \frac{Z}{Z_s} \quad [\text{Vibrar}] \quad (3.5)$$

where Z is given by equation (3.1) and $Z_s = 0.1 \text{ cm}^2/\text{s}^3$ is a reference value.

TABLE 3.4 *Limits of Displacement Amplitudes Produced by Blasting*

Criterion	Frequency, Hz					
	5	10	20	30	40	50
	Displacement amplitude, μm					
Crandell $\dfrac{a^2}{f^2} = 0.28 \left(\dfrac{m}{s}\right)^2$	2670	1350	660	460	330	280
Edwards-Northwood $v = 11.5 \dfrac{\text{cm}}{\text{s}}$	3650	1820	910	610	460	350

3.3 Effects of vibrations on machinery

In Figure 3.7, the effects of machinery vibrations are classified as a function of frequency and displacement amplitude.

Figure 3.8 shows the Rathbone criteria [5] limited to turbines on individual foundations, running at speeds less than 5000 rpm

3.3 *Effects on machinery*

TABLE 3.5 *Classification of Vibrations According to Their Effect on Buildings*

Vibration intensity, Vibrar	Vibration class	Effect on buildings
10...20	weak	no danger
20...30	moderate	no danger
30...40	strong	small damages, cracking of walls
40...50	rough	cracking of resistance walls
50...60	very rough	building demolition

and with small ratios of shaft vibration to bearing housing or pedestal. Similar vibration tolerances have been published by Schenck-Darmstadt (Fig. 3.9) for electric motors.

The V.D.I. —2056 Richtlinien [7] are some of the most exhaustive recommendations in present use, referring to active vibration sources (various machines) and to instruments to be isolated passively.

The measured quantities are:

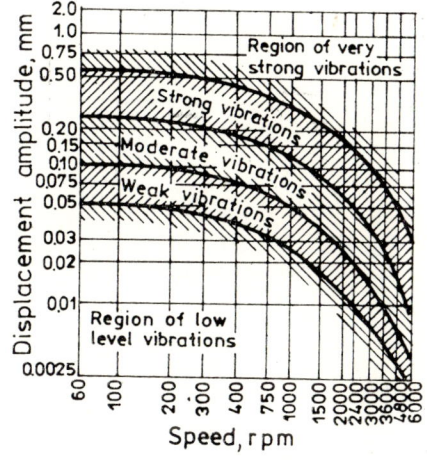

Figure 3.7

— displacement, velocity, acceleration, vibration severity, dynamic force;
— amplitudes, peak-to-peak displacements, rms values;
— absolute and relative values.

Guidelines for measurements include:

— machine operating conditions: on the test stand; idling; full load;
— measurement scope: production reception; fault detection; diagnosis;
— vibration generation: free vibration; produced by impact; shaker excitation; generated during machine operation.

The vibratory state of a machine is characterized by the *vibration severity* defined as the maximum root-mean-square value of

3 Effects of vibrations

the vibration velocity measured at significant points of a machine, such as a bearing, a mounting point, etc. This is calculated using equation (2.5) written in terms of velocity.

Starting from the fact that increasing the velocity 1.6 times leads to a different degree of perception, vibration severity levels

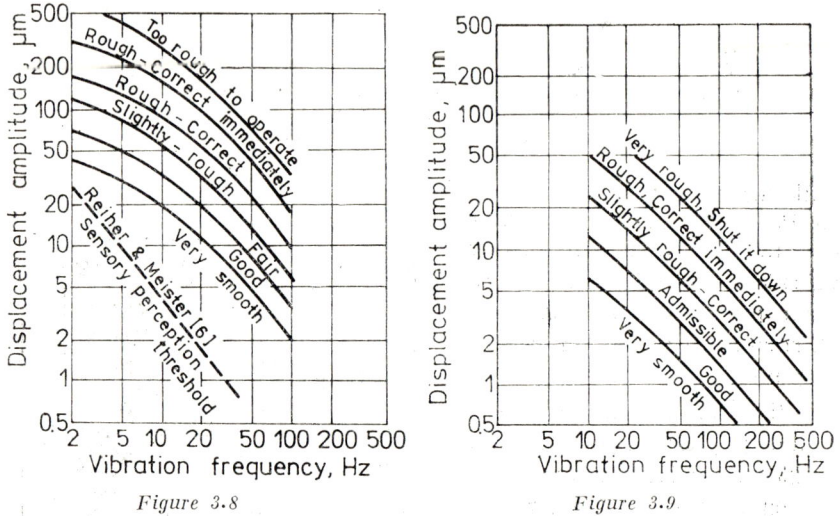

Figure 3.8   Figure 3.9

are established so that succeeding ranges have a 4 dB spacing (1.6 :1 ratio).

Constant rms velocity lines are plotted in Figure 3.10, defining severity levels from 0.071 to 71 mm/s. Based on experience, vibrations with the same r.m.s. velocity anywhere in the frequency range from 10 to 1000 Hz are considered to be of equal severity. As studies have been limited down to 10 Hz, below this frequency, lines of constant rms velocity are replaced by (horizontal) lines of equal displacement amplitude.

Specifications are given for the following classes of machines :
— group K (Fig. 3.11) : individual parts of engines and machines, integrally connected with the complete machine in its normal operating condition (production electrical motors of up to 15 kW);
— group M (Fig. 3.12) : medium sized machines (electrical motors with 15 to 75 kW output) without special foundations, rigidly mounted engines or machines (up to 300 kW) on special foundations ;
— group G (Fig. 3.13) : machines for power generation and process machinery on heavy rigid foundations whose natural frequency exceeds machine speed ;

3.3 Effects on machinery

— group T (Fig. 3.14) : machines for power generation and process machinery with unbalanced rotating masses, operating at speeds above foundation natural frequency (e.g. turbogenerator sets, especially those with light weight substructures).

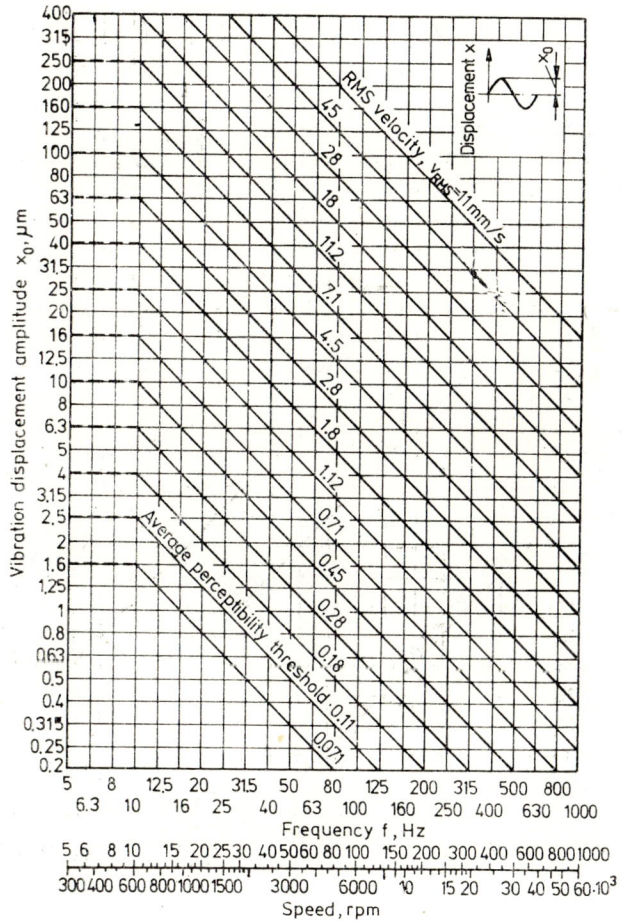

Figure 3.10

Figure 3.15 summarizes the vibration severity criteria presented in Figures 3.11—3.14. It can be seen that, regardless of the machine group, there is a 8 dB separation between ranges of different vibration severity and a 20 dB change of the vibration level determines a jump from "good" to "not permissible".

59

3 Effects of vibrations

This is used for setting warning and shut-down limits when designing surveillance systems.

V.D.I.-2056 Richtlinien give specifications for other two classes of machines:

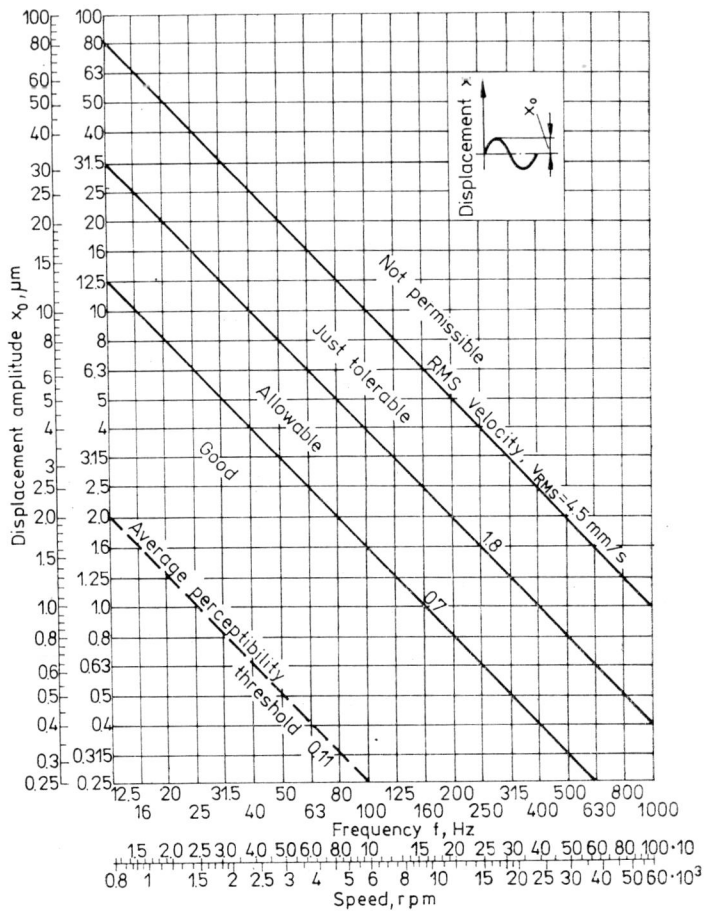

Figure 3.11

— group D, containing reciprocating machines (with unbalanceable masses) on rigid foundations, whose vibration severity level can reach 20—30 mm/s;

— group S, containing machines with unbalanceable rotating masses, supported by a resilient system (soft-mounted), with

3.3 Effects on machinery

allowable levels of 50 mm/s and short term levels up to 500 mm/s while passing through resonance. This class also contains machines with rotating slack-coupled masses such as beater shafts in grinding mills, centrifugal machines, vibrating screens, dynamic fatigue-

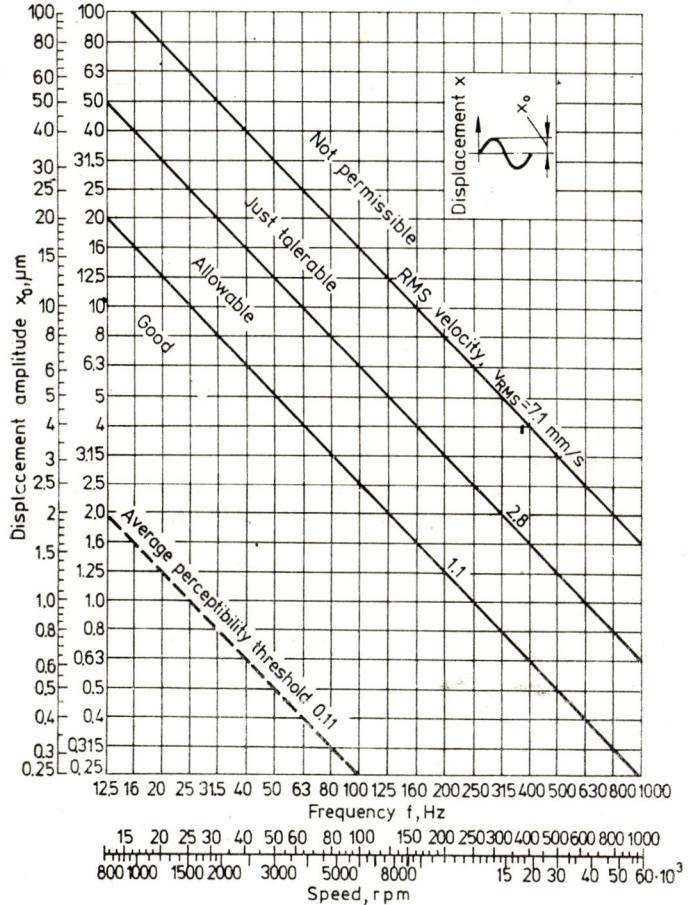

Figure 3.12

testing machines and vibration exciters used in processing plants.

Similar recommendations are contained by the international standard ISO-2372 [8], the British Standard 4675 [9] and the French Norm AFNOR E 90−300 [10].

It must be remembered that the above specifications refer to absolute measurements on the bearing cap or pedestal.

61

3 *Effects of vibrations*

Figure 3.16 shows the general severity chart for bearing cap measurement (filtered readings) developed first by Yates and reworked by IRD Mechanalysis [11]. Blake's severity chart [12] is also shown for comparison in Figure 3.17.

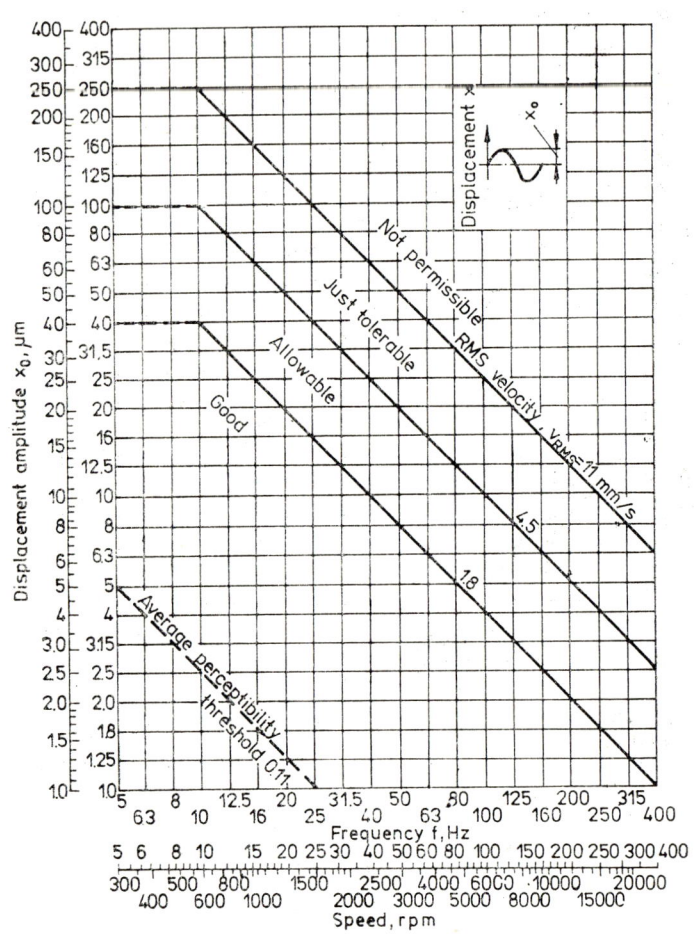

Figure 3.13

All these severity charts must be used only as guides in judging vibration and as a warning of impending trouble.

Blake also introduced the concept of *service factor* (see Table 3.6) taking into account the importance of the particular machine

3.3 *Effects on machinery*

to the prime function of the plant. The vibration to be evaluated is obtained by multiplying the measured single amplitude vibration by the service factor.

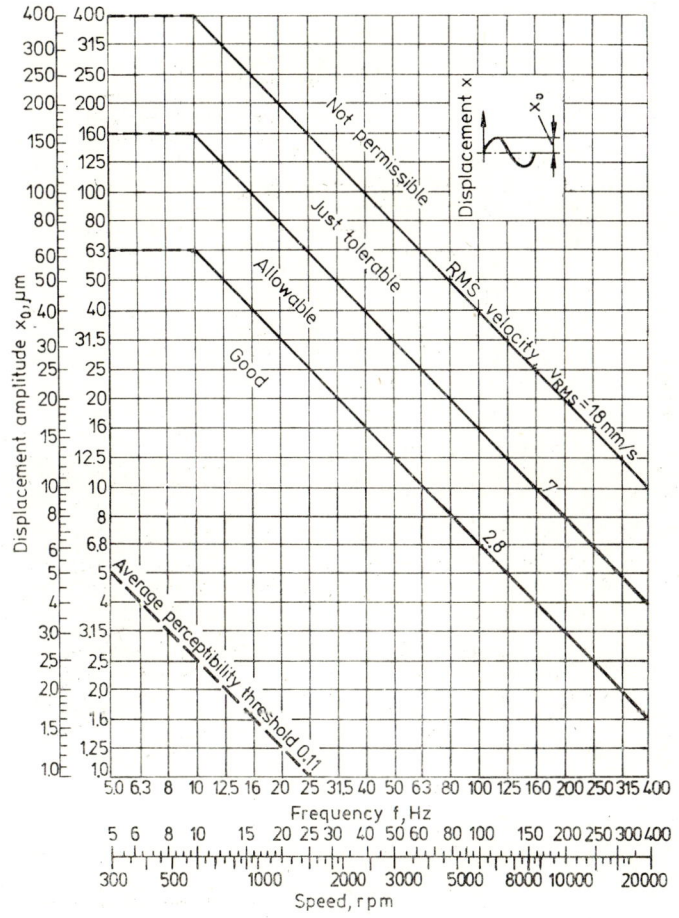

Figure 3.14

Casing or *bearing* (absolute) *vibration measurements* are recommended when the support structure is flexible, bearings are stiff, and the ratio of casing weight to rotor weight is low.

Shaft-motion measurements (relative to some part of the machine structure, usually the bearing) are recommended when a large ratio of casing weight to rotor weight and stiff bearing

3 Effects of vibrations

supports cause most of the energy generated by the rotor to be expended as relative motion between the shaft and the bearing.

In V.D.I.-2059 Richtlinien [13] limit values are given for shaft relative displacement during long-term machinery surveillance.

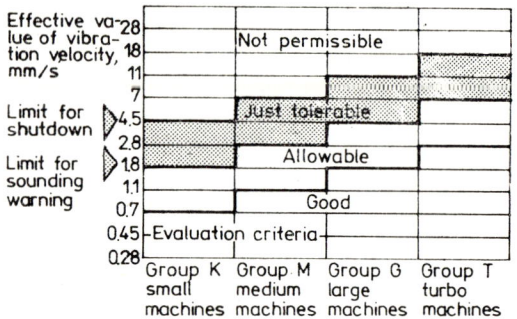

Figure 3.15

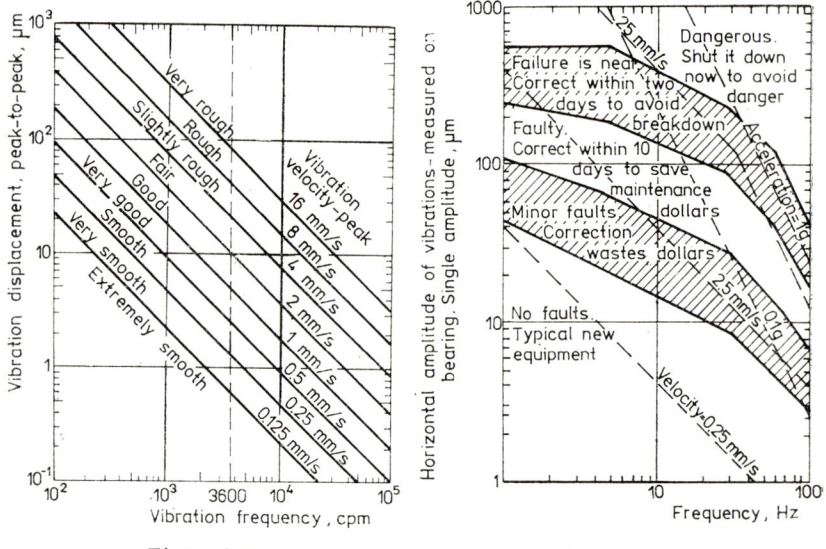

Figure 3.16   Figure 3.17

For turbosets of 60—350 MW with steam turbines, operating at speeds of 3000 or 3600 rpm, with a minimum bearing diameter of 180 mm and rotor vibration main components between 20 and 60 Hz, the following limits are recommended for the shaft maximum displacement: 120 μm — during the long-term normal

3.3 Effects on machinery

TABLE 3.6 *Service Factors (After Blake [12])*

Machine	Service factor
Single-stage centrifugal pump, electric motor, fan	1
Typical chemical processing equipment, noncritical	1
Turbine, generator, centrifugal compressor	1.6
Centrifuge, stiff-shaft (horizontal displacement on basket housing); multistage centrifugal pump	2
Miscellaneous equipment, characteristics unknown	2
Centrifuge, shaft-suspended, on shaft near basket	0.5
Centrifuge, link-suspended, slung	0.3

Machine tools are excluded. Values are for bolted-down equipment; when not bolted, multiply the service factor by 0.4 and use the product as a service factor. Vibration is measured on bearing housing, except as stated.

operating regime; 180 μm — several hours up to several days, e.g. during the analysis of some faults; 360 μm — very short time, e.g. when passing through resonance. The alarm level is set at maximum 30 μm above the normal operating level. Figure 3.18 shows the specified limits for industrial turbine shafts and figure 3.19 — for gas turbines.

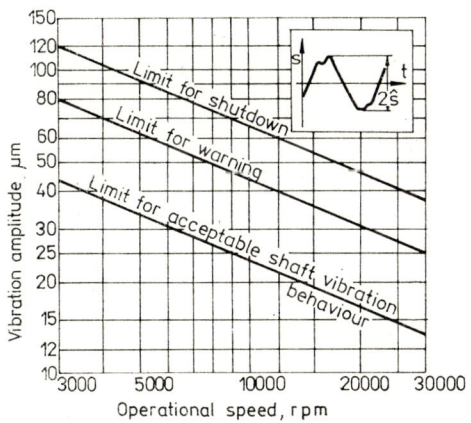

Figure 3.18

For centrifugal compressors, API Std 617 recommends shaft vibrations less than either 50 μm or the value given by *peak-to-peak*

$$displacement = 25 \sqrt{\frac{12{,}000}{rpm}} \; [\mu m].$$

65

3 Effects of vibrations

In the technical literature, allowable limits are specified for machinery vibrations in tabular form:

a) After Savinov [14], the upper limits of machine foundation vibrations are given in Table 3.7.

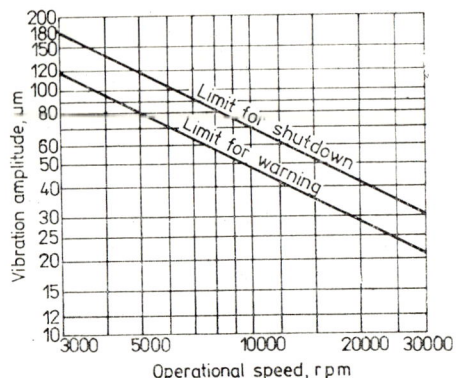

Figure 3.19

TABLE 3.7 *Acceptable Vibration Limits For Rotating Machines*

Direction of vibrations	Speed, rpm						
	<500	500	750	1000	1500	3000	5000	10000
	Vibration Displacement Amplitude, mm						
Vertical	0.15	0.12	0.09	0.075	0.06	0.03	0.015	0.005
Horizontal	0.20	0.16	0.13	0.11	0.09	0.05	0.02	0.0075

b) Norms used by Teploelektroproyekt (USSR) recommend the acceptable vibration levels for turboset foundations:

for $n \geq 3000$ rpm   0.02...0.03 mm;
for $n \geq 1500$ rpm   0.04...0.06 mm;
for $n \geq 750$ rpm   0.08...0.12 mm.

c) At *forge hammers*, acceptable levels for *foundation* vibrations are 1.0...1.2 mm. For the anvil, acceptable levels are dependent on the hammer weight:

for 1 ton hammer   1 mm displacement amplitude;
for 2 tons hammer   2 mm displacement amplitude;
for hammers above 3 tons   3...4 mm displacement amplitude.

Liubimov recommends allowable displacement amplitudes up to 2.5 mm for the foundation block. German Norms DIN 4025

3.3 Effects on machinery

recommend a maximum velocity level of 3.8 mm/s for the vertical vibrations of the foundation.

d) For *machine tools*, several acceptable levels for vertical displacements are given in Table 3.8.

TABLE 3.8 *Acceptable Vibration Levels For Machine Tool Vibrations*

Machine type	Acceptable displacement amplitude, mm
Planing machines	0.35
Lathes, milling machines, boring machines	0.03
Grinders and machines for ball-bearings	0.03
Internal grinders	0.05
Presses	0.09

e) For measurements-standards laboratories [15] the following limit values are prescribed for the floor acceleration: $0.001\ g$ — in laboratories for reference standards (dimensional, electrical, physical); $0.002\ g$ — in laboratories for working standards (dimensional); and $0.003\ g$ — in laboratories for working standards (electrical, physical).

f) For *rooms* with different destinations, the acceptable levels for vibration displacements are given in Table 3.9.

g) For *electrical machines*, limits for vibration severity are given in Table 3.10 according to DIN 45 655 and ISO Standard 2373.

TABLE 3.9 *Acceptable Vibration Levels in Different Rooms*

Room destination	Acceptable level, μm
Laboratories with precision instruments	10...30
Workshops with precision machines and testing stands	20...40
Electrical powerplants with steam turbines	20
Foundries	30...50
Offices and flats	50...70

3 Effects of vibrations

TABLE 3.10 Limits for Vibration Severity for Electrical Machines

Quality grade	Speed range, rpm	Shaft height, mm		
		80...132	160...225	250...315
		Limits for rms velocity, mm/s		
N (normal)	600—1800	1.8	2.8	4.5
	1800—3600			
R (reduced)	600—1800	0.71	1.12	1.8
	1800—3600	1.12	1.8	2.8
S (special)	600—1800	0.45	0.71	1.12
	1800—3600	0.71	1.12	1.8

For large rotating machines with a speed range from 10 to 200 rev/s specifications are given in ISO Standard 3945. For induction motors, allowable vibration levels are specified in NEMA Std MG1—20.52 and API Std 541 [18].

h) For rotating machines, Table 3.11 presents the limits recommended by Steve Maten [16] in 1970 and used in the petroleum industry. Jackson [17] adopts Maten's values for pumps, turbines, centrifugal and rotary compressors, motors, expanders, blowers, fans, centrifuges and dryers. For gearing and reciprocating compressors he recommends values 2.5 mm/s higher; for hammer or ball mills — values 2.5 to 5 mm/s higher.

For medium and high speed centrifugal compressors, turbines and expanders, Jackson recommends proximity shaft vibration measurement, setting the alarm at 55 μm (peak-to-peak) and shut-downs at 105 μm (plus 5 μm allowed for total runout tolerance). In the American Petroleum Institute Standards API-612, API-616, API-619 specifications are given for limit values of vibrations occurring during machinery reception tests. Other permissible values for compressors are given in [12].

i) For *gear units*, specifications for acceptable levels are given in the American Gear Manufacturers Association Standard AGMA 426.01. Values referring to lateral vibrations of high speed helical and herringbone gear units are given in Figure 3.20.

Other vibration standards are presented in References [18] and [19].

3.3 Effects on machinery

TABLE 3.11 Velocity Standards (After S. Maten [16])

Overall velocity, mm/s	Classification	Severity rating	Approximate interpretation
above 12.5	AA	*Extremely rough* danger, consider shutdown	Oil film destroyed. Metal-to-metal contact, seizure, breakage
7.5—12.5	A	*Very rough,* correct soon major damage may occur	Oil film breaks if viscosity or temperature are not controlled, rapid wear
5.0—7.5	B	*Rough,* correct to save wear	Gradual wear over period of time expected
2.5—5.0	C	*Fair,* minor fault, correction uneconomical	Little or no wear expected
up to 2.5	D	*Smooth,* well balanced, well aligned equipment	Normal trouble-free installation Components will last several years

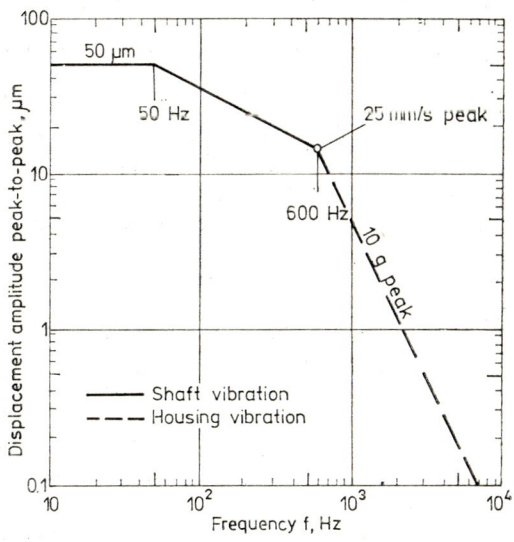

Figure 3.20

3 Effects of vibrations

References for Chapter 3

1. Dieckmann, D., Über die Einwirkung mechanischer Schwingungen bis 100 Hz auf den Menschen, *VDI-Berichte*, **24**, 117–120 (1958).
2. Zeller, W., Vorschlag für ein Mass der Schwingungsstärke, *Zeitschrift V.D.I.*, **77**, *12*, p. 323 (1933).
3. Bobbert, G., Dieckmann, D., Federn, K., Lübcke, E., Beurteilungsmaßstäbe für mechanische Schwingungen, *Forschungsberichte des Landes Nordrhein-Westfalen*, Westdeutscher Verlag, Köln und Opladen, 1967.
4. Crandell, F. J., Ground vibration due to blasting and its effect on structures, *J. Boston Society of Civil Engineers* (April 1949).
5. Rathbone, T. C., Vibration tolerance, *Power Plant Engineering*, **43**, 721 (Nov. 1939).
6. Reiher, H. and Meister, F. J., Die Empfindlichkeit der Menschen gegen Erschütterungen, *Forsch. auf dem Gebiete des Ingenieurwesens*, **2**, *11*, 381–386 (1931).
7. *V.D.I.-2056 Richtlinien*, "Beurteilungsmaßstäbe für mechanische Schwingungen von Maschinen", Okt. 1964.
8. *ISO-Standard 2372*, "Mechanical Vibration of Machines With Operating Speeds From 10 to 200 rev/s, Basis for Specifying Evaluation Standards", First Edition, 1974.
9. *British Standard 4675 : 1971* "A Basis for Comparative Evaluation of Vibration in Machinery".
10. *AFNOR E 90–300* : "Vibrations mécaniques des machines ayant une vitesse de fonctionnement comprise entre 10 et 200 tours par seconde".
11. *IRD Training Manual*, General Machinery Vibration Severity Chart, Form No. 305 D, IRD Mechanalysis, Columbus, Ohio.
12. Blake, M. P., *Standards and Tables : Vibration and Acoustic Measurement Handbook*, Spartan Books, New York, 1972.
13. *V.D.I.-2059 Richtlinien* (Entwurf 01.79, Blatt 2) "Wellenschwingungen von Dampfturbosätzen fur Kraftwerke"; (Entwurf 02.76, Blatt 3) "Wellenschwingungen von Industrieturbosätzen"; (Entwurf 02.76, Blatt 4) "Wellenschwingungen von Gasturbinensätzen".
14. Savinov, O. A., *Fundamenty pod mashin*, GILSA, Leningrad, 1955 (In Russian).
15. * * * Recommended Environments for Standards Laboratories, *I.S.A. Journal*, **8**, *2*, 58–62 (1961).
16. Maten, S., Program machine maintenance by measuring vibration velocity, *Hydrocarbon Processing*, **49**, *9*, 291–296 (Sept. 1970).
17. Jackson, Ch., A practical vibration primer (Part 8), *Hydrocarbon Processing*, **57**, *4*, 209–215 (April 1978).
18. Eshleman, R. L., Vibration Standards, Ch. 19 of *"Shock and Vibration Handbook"*, 2nd Ed., Harris C.M. & Crede C.E.(Eds), McGraw-Hill Book Comp., N.Y., 1976.
19. Eshleman, R. L., Vibration standards for power transmission equipment, *Power Transmission Design*, **22**, *4*, 26–29 (1980); **22**, *5*, 45–46 (1980).

4

Transducers and pickups for vibration measurement

4.1 Transducers for electrical measurement of vibrations

The transducer is a device which converts changes of mechanical quantities into changes of other physical quantities, most commonly to an electrical signal proportional to a parameter of the experienced motion. Mounted in a more complicated device, the transducer becomes a pickup (velocity pickup, displacement pickup, etc.).

Transducers can be classified into two large categories (Fig. 4.1):

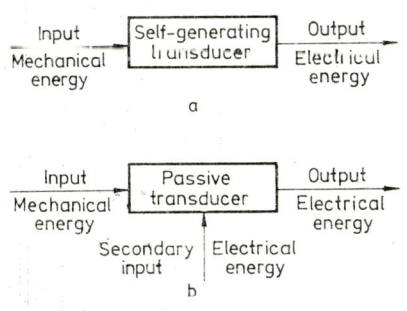

Figure 4.1

— *self-generating* (active) transducers, which convert the input mechanical energy into electrical energy (Fig. 4.1 a);

— *passive* transducers, which are supplied (by a secondary input) with electrical energy, so that changes of the mechanical

4 Transducers and pickups

energy of the system on which measurements are carried out produce changes of the electrical energy (Fig. 4.1 b).

When the transducer converts the input mechanical energy directly into electrical energy it is called *with direct conversion*; when another quantity — acoustical, optical etc. — intervenes in the conversion process, the transducer is *with indirect conversion*. A more detailed classification of transducers is given in Table 4.1.

TABLE 4.1

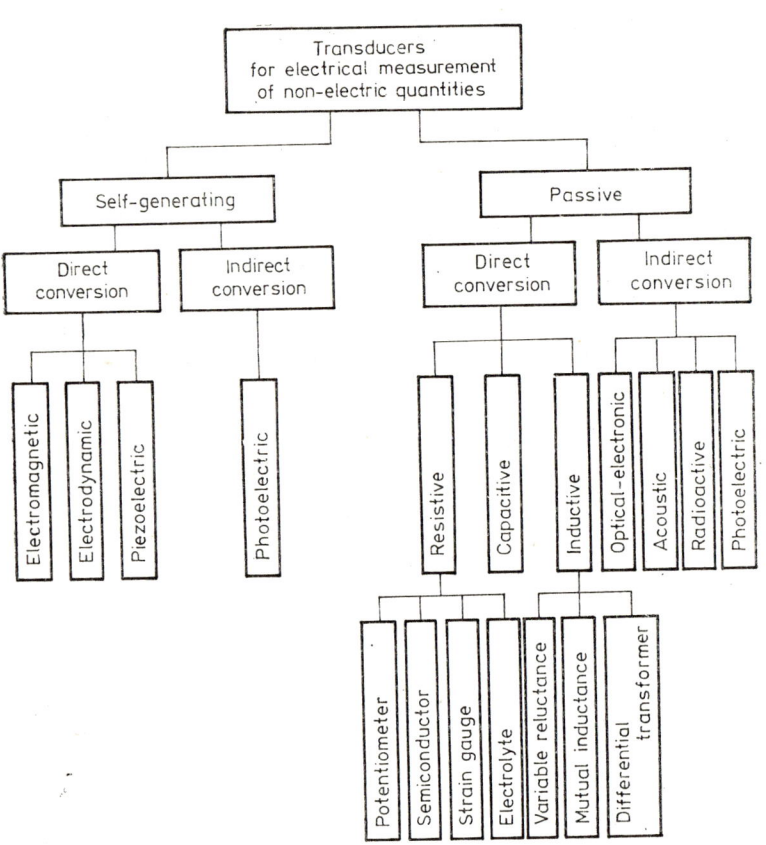

Not all transducers mentioned in the table are suitable for vibration measurement, due to either non-linearity or the slow response to the measurand. Therefore, in the following, the operating principle and the characteristics will be presented only for the transducers most often used in vibration measurement [1].

4.1 Vibration transducers

4.1.1 Passive transducers

From this category, only resistive, capacitive, inductive and eddy-current transducers are described in the following.

4.1.1.1. *Variable resistance transducers.* With resistance transducers, the mechanical motion produces a change of the electrical resistance which, in turn, produces a change of the output current or voltage. The resistance, which may consist of a rheostat, a semiconductor, a strain gauge or an electrolyte (see Table 4.1) is included in a circuit energized by an external supply.

a) *Potentiometer transducers.* Relatively large displacements can be measured using a potentiometer transducer in which a slider moves over a resistor and taps a voltage proportional to the displacement. In the translatory devices from Figures 4.2, the resistance R_x changes proportionally with the measured displacement x, R_i is the internal resistance of the "meter", E — energizing d.c. voltage, I — current, U_1 — output voltage. Let $U = E - Ir$, where r is the source internal resistance.

Analysis of the circuit from Figure 4.2a yields

$$I = \frac{U}{R_x + R_i}. \qquad (4.1)$$

The current has a hyperbolic variation with R_x and the meter scale is divided accordingly. For the circuit from Figure 4.2 b,

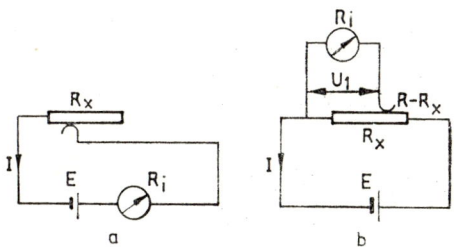

Figure 4.2

the voltage at the meter input is

$$U_1 = U \frac{R_x}{\frac{R_x(R - R_x)}{R_i} + R}, \qquad (4.2)$$

4 Transducers and pickups

where R is the total resistance of the rheostat. When $R \ll R_i$, and also $R_x \ll R_i$, equation (4.2) becomes

$$U_1 = U \frac{R_x}{R}, \qquad (4.3)$$

so that the voltage U_1 is proportional to R_x (and to x) and a linear scale may be used. It can be seen that in order to achieve linearity, the potentiometer resistance should be relatively small compared to the meter resistance.

Potentiometer transducers are currently used for measuring low frequency and large amplitude (of the order of centimeters) vibrations.

b) *Resistance strain gauges*. Strain gauges consist of a conducting element whose resistance changes when it is strained. They usually consist of a grid of fine wire (or suitably etched foil) bonded onto an insulated backing sheet which is then cemented to the surface where the strain is to be measured. The most common are wire gauges (Fig. 4.3) which consist of : 1 — sensitive wire, 2 — electrical leads, 3 — backing material. When properly cemented, the gauges undergo the same strain as the surface to which they are fastened.

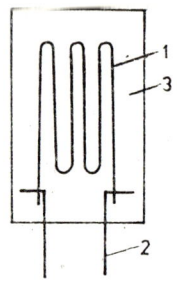

Figure 4.3

Let the wire resistance be

$$R = \rho \frac{l}{A}, \qquad (4.4)$$

where ρ is the material resistivity, l — sensitive wire length, A — wire cross sectional area. If the wire is stretched or compressed, a resistance change ΔR will occur. The fractional change in resistance is given by

$$\frac{\Delta R}{R} = \left(1 + 2\nu + \frac{\Delta \rho}{\rho} \frac{l}{\Delta l}\right) \frac{\Delta l}{l}, \qquad (4.5)$$

where $\Delta \rho$ is the change of resistivity (due to piezoresistance effect) and ν is Poisson's ratio. Bonded metallic strain gauges are based on materials with $\Delta \rho \cong 0$. (e.g. : constantan, Advance, Isoelastic). Denoting $\varepsilon = \frac{\Delta l}{l}$, equation (4.5) becomes

$$\frac{\Delta R}{R} = (1 + 2\nu)\varepsilon. \qquad (4.6)$$

4.1 Vibration transducers

The ratio

$$k = \frac{\Delta R/R}{\varepsilon} = 1 + 2\nu \qquad (4.7)$$

is called the *gauge factor*. Given the value of k (experimentally determined by the manufacturer on similar gauges) and measuring

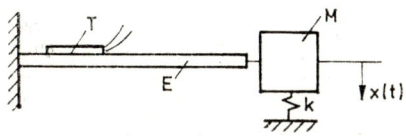

Figure 4.4

ΔR, the strain ε can be determined. These days, almost all strain gauges are foil, usually epoxy backed.

In semiconductor (piezoresistive) gauges, most of the resistance change comes from piezoresistance effects, resulting in gauge factors of about 150. Typical metallic strain gauges have factors from 1.9 to 2.6 and nominal resistances from 100 to 600 ohms. At normal temperatures, wire gauges usually dissipate a power of about 1 Watt/cm^2.

Bonding the gauge T to the flexible element E of a mass-spring system, the simplest vibration pickup is obtained (Fig. 4.4). The deflection of the cantilever beam, which equals the displacement $x(t)$ to be measured, is proportional to the strain at every point. This is proportional to the resistance increment ΔR, which can be measured using a potentiometer circuit or a Wheatstone bridge circuit.

The strain gauge behaves linearly up to strains corresponding to the elastic limit of the strain wires (about 0.01 mm/mm). Details on the various configurations of strain gauges together with their characteristics, bonding techniques and determination of gauge factor can be found in specialized monographs [2].

Its qualities make the strain gauge to be successfully used as transducing element in vibration pickups. In this case, steps should be taken for protection against adverse agents which might modify its behaviour (humidity, corrosion, temperature etc.).

4.1.1.2 *Capacitive transducers*. The capacitance of a parallel-plate capacitor (Fig. 4.5) is

$$C = \varepsilon_0 \varepsilon_r \frac{A}{\delta} \quad [\text{Farads}] \qquad (4.8)$$

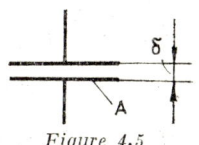

Figure 4.5

4 Transducers and pickups

where $\varepsilon_0 = \dfrac{1}{36\pi} \cdot 10^{-9}$ F/m is the absolute permittivity of air, A is the relative permittivity of the medium (see Table 4.2), ε_r — plate area, m², δ — spacing of the plates, m.

TABLE 4.2 *Relative Permittivity Constants*

Medium	Vacuum Air	Paper	Oil	Plastics	Porcelain Quartz	Glass	Water
ε_r	1	1.8—2.6	2.2—2.5	4—6	4.5—5	5—10	80

In a capacitive transducer, mechanically induced changes in the quantities from equation (4.8) determine changes in the electrical capacitance. Usually, the spacing δ between the plates changes, and the relationship between C and δ is hyperbolic. The capacitance change can be converted to an electrical signal by means of either a d.c. polarizing circuit or a feedback circuit. The latter is schematically shown in Figure 4.6. A linear relationship is looked for between the output voltage U_0 and the plate separation x.

The capacitive transducer C_2 is introduced in the feedback loop of an operational amplifier ($A \gg 1$). The carrier voltage U_i is applied at the amplifier input through the capacitor C_1.

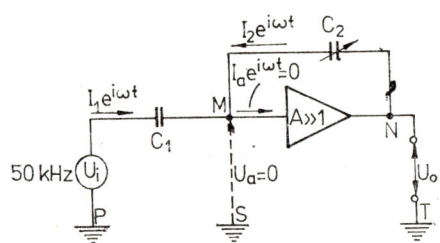

Figure 4.6

The amplifier is designed so that a "virtual earth" should be achieved at point M. For the PMS portion of the circuit, the relationship $U_i = X_1 I_1$ can be established, where the reactance of the capacitor C_1 is $X_1 = \dfrac{1}{i\omega C_1}$. For the TNMS portion, equation $U_0 = X_2 I_2$ can be established, where the reactance of the capacitor

4.1 Vibration transducers

C_2 is $X_2 = \dfrac{1}{i\omega C_2}$. At point M, $I_1 + I_2 = 0$. Taking into account that C_2 is inversely proportional to x,

$$U_0 = -\frac{X_2}{X_1} U_i = -\frac{C_1}{C_2} U_i = -\frac{k}{C_2} = -Kx, \qquad (4.9)$$

where $k = C_1 U_i$. For vibratory displacements, the signal U_0 is an amplitude-modulated sine wave, whose variation about the mean value is proportional to the vibration amplitude. After demodulation, the vibration waveform can be obtained.

Usually, the transducer is a noncontacting probe, requiring a metallic or metallized surface structure.

Capacitive transducers are also built whose operation is based on the variation of the area A the condenser plates are facing. In this case, the

Figure 4.7

plates are either cylindrical surfaces moving relatively to each other in axial direction or toothed surfaces (Fig. 4.7) having a rotational relative movement.

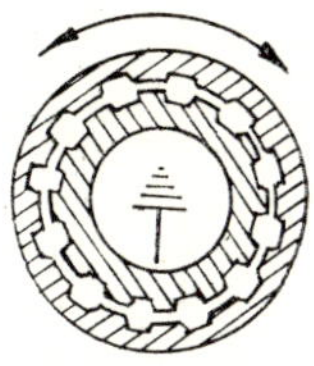

4.1.1.3. *Inductive transducers.* In these devices, the observed motion produces an inductance change of a circuit supplied by an a.c. source, resulting in a corresponding change in the output voltage. There are several types of passive inductive transducers, according to the moving component and the principle of operation (see Table 4.1).

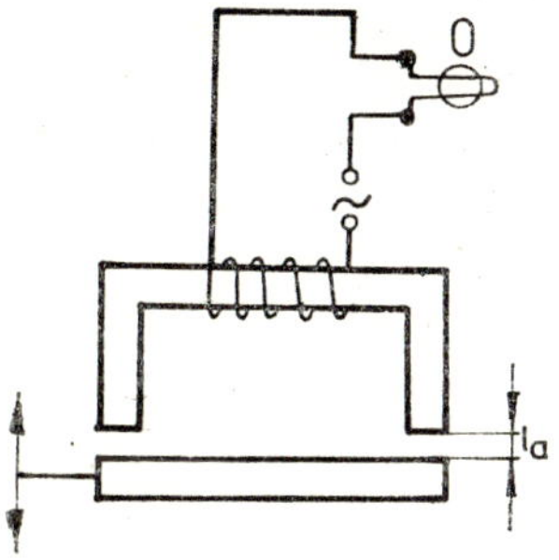

A first category includes variable-reluctance transducers.

Transducers with *variable air gap* correspond to the scheme from Figure 4.8. The inductance of the coil L is

Figure 4.8

$$L = 1.25 \cdot 10^{-8} \frac{N^2}{\dfrac{l_f}{\mu S_f} + \dfrac{l_a}{S_a}}, \quad \text{[Henrys]} \qquad (4.10)$$

where N is the number of coil turns; l_f — length of magnetic flux path in the magnetic core, cm; l_a — air gap thickness, cm; μ — the permeability of the magnetic core, at maximum flux density; S_f — the area of the magnetic core perpendicular to the direction of the flux flow, cm^2; S_a — air gap area, cm^2.

4 Transducers and pickups

Usually the core reluctance is negligible compared to the air reluctance so that expression (4.10) reduces to

$$L \cong \frac{S_a}{l_a} N^2 1.25 \cdot 10^{-8}. \text{ [Henrys]} \tag{4.11}$$

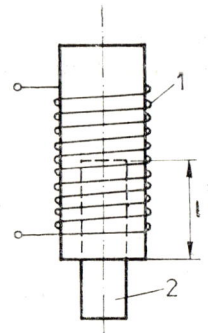

Figure 4.9

Generally, the resistance R contributes little to the total inductance of the coil

$$Z = \sqrt{(L\omega)^2 + R^2},$$

so that practically it becomes $Z = L\omega$. Therefore, the impedance Z and the inductance L of the coil are inversely proportional to the air gap thickness l_a. For small air gaps and small displacements of the movable armature, a nearly linear variation is obtained.

When the magnetic flux path is not closed through a core (Fig. 4.9), as for a coil 1 wound on a cylinder in which a magnetic core 2 can move, the coil inductance is directly proportional to the length l of the core within the coil, hence this *movable-core transducer* is indicated for relatively large amplitude vibrations.

The transducer with (closed magnetic flux path and) variable air gap can be improved designing it as a *differential transducer* (Fig. 4.10). Two coils are used in this case; when the core is moved, the inductance of one coil is increased, while that of the other coil is decreased. The three arrangements with double magnet from Figure 4.10 are used for (a) rectilinear motion, (b) angular motion, and (c) whirling motion of a rotating shaft.

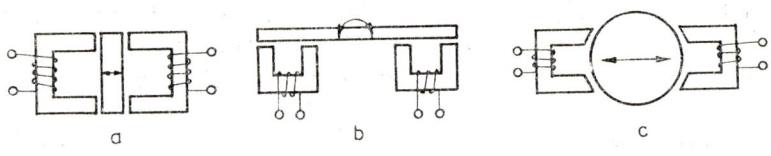

Figure 4.10

Other inductive transducers are based on the *mutual-inductance* principle, having two or more circuits, with energizing (primary) coils and induced current (secondary) coils. Examples are given in Figure 4.11, where (a) and (b) are simple transducers, with double or single core, (c), (d), (e) — differential transducers with two energizing coils A and a single measuring coil M, (f) — differential transducer with movable core.

4.1 Vibration transducers

On a similar principle is based the *linear variable differential transformer transducer* (Fig. 5.19) consisting of a single primary coil, two secondary coils and a movable core of magnetic material. For constant current through the primary coil, the output voltage equals the difference of the voltages induced in the secondary coils which are wound in series opposition.

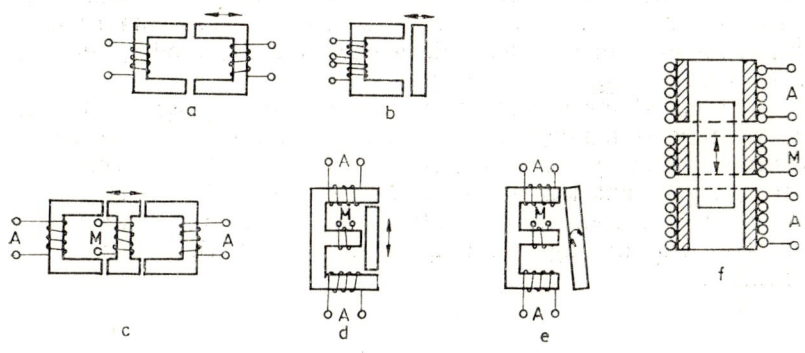

Figure 4.11

Inductive transducers can be made to have very small size and masses down to 1 gram, which makes them suitable for measurements on light weight parts or experiencing severe vibrations (e.g.: propeller blades, at accelerations up to 1000 g). They can be made from materials able to cancel or to compensate the temperature change effects; at the same time, they are essentially not affected by humidity, which places them before the variable-resistance transducers. They are hysteresis free, which permits their use for measuring vibrations up to several hundred Hertzs.

Some inductive transducers are based on the property of ferromagnetic materials to change their permeability under mechanical loading. When the permalloy wire 2 from Figure 4.12 is acted upon by the force F, its permeability changes, determining a corresponding change of the inductance L of the coil 1. Such transducers are especially used for measuring cutting forces at machine tools and pressures in engine cylinders.

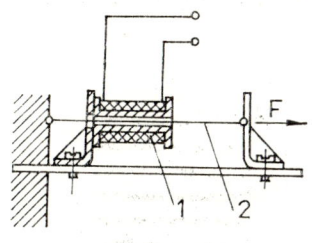

Figure 4.12

Eddy-current transducers are available for contactless measurements of rotating metallic (both ferrous and nonferrous) components, such as the machine shaft vibrations relative to

79

4 Transducers and pickups

the bearings. Some of them contain an oscillator-transformer and a coil inside the sensing head which generates a high-frequency magnetic field that induces eddy-currents in metal targets. Others have a separate proximitor. The eddy currents change the oscillator amplitude as the air gap between the top of the sensing head and the test object changes. *"Killed-oscillator"* probes detect the oscillator amplitude change with conventional demodulator/integrator circuits. *Current-source transducers* change the oscillator amplitude into pulse widths whose change is detected by an active filter which triggers the output.

For speed measurements these transducers register metallic discontinuities in a moving target at rates up to 5 kHz, the sensing distance being typically limited to the diameter of the probe. The eddy-current transducer is the only one that can be used for direct measurement on a rotating shaft, providing information on both the average running position and the peak-to-peak vibration.

4.1.2 Self-generating transducers

From the wide range of self-generating transducers, only piezoelectric, electrodynamic and electromagnetic transducers are presented, being most widely used in vibration measurement.

4.1.2.1. *Piezoelectric transducers.* Certain materials, natural or ceramic, exhibit the property of generating an electrical charge when subjected to mechanical stresses. The effect vanishes when the mechanical loading is canceled. Regardless of whether a natural crystal or a polarized ferroelectric ceramic is used, transducers operating on this principle are referred to as piezoelectric transducers.

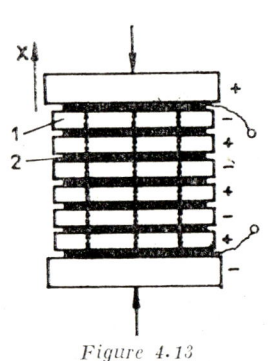

Figure 4.13

A compression-type piezoelectric transducer consists of a set of discs assembled as in Figure 4.13 : the quartz discs 1 alternate with metallic discs 2. Figure 4.14 shows a quartz cut, of dimensions a, b, c, where the largest face is cut parallel to the plane YZ, where Z is the optical axis of the crystal.

The charge generated by the crystal cut, due to a force F_x, parallel to the axis X, is

$$q_x = kF_x. \qquad (4.12)$$

4.1 Vibration transducers

The charge due to a force F_y, parallel to the axis Y, is

$$q_y = k \frac{a}{b} F_y. \qquad (4.13)$$

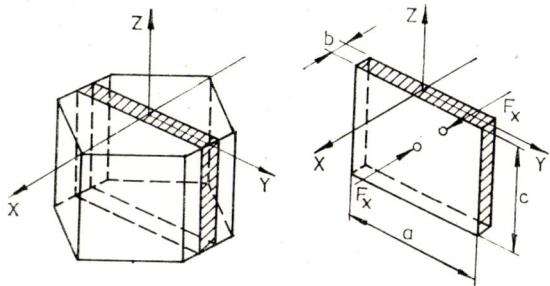

Figure 4.14

The loading on the X direction is preferred, the effect being independent of the crystal cut dimensions. The piezoelectric constant of quartz is

$k = 2.1 \cdot 10^{-12}$ Coulombs/Newton.

Other materials, e.g. Rochelle salt and barium titanate, have much greater piezoelectric constants than the quartz.

In order to better understand the limits between which a piezoelectric transducer can be used, the equivalent electrical circuit of the system consisting of transducer, cable and measuring instrument is presented in Figure 4.15. At the frequencies of interest, the transducer can be represented as a charge generator with a high internal capacity.

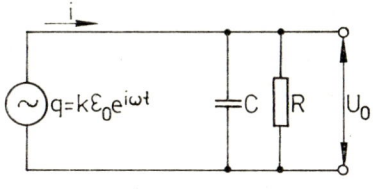

Figure 4.15

The generated charge is proportional to and in-phase with the applied (harmonic) strain $\varepsilon_0 e^{i\omega t}$. The capacitance C includes those of the transducer and of connecting cables. Resistance R includes those of the transducer and cables and the input impedance of the measuring device, connected in parallel.

4 Transducers and pickups

The current i produced by the charge q is

$$i = \frac{dq}{dt} = \frac{u_0}{R} - C\frac{du_0}{dt},\qquad(4.14)$$

where u_0 is the output voltage. For the steady-state regime, substitution of $u_0 = U_0 e^{i\omega t}$ in equation (4.14) yields

$$u_0 = \frac{i\omega q}{\dfrac{1}{R} + i\omega C} = \frac{1}{1 - \dfrac{i}{RC\omega}} \frac{k\varepsilon_0}{C} e^{i\omega t},$$

or

$$U_0 = \frac{e^{i\theta}}{\sqrt{1 + \left(\dfrac{1}{RC\omega}\right)^2}} \frac{k\varepsilon_0}{C},\qquad(4.15)$$

where

$$\theta = \tan^{-1}\frac{1}{RC\omega}.\qquad(4.16)$$

When ω is very large ($\omega \to \infty$), $\theta \to 0$ and $U_0 \to \dfrac{k\varepsilon_0}{C} = U_{\mathrm{HF}}$. Using this notation, equation (4.15) becomes

$$\frac{U_0}{U_{\mathrm{HF}}} = \frac{e^{i\theta}}{\sqrt{1 + \left(\dfrac{1}{RC\omega}\right)^2}}.\qquad(4.17)$$

The modulus $\left|\dfrac{U_0}{U_{\mathrm{HF}}}\right|$ and the phase shift θ are plotted versus ω in Figure 4.16.

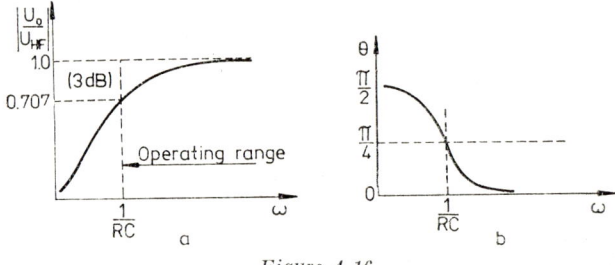

Figure 4.16

The nominal low-frequency limit of a piezoelectric transducer depends on the time constant RC. This is strongly affected by

4.1 Vibration transducers

the connecting cables. At low frequencies, the output voltage U_0 is a function of cable length. The low frequency response can be extended by increasing RC, but as C is fixed, R should be made as high as possible, using amplifiers with high input impedance (e.g.: cathode followers).

4.1.2.2 Electrodynamic transducers.
In an electrodynamic transducer (Fig. 4.17), a coil moves through the magnetic field produced by a permanent magnet or by an electromagnet. The transducer can also be designed with the magnet movable and the coil stationary. The voltage e generated in the conductor is

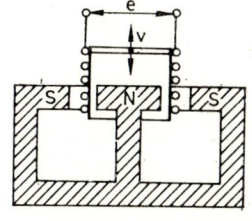

Figure 4.17

$$e = -Blv, \text{ [Volts]} \qquad (4.18)$$

where B is the magnetic flux density, in Teslas, l — the conductor length, in meters and v — the relative velocity between the conductor and magnetic field, in meters per second.

Electromagnetic transducers generate relatively high output voltages, so that they can be used without amplifiers in some cases. As the output voltage is proportional to the velocity, they are currently used in velocity pickups.

Electrodynamic transducers are reversible. If a current i is passed through the coil, it is acted upon by a force

$$F = Bli, \text{ [Newtons]}. \qquad (4.19)$$

Equations (4.18) and (4.19) yield

$$Bl = \left|\frac{e}{v}\right| = \frac{F}{i}, \qquad (4.20)$$

which suggests a procedure for static calibration of the transducer, measuring the force F, in Newtons, at different values of the current i, in Amps, using a force cell and an amperemeter (see Chapter 8).

Reversibility of the electrodynamic transducer makes it suitable for use as a vibration exciter.

4.1.2.3 Electromagnetic transducers.
The simplest electromagnetic transducer consists of a wire coiled around a permanent magnet. When a ferrous object moves relative to the transducer, the magnet flux density changes, generating a voltage at the coil terminals. The induced voltage is

$$e = -n\frac{d\Phi}{dt} = -n\frac{d\Phi}{dx}\frac{dx}{dt},$$

4 Transducers and pickups

where n is the number of coil turns, Φ is the magnetic flux and x is the air gap between the movable object and the transducer pole piece.

When a metal part nears the transducer, the reluctance decreases, thereby increasing the magnetic flux through a pole-piece and inducing a voltage across the coil. When the metal part moves away from the transducer, the reluctance increases, the magnetic flux decreases, inducing a voltage of opposite polarity across the coil. This way, the vibratory motion of whirling shafts can be measured using contactless probes. A spur gear generates a waveform whose pulse frequency is proportional to the shaft speed.

4.2 Vibration pickups

In the following, the most often used types of vibration pickups for measuring displacements, velocities and accelerations are described, as well as those for measuring dynamic forces, torques and pressures.

4.2.1 Theory of seismic instruments

There are two basically different instruments for vibration measurement, hence two types of vibration pickups:
— *fixed reference instruments* or *quasistatic devices*, in whic- the vibratory motion is measured relative to some fixed reference point;
— *seismic instruments*, in which the vibratory motion is measured relative to the mass of a mass-spring system attached to the vibrating structure.

With fixed reference instruments, the moving part follows the motion of the vibrating structure. The main errors arise from deformations produced by inertia forces in the transmission linkage.

The principle of operation of seismic instruments is described below.

4.2.1.1 *Principle of operation*. The seismic instrument (Fig. 4.18) consists of the frame S, rigidly attached to the vibrating system, the *mass m*, supported on a *linear spring k* and a *dashpot c*, which are fastened to the frame of the instrument, and the *transducer T*, that measures the relative motion of the mass and the frame.

4.2 Vibration pickups

It is assumed that the vibrating system, hence the instrument base, experiences a harmonic motion

$$x_1 = X_1 \sin \omega t. \tag{4.21}$$

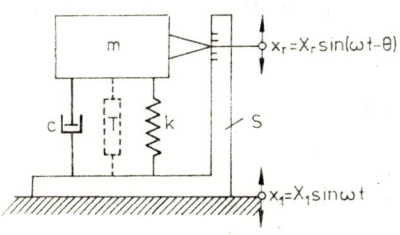

Figure 4.18

Neglecting transient terms, the relative displacement between the mass m and the frame S can be defined by

$$x_r = X_r \sin(\omega t - \theta). \tag{4.22}$$

The absolute displacement of the mass m, relative to a fixed reference point, is

$$x = x_1 + x_r$$

and the absolute acceleration is

$$\ddot{x} = \ddot{x}_1 + \ddot{x}_r.$$

The equation of motion of the mass m can be written

$$m(\ddot{x}_1 + \ddot{x}_r) + c\dot{x}_r + kx_r = 0 \tag{4.23}$$

or

$$m\ddot{x}_r + c\dot{x}_r + kx_r = -m\ddot{x}_1 = mX_1\omega^2 \sin \omega t. \tag{4.24}$$

Equation (4.24) has a steady-state solution of the form (2.52) which can be written

$$\frac{X_r}{X_1} = \frac{\left(\dfrac{\omega}{p}\right)^2}{\sqrt{\left(1 - \dfrac{\omega^2}{p^2}\right)^2 + \left(2\zeta\dfrac{\omega}{p}\right)^2}} = \frac{\eta^2}{\sqrt{(1-\eta^2)^2 + (2\zeta\eta)^2}}, \tag{4.25}$$

$$\theta = \tan^{-1} \frac{2\zeta \dfrac{\omega}{p}}{1 - \dfrac{\omega^2}{p^2}} = \tan^{-1} \frac{2\zeta\eta}{1 - \eta^2}, \tag{4.26}$$

4 Transducers and pickups

where

$$p = \sqrt{\frac{k}{m}}, \quad \zeta = \frac{c}{c_c}, \quad c_c = 2\sqrt{km}, \quad \eta = \frac{\omega}{p},$$

and p is the natural angular frequency of the mass-spring system.

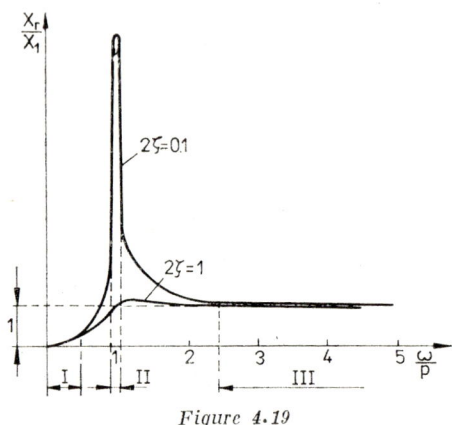

Figure 4.19

Figure 4.19 shows the variation of the amplitude ratio X_r/X_1 plotted against ω/p for two damping ratios. Figure 4.20 shows the variation of the phase shift θ plotted against ω/p.

Three ranges can be distinguished in Figure 4.19. Within the range III, when $\omega \gg p$, it can be seen that $X_r \cong X_1$, so

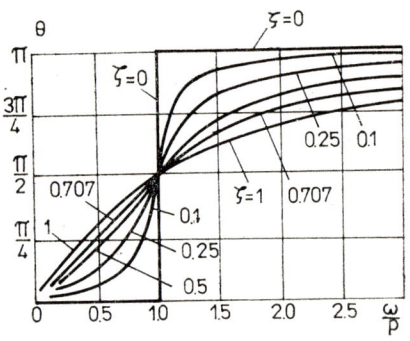

Figure 4.20

that the relative motion X_r between mass and frame, sensed by the transducer, is essentially the same as the displacement X_1 of the structure being measured. Figure 4.20 shows that, within

4.2 Vibration pickups

this frequency range, the phase shift is $\theta = \pi$ for light damping ($\zeta \to 0$), so that the frame and the mass m are vibrating 180° out of phase. Relative to an inertial space (fixed reference point) the mass m remains nearly stationary (becomes a fixed point) and the frame motion is measured with respect to it.

When T is a displacement transducer, the instrument is a seismic absolute *displacement pickup*. When T is a velocity transducer, the instrument becomes a *velocity pickup*. This can be demonstrated by differentiating equations (4.21) and (4.22) with respect to time

$$\dot{x}_1 = X_1\omega \cos \omega t, \quad \dot{x}_r = X_r\omega \cos(\omega t - \theta)$$

and calculating the ratio

$$\left|\frac{\dot{x}_r}{\dot{x}_1}\right| = \frac{X_r}{X_1}. \qquad (4.27)$$

Seismic displacement-measuring instruments should have very low natural frequencies (of the order of 1 to 5 Hz) which are obtained with low values of k, hence with a *soft suspension* of the seismic mass, respectively with relatively large masses m. These facts tend to make the displacement measuring instruments bulky and heavy, compared to the accelerometers.

Within the range I, for $\omega \ll p$, equation (4.25) becomes

$$\frac{X_r}{X_1} \cong \left(\frac{\omega}{p}\right)^2$$

or

$$X_r \cong \frac{1}{p^2}(X_1\omega^2), \qquad (4.28)$$

where $X_1\omega^2$ is the acceleration of the structure which is being measured.

In this case, the instrument measures a quantity proportional to the absolute acceleration of the structure (the factor of proportionality being the constant of the instrument) and is called an *accelerometer*. It has a high natural frequency, obtained with a small seismic mass and a hard spring (high stiffness k). Consequently, accelerometers can have small size, which is of interest for model testing and measurements on light-weight structures.

In the frequency range II, at $\omega \cong p$, the mass-spring system exhibits large amplitude vibrations, property used in the design of reed-type frequency-indicating vibrometers. The most used are multiple-reed gauges which employ a set of reeds of different

4 Transducers and pickups

natural frequencies that are closely spaced. The reed whose natural frequency corresponds most nearly to the excitation frequency has the greatest displacement amplitude. A frequency scale indicates the natural frequency of each reed.

4.2.1.2 *Amplitude distortions.* Beyond the linear operating range, the pickup output is distorted because the amplitude and the phase shift are non-linear with frequency. This factor should be taken into account especially for complex waveforms or for periodic vibrations consisting of several harmonics. When different harmonics are distorted non-uniformly, the output may significantly differ from the input, producing erroneous results. Consequently, frequency response ranges are given for each pickup so that the distortions might remain within prescribed limits.

Let consider a seismic instrument, working as a displacement-measuring device, with a maximum distortion of 5%. The output should be comprised within $X_r = (1 \pm 0.05)X_1$. Figure 4.21 shows an augmented part of Figure 4.19, bounded by 1 ± 0.05, where the curves have been plotted for four damping ratios: $\zeta = 0$; 0.6; 0.65 and 0.7. When the damping is zero, the amplitude distortion remains within $\pm 5\%$ for $\dfrac{\omega}{p} \geqslant 5$. For instruments with a damping ratio of 0.6 it is possible to operate down to $\dfrac{\omega}{p} = 1.2$ and for $\zeta = 0.7$ down to $\dfrac{\omega}{p} = 1.75$.

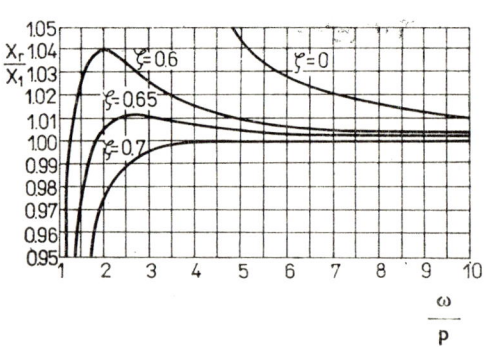

Figure 4.21

For an *accelerometer*, part of the plot of $\dfrac{p^2 X_r}{\omega^2 X_1}$ versus $\dfrac{\omega}{p}$ is depicted in Figure 4.22, for distortions of $\pm 5\%$. It can be seen that an accelerometer with zero damping can operate within the range $0 \leqslant \dfrac{\omega}{p} \leqslant 0.2$, whereas for a damping ratio of 0.6 it can operate up to $\dfrac{\omega}{p} = 0.84$.

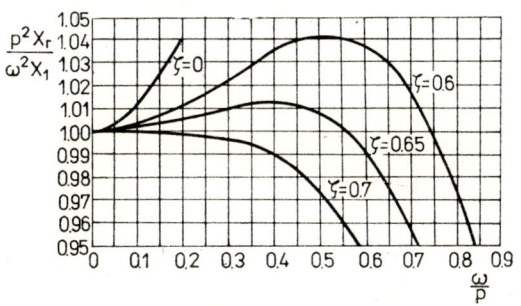

Figure 4.22

In conclusion, amplitude distortions set an upper frequency limit to the accelerometer and a lower frequency limit to the displacement-measuring instrument.

4.2.1.3 *Phase distortions.* Figure 4.23 presents a magnified portion of Figure 4.20, corresponding to the *displacement-measuring instrument.* When the damping is zero, $\theta = \pi$ for $\dfrac{\omega}{p} > 1$, so that no phase distortion occurs. For damping ratios between 0.6 and 0.7, phase distortions are fairly close to each other and for $\dfrac{\omega}{p} \geqslant 5$ have a practically constant value approaching π. For other values these instruments can never exactly reproduce time records of complex displacements because components

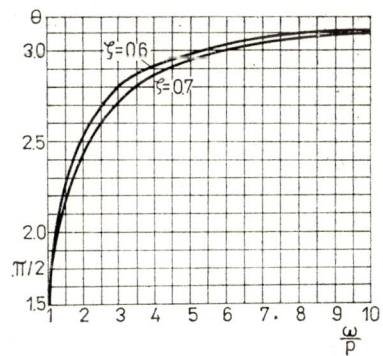

Figure 4.23

of various frequencies are delayed by different amounts. Hence, the lower operating frequency limit of a displacement pickup is set by phase distortions.

4 Transducers and pickups

For an *accelerometer* (Fig. 4.24), phase distortions are zero when damping is zero, but such an instrument is unsuitable due to amplitude distortions. When the damping ratio is about 0.7 a nearly linear relationship exists between phase angle and frequency

$$\theta = \frac{\pi}{2}\frac{\omega}{p}. \qquad (4.29)$$

Measuring the i-th frequency component of a periodic vibration, for which equation (4.22) becomes

$$x_{r_i} = X_{r_i} \sin \omega_i \left(t - \frac{\theta}{\omega_i}\right) \qquad (4.30)$$

and substituting the expression (4.29) for θ, one obtains

$$x_{r_i} = X_{r_i} \sin \omega_i \left(t - \frac{\pi}{2p}\right). \qquad (4.31)$$

In this case, for each frequency component, the time delay $\Delta t = \frac{\pi}{2p}$ is the same, so that the damping ratio $\zeta = 0.7$ is optimum, because phase distortions are very small and amplitude distortions remain within $\pm 5\%$ up to $\omega = 0.58\, p$.

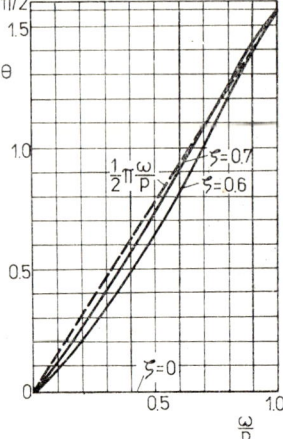

Figure 4.24

4.2.2 Displacement and acceleration pickups

Displacement and acceleration pickups are either seismic or fixed reference instruments. Seismic devices correspond to the scheme from Figure 4.18, their operating principle being described in the previous paragraph. The *elastic element* can be: a rod loaded in tension-compression, bending or torsion, a cylindrical tube, a helical spring, an elastic membrane etc. The transducer is connected either to the seismic mass or to the elastic element, sensing the relative movement of the moving element. Depending on the type of transducer, seismic pickups can be: inductive, resistive, piezoelectric, capacitive, piezoresistive, eddy-current or electrodynamic. Depending on the measured quantity, pickups can be classified as *motion pickups* — for displacement, velocity and acceleration, and *load transducers* — for dynamic forces, torques and pressures. At the design of a seismic pickup, the natural frequency is determined using the formulae given in Chapter 2. Fixed reference instruments are also mentioned.

4.2 Vibration pickups

4.2.2.1 Displacement-measuring pickups

a) *Inductive pickups.* The seismic pickup Type B3, manufactured by Hottinger-Baldwin, is schematically depicted in Figure 4.25, having the basic elements shown in Figure 4.18. The pickup contains an inductive differential transducer Type W1, having two coils with terminals 1, 2, and 1, 3, respectively, which are connected to form two arms of an a. c. *bridge circuit.* Movement of the ferromagnetic core from the position at which the bridge was initially balanced will produce an output proportional to the displacement.

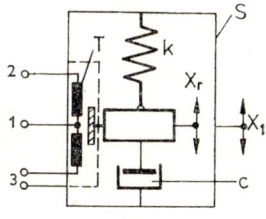

Figure 4.25

The pickup has a natural frequency of 5 Hz, an amplitude range of ± 2 mm and a frequency range from 7 to 100 Hz. Like any other seismic instrument it can operate as an accelerometer at frequencies between 0 and 3 Hz. The total mass of the pickup is 1.2 kg. The damping is achieved by generating dissipative eddy currents in a piece of copper which moves in a magnetic field.

Figure 4.26 illustrates this pickup. Using a "displacement converter" between pickup and the measuring bridge, the linear operating range is extended down to 0.7 Hz, that is below its natural frequency. Hottinger-Baldwin Series B pickups have low natural frequencies, i.e. between 0.7 and 5 Hz, for displacement instruments, and between 170 and 3000 Hz, for acceleration-measuring instruments. They contain Hottinger inductive transducers of the Series W.

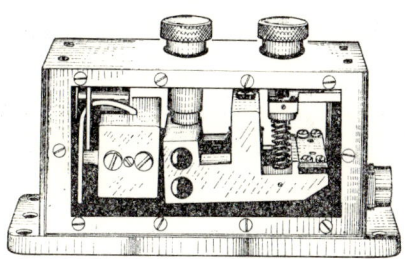

Figure 4.26

Fixed-reference inductive transducers of the Type IWT, manufactured by VEB Funkwerk Dresden, are moving-core differential transducers. The model IWT-201 is shown in Figure 4.27: the housing containing the coils can be mounted on a support,

91

4 Transducers and pickups

whereas the sensing rod with the moving core is attached to the vibrating structure. Table 4.3 lists some technical data of Series IWT transducers. Similar models covering a larger amplitude

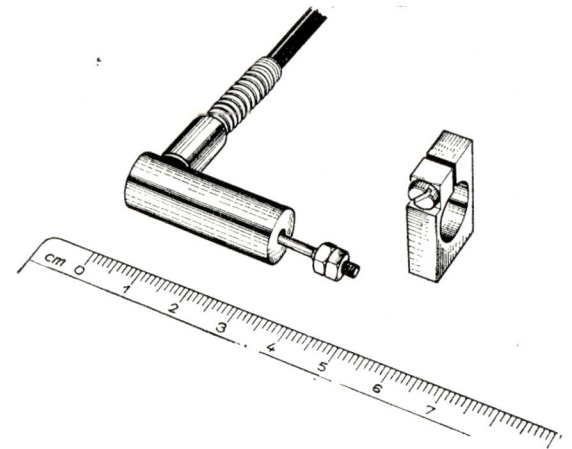

Figure 4.27

range are Hottinger-Baldwin Series W transducers (Table 4.4). Contactless transducers of Series IWB can be used as proximity pickups. They can have the two coils either in two housings

TABLE 4.3 *Characteristics of Displacement Inductive Transducers Manufactured by VEB Funkwerk Dresden*

Technical Data	Moving-core inductive displacement pickups, IWT Series				Proximity pickups IWB Series	
	IWT 102	IWT 202	IWT 302	IWT 402	IWB 102	IWB 202
Frequency range, Hz	0−1000	0−1000	0−1000	0−400	0−1000	0−1000
Minimum displacement, μm	0.15	0.15	1.5	15	0.1	0.2
Maximum displacement, mm	1	1	10	100		
Total mass, g	65	20	90	1500	20	40

4.2 Vibration pickups

TABLE 4.4 Technical Data of Hottinger-Baldwin Series W Inductive Transducers

Configuration	Model number									
Lateral plug, sensitivity adjustment			W1E			W10	W20	W50	W100	W200
Axial plug, sensitivity trimmer	W0.5T	W1T3		W2K	W5K	W10K	W20K	W50K	W100K	W200K
With sensing rod			W1TM	W2TK	W5TK	W10TK W10TS	W20TK W20TS	W50TS		
Amplitude range, mm	0.5	1	1	2	5	10	20	50	100	200

4 Transducers and pickups

(Fig. 4.28) or in a single one. For absolute measurements they should be calibrated taking into account the distance to and the size of the vibrating surface.

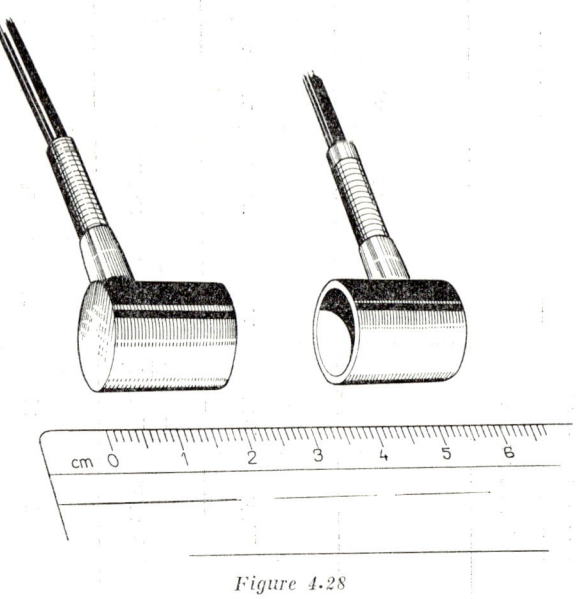

Figure 4.28

Characteristics of the Philips inductive displacement transducers PR 9314 Series are listed in Table 4.5. The two coils forming the half-bridge circuit are energized by a carrier-wave voltage of 4 to 6 kHz from the measuring bridge. The position of a ferroxcube core to the two coils determines the inductive unbalance which is detected by the measuring bridge. Philips LD 5000 Series transducers are available for displacements from ± 12 mm to ± 150 mm.

Philips' inductive displacement transducer Type PR 9310 is a differential transformer having a primary and two secondary coils wound on a non-magnetic tube within which a core of ferroxcube moves freely. The transducer can detect static or dynamic relative displacements from ± 0.1 μm to ± 1 mm having a sensitivity of 0.25 mV/μm at 2.75 V and 4 kHz, and a frequency range from 0 to 1000 Hz.

A.C.B. (Ateliers de Construction de Bagneux, France) manufactures four types of transformer type inductive transducers (corresponding to the configuration from Figure 4.11 a), weighing from 50 to 70 grams, having a frequency range up to 200 Hz and a linear amplitude range from 25 μm to 1 mm.

4.2 Vibration pickups

b) *Resistive pickups* contain strain gauges bonded either on the spring of the seismic instrument or on other elastic element.

TABLE 4.5 *Characteristics of Linear Displacement Inductive Transducers Manufactured by Philips*

Technical data	PR 9314 Series					LD 5000 Series		
	Model number							
	PR 9314/01	PR 9314/02	PR 9314/05	PR 9314/10	PR 9314/20	LD 5004 M	LD 5008 M	LD 5012 M
Maximum linear measuring length, mm	1	2	5	10	20	50	100	150
Nominal sensitivity, mV/mm per V	55	45	37.5	16.5	10.5	270	270	270
Overall length including sensing rod, mm	65	65	78	92	119	326	649	798
Total mass, g	35	35	45	45	60	370	660	880

At the model from Figure 4.29, the following parts can be seen:
1 — frame, 2 — leveling screws; 3 — strain-gauge instrumented elastic member (flat spring); 4 — calibrating screw; 5 — rigid

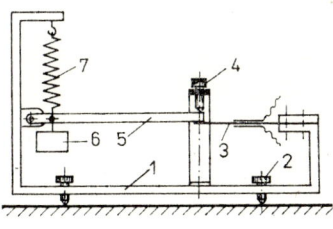

Figure 4.29

arm; 6 — seismic mass; 7 — helical spring. Pickups of this model have been manufactured at the Strength of Materials Laboratory of the Polytechnic Institute of Bucharest, Romania.

4   *Transducers and pickups*

The pickup of the vibrometer manufactured by Sexta (France) is shown in Figure 4.30. The seismic mass 1 is suspended on the helical spring 2. The seismic mass has a conical end that is in contact with the flat springs 3 and 4, which are bent during vibration. Four strain gauges are bonded on these flat springs and connected to form the arms of a Wheatstone bridge circuit. The tube 5 constrains the seismic mass to have a single degree of freedom motion (linear translation) and provides the necessary damping (by friction). The natural frequency is 5.4 Hz.

c) The *capacitive transducer* Type MM 0004, manufactured by Brüel & Kjaer, is schematically shown in Figure 4.31. It consists of a gold plated electrode shielded by the housing to prevent stray capacitance from influencing the measurements. Its sensitivity is inversely proportional to the square of the distance d_0 between the transducer electrode and the test specimen; for $d_0 = 0.5$ mm and a peak-to-peak vibration amplitude of 0.01 cm, the output voltage is 0.9 V r.m.s. The transducer is used in contactless relative measurements, where the absolute displacement amplitude is not required.

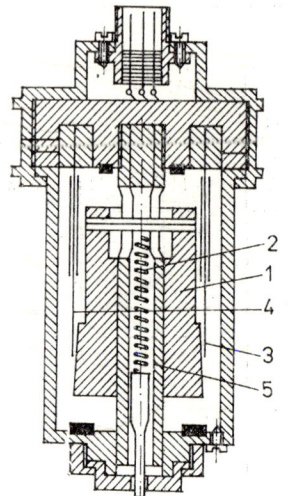

Figure 4.30

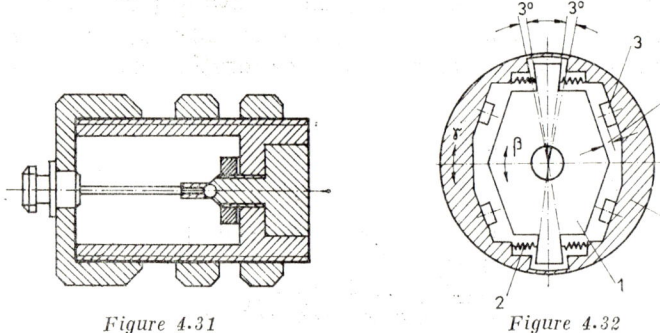

Figure 4.31   Figure 4.32

d) *Angular vibrations* are measured using inductive or resistive rotational seismic pickups. Figure 4.32 is a schematic cross section of the Hottinger-Baldwin differential inductive rotational pickup Type BD. The following parts can be seen: 1 — rotary seismic mass; 2 — helical springs; 3 — contactless inductive transducer; 4 — variable air gap; 5 — housing. The pickup senses the relative angular motion $\alpha = \beta - \gamma$ between housing and seismic mass. The instrument is attached to the end of the vibrat-

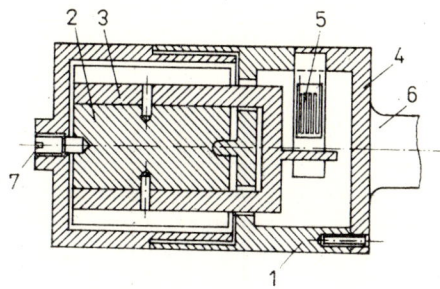

Figure 4.33

ing shaft and electrical connections are made through slip rings. The pickup has a natural frequency of 3 Hz, operating as an angular displacement pickup — in the frequency range from 5 to 1000 Hz and as an angular accelerometer — in the range from 0 to 2 Hz. Maximum amplitudes are 3 degrees for angular displacements and 18.6 rad/s^2 for angular accelerations.

The pickup with strain gauges from Figure 4.33 consists of: 1 — case; 2 — cylindrical seismic mass; 3 — hollow cylindrical element attached to the seismic mass and acting on the flat spring 5 on which two strain gauges are bonded; 4 — flat spring support; 6 — instrument shaft; 7 — adjusting screws of the seismic mass pivots. The pickup is attached with the shaft 6 at the end of the vibrating system. The relative motion between the seismic mass and the base 4 is sensed by strain gauges.

e) *Eddy-current* proximity transducers are mostly used with rotating shafts. They are non-contacting gap-to-voltage systems measuring both static and dynamic distances. The complete system consists of a probe of 5 to 8 mm tip diameter, an oscillator-demodulator unit (called proximitor) and an interconnecting extension cable. Most eddy-current proximity transducer systems are designed according to American Petroleum Institute API-670 Standard. The characteristics of some commercially available systems are listed in Table 4.6.

Generally, a 24 V d.c. supply is used, the frequency range is 0 to 10 kHz, the oscillator frequency is 2 MHz, the characteristics being determined for an AISI E 4140 steel target and a 10 kiloohm load.

TABLE 4.6 *Characteristics of Eddy-Current Proximity Transducer Systems*

Manufacturer	Transducer type	Oscillator type	Static measuring range, mm	Nominal operating position, mm	Probe tip diameter, mm	Sensitivity, mV/mm	Probe temperature range, °C
Bently Nevada	22810 22811 22812 22813	18745	2.03		7.62 7.62 5.08 5.08	8	−34 to +121
Schenck	SD-051 SD-052 SD-053	OD-051	2.0	1.2−1.5	5.0 7.5 7.5	8	−34 to +177
	SV-101	SV-200 SV-202 SV-204	0.6 1.0 0.6	1.2−1.8 2 1.2−1.8	8	5 5 8	−34 to +200
Vibro-Meter	TQ 101 TQ 102	IQS 603 IQS 604	0.2−2.05 0.1−3.5		8	8	−35 to +180
Dymac	M60 M61 M62	M600 M606 M620−1	0−2.5 0−2.15 1.5−7.5		8 5 19	8	
Philips	PR 6422 PR 6423 PR 6424	CON 010	1.0 1.0 2.0	1.5 1.5 3.0		8 4	−35 to +180

4.2 Vibration pickups

4.2.2.2. Accelerometers. Due to their advantages — lightweight, ruggedness, wide frequency response, good temperature resistance and moderate pricing — acceleration pickups are the most often used vibration sensing instruments today. The majority of accelerometers are of the seismic-type.

A compression-type seismic accelerometer with a piezoelectric transducer is shown schematically in Figure 4.34, where:
1 — preloading spring; 2 — seismic mass; 3 — piezoelectric discs; 4 — base; 5 — lead wires attached to the plated surfaces of discs; 6 — housing.

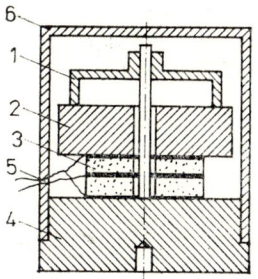

Figure 4.34

When the accelerometer is subjected to vibrations, the mass will exert a variable force on the piezoelectric discs. The charge developed across the piezoelectric discs is proportional to the force, which in turn is proportional to the acceleration of the mass. For frequencies much lower than the resonance frequency of the accelerometer assembly, the acceleration of the seismic mass is equal to the acceleration of the whole pickup.

The frequency response curve of an accelerometer is presented in Figure 4.35. The frequency range over which one can get a true and accurate signal is limited. The upper limit f_u is determined by the mechanical resonance of the mass-spring system of the pickup, whereas the lower limit f_l is determined by cable and preamplifier. The lower limit is set by the effect of ambient

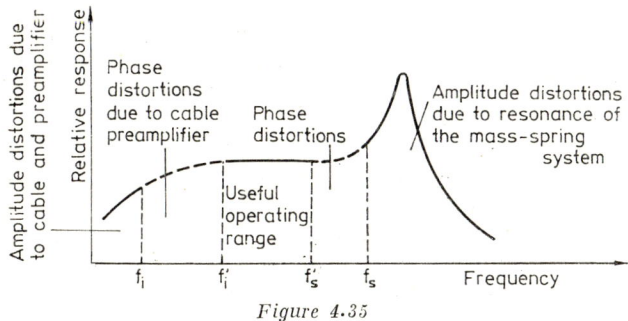

Figure 4.35

temperature fluctuations, the low frequency cut-off of the preamplifier being well below f_l. In order to completely avoid the phase distortions, the lower limit can be set to $f'_l \cong 10 f_l$ and the upper limit to $f'_u \cong 0.2 f_u$ obtaining the "range of ideal operation" shown in the figure. Most accelerometers do not incorporate a significant amount of damping, having less than 0.1 of critical. Thus, it is better to operate at frequencies much lower than the resonance frequency.

When selecting a piezoelectric accelerometer, the fundamental criterion is the *sensitivity*, defined as the ratio between its electrical output and the acceleration causing it. It can be expressed either as a *charge sensitivity*

$$S_q = \frac{q}{\ddot{x}}, \ [\text{pC}/g] \qquad (4.32)$$

where q is the charge developed at acceleration $\ddot{x}$, or as a *voltage sensitivity*

$$S_e = \frac{u}{\ddot{x}}, \ [\text{mV}/g] \qquad (4.33)$$

where u is the transducer output voltage.

The two sensitivities are related by equation

$$S_e = S_q/C, \qquad (4.34)$$

where C is the total capacitance of the transducer, cable and convertors.

Sensitivity depends on the properties of the piezoelectric material and on the loading mass. Accelerometers of larger size have increased sensitivity for a given piezoelectric material, but have a lower high frequency limit, being more sensitive to acoustic noise.

For a good accelerometer, the *transverse* (cross-axis) *sensitivity*, i.e. the sensitivity to motions perpendicular to the main accelerometer axis, should be less than 5% of the main axis sensitivity and the sensitivity to environmental factors (temperature, humidity, acoustic noise, magnetic fields, nuclear radiation, base strains) should be known.

When selecting an accelerometer, the following should be considered:

— the weight of the accelerometer should be at least ten times less than the effective weight of the vibrating structure;
— the frequency range of the accelerometer should correspond to the frequency range required for measurement;
— the dynamic range (see Section 5.1) of the accelerometer should be adequate; the maximum vibration level expected during measurement should not exceed one third of the maximum shock rating of the accelerometer;
— the maximum operating temperature should not be overpassed;

4.2 Vibration pickups

— the accelerometer should stand the other environmental conditions : humidity, noise, magnetic fields, radiation.

Piezoelectric accelerometers are built in different mechanical designs. Some configurations are shown in Figure 4.36 [3];

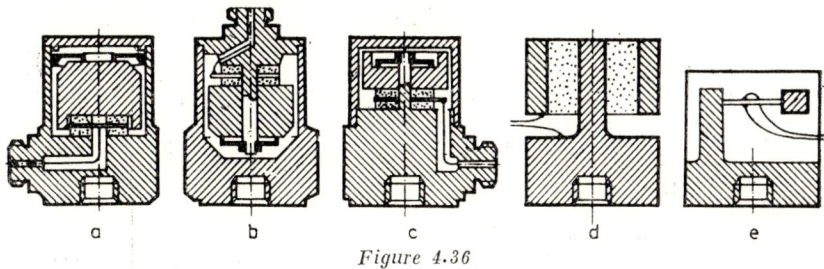

Figure 4.36

a) The peripheral mounted compression-type has a rugged construction, a fairly high resonance frequency, but is sensitive to external agents like acoustic noise, temperature transients and base strain.

b) The inverted centre-mounted compression type is better isolated from base strain, but has a somewhat lower resonance frequency.

c) The single-ended centre-mounted compression type is rugged, has a very high resonance frequency and reduced sensitivity to base strains, being the most widely used and having the best overall characteristics.

d) The shear type provides good isolation of piezoelectric elements and is adequate for miniaturisation.

e) The bending type has very high sensitivity but a relatively low resonance frequency.

Specifications for some of the piezoelectric accelerometers manufactured by Endevco and Brüel & Kjaer are listed in Table 4.7. These are versions of the compression and shear modes of operation (Fig. 4.36 c, d).

Figure 4.37 shows the Brüel & Kjaer standard accelerometer Type 8305, having mounting threads at both ends, one in the base, for mounting it on a vibrator table, and one in the top, for the attachment of the accelerometer to be tested. Numbers have the same meaning as for Figure 4.34.

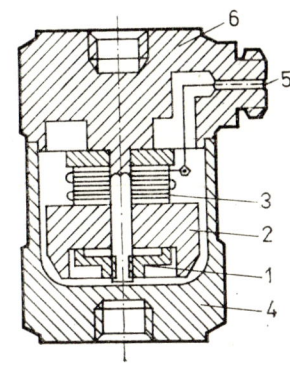

Figure 4.37

101

4 Transducers and pickups

TABLE 4.7 *Specifications for Some*

Model number	Sensitivity		Frequency range, Hz	g Range (±)	Transducer capacitance, pF
	mV/g	pC/g			

Ende

2219 E	370	85	2—3000	0—100	135
2271 A	5	11.5	2—5500	0—10000	2000
22 PICOMIN™	1.0	0.4	5—10000	0—2500	250
213E, 233E	45	60	4—8000	0—1000	1000
2229 C	5.5	3.2	3—5000	0—2000	400
236 ISOBASE®	26	60	3—8000	0—3000	2000
224 C	10.5	11	4—6000	0—1000	750
215 E	16	170	1—8000	0—1000	10000
2292	0.4	0.14	20—15000	0—20000	80
2223 D	11	12	3—4000	0—1000	750
2285	9	2.5	20—3000	0—2000	315

Brüel &

4366, 4368	40	45	0.2—9000		1200
4367, 4369	15	20	0.2—10600		1200
4370	100	100	0.2—6000		1200
4374	2	1	1—27000		500
8309	0.3	0.04	1—60000		90
4321	8	10	1—12000		1200
8305	—	1.2	0.2—4400		180
8306	1000	1000	0.2—1000	0—100	1000
8308	10	10	1—10000		1100

4.2 Vibration pickups

Piezoelectric Accelerometers

Maximum shock, g	Maximum vibration, g	Mounted resonant frequency, kHz	Total mass, grams	Application

vco

Maximum shock, g	Maximum vibration, g	Mounted resonant frequency, kHz	Total mass, grams	Application
1000	200	16	72	Very high sensitivity
10000	1000	27	27	Precision Isobase ®
10000	1000	54	0.14	Small size, lightweight, shear
2000	1000	32	32	High sensitivity
2000	1000	24	4.9	Small size, lightweight
10000	1000	29	28	Low acceleration
2000	1000	32	16	General purpose
2000	1000	32	32	High capacitance
20000	1000	125	1.3	Shock measurement
2000	1000	17	41	Triaxial
2000	500	30	20	Very high temperature (760°C)

Kjaer

Maximum shock, g	Maximum vibration, g	Mounted resonant frequency, kHz	Total mass, grams	Application
5000	2000	27	28, 30	General purpose
10000	3000	32	13, 14	General purpose
2000	2000	18	54	High sensitivity
25000	5000	90	0.65	High frequency, high level
100000	30000	180	3	Shock measurement
1000	500	40	55	Triaxial
1000	1000	30	40	Accelerometer calibration
1	1	4.5	500	High sensitivity
2000	2000	30	100	High temperature (400°C)

4 Transducers and pickups

Brüel & Kjaer DELTA SHEAR® design employs three piezoelectric elements, each with its own mass, arranged in the shear mode around a triangular centre post. Endevco's ISOSHEAR™ consists of two stacks of flat plate crystals, with seismic masses at the end, bolted to opposite sides of the transducer centre post. The crystals are oriented perpendicular to, and isolated from the transducer base.

PCB Piezotronics manufactures quartz accelerometers with built-in microelectronic unity-gain amplifiers. Models 302A (general purpose), 308A (high sensitivity), 305A (high shock) and 303A (miniature) are mostly used. Quartz accelerometers with similar performances are manufactured by Kistler.

Strain gauge instrumented *accelerometers* are available in a large range of configurations. For instance, acceleration pickups can be constructed by bonding strain gauges on the flexural element of the seismic instruments from figure 4.38 and designing them to have relatively high natural frequencies.

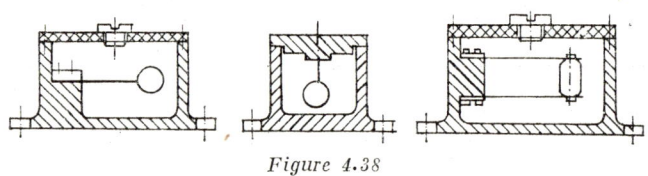

Figure 4.38

Figure 4.39 presents a schematic view of the Type BWH accelerometers manufactured by VEB Schwingungstechnik und Akustic, Dresden. On the figure: 1 — seismic mass; 2 — cantilever spring; 3 — semiconductor transducers; 4 — wire leads; 5 — space filled with silicon oil; 6 — measuring axis. These

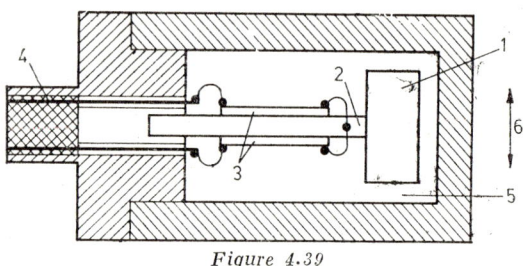

Figure 4.39

pickups (built in four models) can be used for measuring maximum accelerations from 20 to 10000 m/s^2 in the frequency range from 0 to 30...900 Hz, weighing from 35 to 125 grams.

4.2 Vibration pickups

SFIM-Massy (France) manufactures potentiometric resistive accelerometers, for maximum accelerations from 0.6 g to 24 g and frequencies up to 90 Hz.

Philips linear acceleration pickups of PR 9367 Series are used for measurement ranges from $\pm 10\ g$ to $\pm 1000\ g$ and frequency response ranges from $0 - 300$ Hz to $0 - 5000$ Hz,

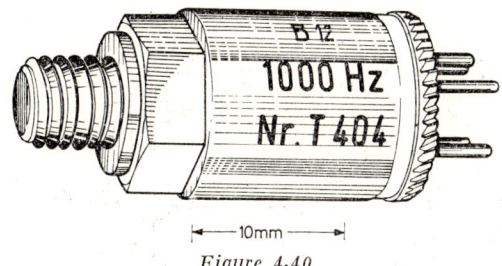

Figure 4.40

weighing 8 or 12 grams without cable. Philips three-directional acceleration pickups of the PR 9368 and PR 9369 Series have measurement ranges from $\pm 2\ g$ to $\pm 1000\ g$ and frequency response ranges from $0 - 31$ Hz to $0 - 5000$ Hz.

Piezoresistive accelerometers manufactured by Endevco include models 2260A-250 (general purpose, $\pm 250\ g$), 2261A-10K (shock measurements, $\pm 10,000\ g$), 2262-25 (low g, $\pm 25\ g$, 20 mV/g), 2264-150 (miniature, one gram, $\pm 150\ g$), 2267C-750 (triaxial, $\pm 750\ g$).

Inductive transducers are also used with accelerometers. Hottinger-Baldwin accelerometers of B12 Series correspond approximately to the arrangement from Figure 4.25. They have frequency response ranges of $0 - 100$ Hz, $0 - 250$ Hz, $0 - 500$ Hz and $0 - 1000$ Hz, respectively maximum accelerations of ± 200, ± 1000, ± 2000 and ± 2500 m/s^2, weighing 17 grams. Figure 4.40 illustrates such an accelerometer.

4.2.3 Velocity pickups

Velocity sensing devices are either seismic pickups or fixed reference devices including electrodynamic transducers.

Philips seismic velocity pickup PR 9266 is shown in Figure 4.41, where: 1 — permanent magnet; 2 — correction

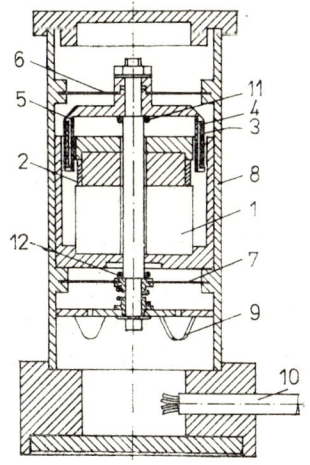

Figure 4.41

coil; 3 — measuring coil; 4 — additional damping coil; 5 — damping cylinder; 6 and 7 — membranes; 8 — casing; 9 — output leads; 10 — three-core screened cable; 11 and 12 — limit stops. It is used for measuring vibrations from 10 to 1000 Hz, with displacement amplitudes up to 1 mm or acceleration amplitudes up to 10 g.

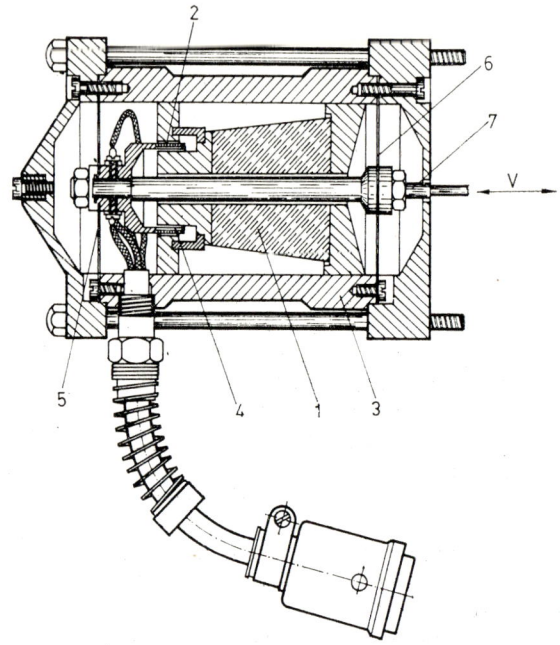

Figure 4.42

Figure 4.42 shows the fixed reference electrodynamic transducer manufactured by Schenck. One can see: 1 — permanent magnet; 2 — air gap; 3 — casing; 4 — moving coil; 5 and 6 — membranes; 7 — sensing rod. The instrument is attached to a support and the sensing rod is held against the vibrating body.

Velocity pickups are omnidirectional, generate a strong signal, are stable for accelerations below 30 g, having a transverse sensitivity less than 5 % up to 1 kHz. As disadvantages one could mention their large size and weight, the exponential decay of the output signal below 10 Hz (requiring correction), the decreased accuracy beyond 1 kHz. One should understand that velocity is the best single medium-frequency parameter to measure but it is not necessarily the best type of sensor.

4.2.4 Force and torque gauges

It is usual to call "load cells" those force transducers used primarily to measure static forces (or weight carried), while those used for the measurement of alternating forces are more often called "force cells" or "force gauges".

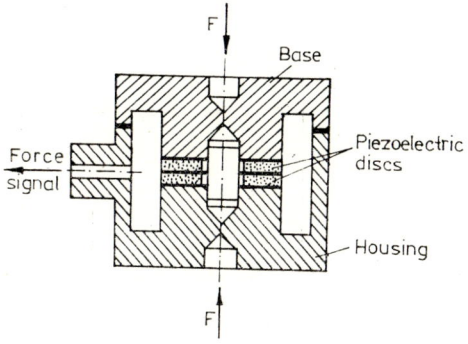

Figure 4.43

Force cells of the piezoelectric type are made with tubes or discs of piezoelectric material, preloaded to a high compressive strain so as to avoid the need for bonding under tension conditions. The construction is diagrammatically shown in Figure 4.43.

The force cell is mounted in the force transmission path so that it is subjected to the forces to be measured. In order to have minimal disturbing effect due to deformation, force cells must have high overall stiffness, so that quartz transducers are preferred. Table 4.8 lists some characteristics of commercially available piezoelectric force transducers.

Impedance heads (Fig. 4.44) contain in the same housing two transducers, a *force transducer* which measures the force applied

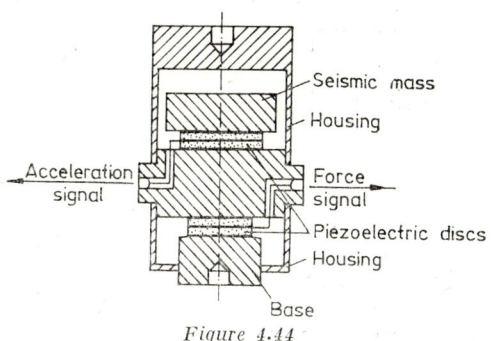

Figure 4.44

4 Transducers and pickups

TABLE 4.8 Specifications of Piezoelectric Force Transducers

Model number	Range, N	Sensitivity mV/N	Sensitivity pC/N	Resonant frequency, kHz	Weight, grams	Application
Endevco						
2103−100	±445	2250	667	15	114	Wide frequency response
2103−500	±2225	667	200	20	57	
2104−1000	±4450	490	155	15	160	Very high gauge stiffness
2104−5000	±22250	147	45	13	292	
2106 E	±22250	17	89	25	200	High forces
Brüel & Kjaer						
8200	−1000 to +5000		4	35	21	Wide force ranges
8201	−4000 to +20000		4	20	112	
Kistler						
9201, 9203	±500	50−5000	4800	27	13	Quartz force links
9311	±5000		400	75	25	
9321	±10000		400	57	95	
9331	±20000		400	50	170	
9341	±30000		200	42	335	
9351	±40000		200	37	485	
9361	±60000		200	29	1040	
9211, 9213	±2500		4.4	∼200	20, 21	Miniature
PCB Piezotronics						
208A	±45	2225		70		General purpose
208A 03	±2225	45		70		
208A 05	−2225 to +22250	4.5		70		
218A	−2225 to +22250		89	70		

to the structure and an *accelerometer* which measures the driving point acceleration. Characteristics of some commercially available impedance heads are given in Table 4.9.

The more usual construction of load cells is a straingauged tube carrying one or more full strain gauge bridges. These are arranged to eliminate errors due to temperature change and

4.2 *Vibration pickups*

TABLE 4.9 *Specifications of Some Impedance Heads*

Model	Part	Range		Sensitivity				Frequency response, Hz	Mass, grams
		g	N	mV/g	pC/g	mV/N	pC/N		
PCB Piezotronics 288A	accelerometer	±500		10				1–10000	140
	force sensor		±2225			45		1–20000	
Endevco 2110E	accelerometer	±500		65	150			2–5000	230
	force sensor		±22250			17	89	2–5000	
Brüel & Kjaer 8000, 8001	accelerometer	±400		30	30			1–6000	29
	force sensor		−300 to +2000			370	370	1–6000	

109

asymmetric loading. In the diagrammatic sketch from Figure 4.45 one can see: 1 — elastic member (tube); 2 — strain gauges; 3 — membrane supporting transverse forces F_q.

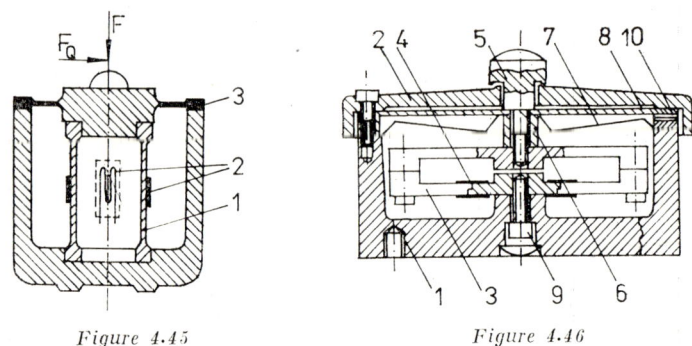

Figure 4.45   Figure 4.46

Figure 4.46 shows the force cell type HLW manufactured by VEB RFT Messelektronik Dresden, where: 1 — casing; 2 — cover; 3 — elastic member; 4 — semiconductor transducers; 5 — force application point; 6 — force transmission ring; 7 — plastic membrane; 8 — steel membrane; 9 — mounting stud; 10 — groove for damping oil pressure control.

Torque transducers are used for monitoring both static and varying torques. Philips transducers of MMS 9372 Series consist of a rotating measuring shaft onto which strain gauges are bonded (arranged in a Wheatstone bridge circuit) and a contactless transmission system is mounted. This consists of bridge controlled oscillator units that convert resistance variations, due to torque changes, into frequency variations in the output signal. A discriminator demodulates the transducer output signal.

Hottinger-Baldwin torque transducers of T30FN Series are used for measuring steady-state torques from 50 to 10,000 Nm and peak-to-peak dynamic torques from 35 to 7000 Nm at speeds from 3000 to 10,000 rpm, without slip rings. Other strain gauge torque transducers with slip rings are Hottinger Baldwin T1, T2, T4 and T4W transducers.

4.2.5 Dynamic pressure transducers

Almost all basic transduction elements can be used in pressure transducers. All models are basically variants of the Bourdon tube or of the diaphragm.

4.2 Vibration pickups

Resistive transducers have either a Bourdon tube moving an arm over a rheostat (limited to about 30 Hz) or strain gauges bonded directly on a diaphragm, as Philips PR 9362, Dynisco

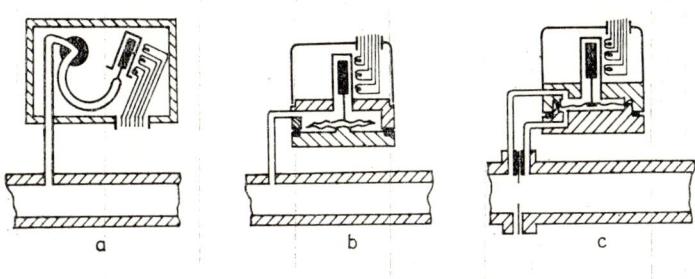

Figure 4.47

300 Series and Hottinger-Baldwin P3M, P4, P5 pressure transducers.

In inductive transducers, the diaphragm or the Bourdon tube are connected to the moving core. In the three sketches from Figure 4.47, illustrating transducers manufactured by ACB (France) : (a) is a pressure transducer with Bourdon tube; (b) — a pressure transducer with elastic diaphragm; (c) — a differential pressure transducer with diaphragm.

A pressure gauge with a capacitive transducer is presented in Figure 4.48, where :
1 — outer membrane; 2 — inner membrane; 3 — isolated condenser plate; 4 — air gap; 5 — mica plate; 6 — insulation; 7 — rubber gasket; 8 — cooling air inlet. The two membranes are connected by a rod that forms the moving element of the condenser whose plates are 2 and 3. A pressure transducer of this type is based on the accelerometer principle, operating at frequencies up to 30% of its natural frequency. The seismic mass consists of membranes 1, 2 and the connecting rod.

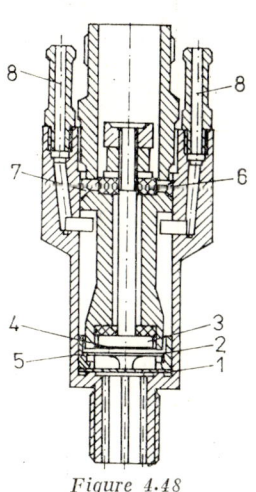

Figure 4.48

Characteristics of several piezoelectric dynamic pressure transducers are listed in Table 4.10.*[)]

---

*[)] See also The Bell & Howell *Pressure Transducer Handbook*, 1974.

TABLE 4.10 Specifications of Some Dynamic Pressure Transducers

Specification	Endevco		Kistler			PCB Piezotronics	
Model	2501-500	2501-2000	7005	701H	7261	111A	112A
Dynamic pressure range, bar*)	±34.5	±138	0–600	0–600	−1 to 10	0.007–207	0.007–207
Sensitivity, pC/bar	1.72	0.96	50	80	2200	0.024	0.07
Transducer capacitance, pF	300	300	9	9	24		
Maximum pressure, bar	69	207	1000	750	12		
Resonant frequency, kHz	45	50	80	65	13	400	250
Mass, grams	20	21	10	8.5	180		

*) 1 bar = $10^5 \, N/m^2$

4.2 Vibration pickups

4.2.6 Limits of vibration pickup performance

It is customary to specify the operating range of a vibration pickup by an operating envelope, as shown in Figure 4.49. This is a graph of velocity amplitude plotted against frequency, upon which are superimposed lines of constant displacement amplitude ($+$ 6 dB/octave) and lines of constant acceleration amplitude ($-$ 6dB/octave). Limit lines are then drawn for the pickup, closing an area inside which the *useful range of the instrument* can be defined.

The upper operational limits are imposed by the mounted resonance frequency and mechanical strength, while the lower limits are set by electronics (linear range and signal-to-noise ratio), being generally resolution limits. Measurement of shock motions requires a pickup response down to zero frequency and considerable efforts are made to attain linear response down to zero frequency.

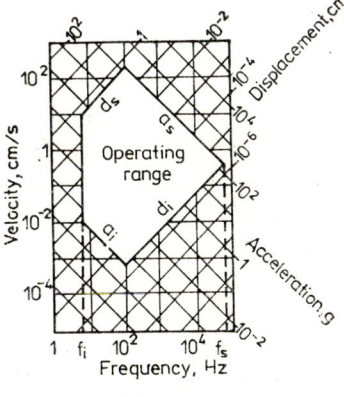

Figure 4.49

4.2.7 Mounting

Several mechanical and electrical factors should be taken into account when mounting a vibration pickup. From the mechanical point of view, the pickup must follow the motion of the vibrating structure without altering its response; from the electrical point of view, the output must not be influenced by the motion of connecting cables or by ground loops (due to improper grounding). As a general guideline, pickups should be attached as rigidly as possible to the structure under test.

Figure 4.50 shows some methods of mounting piezoelectric accelerometers [4]. The steel stud mounting (Fig. 4.50 a) is most reliable for high frequency measurements whereas the magnet attachment is suitable at low frequencies (Fig. 4.50 b). In between these extremes, good results are obtained using a thin layer of bees wax (Fig. 4.50 c), a quick-set cement or epoxy resin glue, and a double-sided adhesive tape. An isolated stud and a mica washer are recommended when electrical isolation between accelerometer and vibrating body is required. A thin film of oil applied

4 Transducers and pickups

before screwing improves the coupling of mating surfaces at high frequencies and is recommended above 4000 Hz. Regardless the mounting method, high frequency measurement accuracy depends on the stiffness of the mounting.

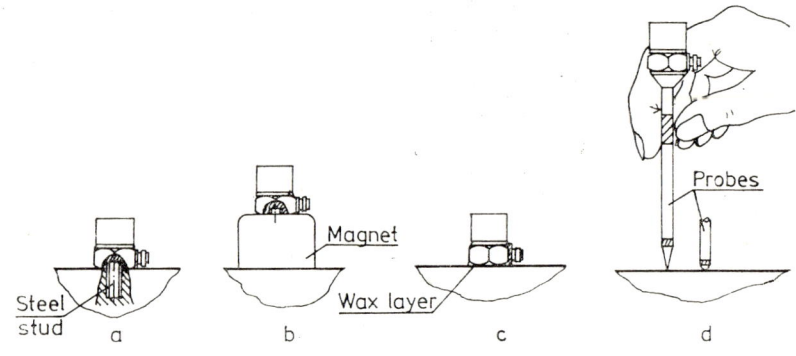

Figure 4.50

Hand-held probes with interchangeable pointed or round tips (Fig. 4.50 d) are used for short duration or prospective measurements, in order to establish the order of magnitude of the vibration and to determine "modal maps" (consisting of equal vibration level lines).

Some accelerometers are highly sensitive to base strain. A piezoelectric accelerometer placed on a node in a vibrating structure might indicate a relatively large acceleration, although no significant motion is occurring. Likewise, accelerometer attachment on an anti-none could result in erroneous measurement values. Care must be taken when screwing down an accelerometer, because a too high mounting torque might produce bending of the base.

The motion to be measured can be altered by mounting a seismic pickup at the point of measurement. Mass loading errors can be avoided if the mass of the pickup is much less than the dynamic mass of the structure at the point of attachment. The stiffening effect of transducers, when mounted on structures with extremely thin sections, can be avoided using proximity pickups. When the motion is not purely translational (e.g. : vibration of the tip of a cantilever), the transducer output is influenced by its transverse sensitivity.

Special attention must be paid to ensure that tribo-electric effects, originating from local capacity changes due to cable dynamic bending or whip, are minimised by proper clamping or by limiting cable displacement. Accelerometer mechanical

4.2 *Vibration pickups*

loading due to connecting cable inertia or to too tied clamping should be avoided.
Another source of noise are the ground loops whose formation is illustrated in Figure 4.51 a. The voltage drop ΔV produces

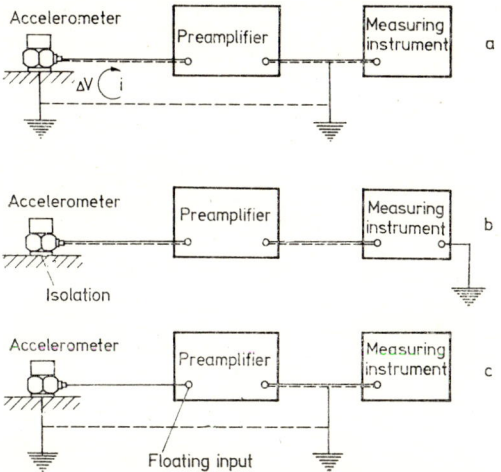

Figure 4.51

an electrical hum at the mains frequency which adds directly to the signal from the accelerometer. The ground loop is eliminated by grounding the whole installation in one point only (Fig. 4.51 b) and electrically isolating the accelerometer from the structure. Grounding should preferably be made at one of the measuring instruments in the system. An alternative solution is use of a preamplifier having a differential (or floating) input, without direct coupling between input and output (Fig. 4.51 c), isolated from ground. In this case it is not necessary to isolate the accelerometer from the vibrating structure.

References for Chapter 4

1. Neubert, H. K. P., *Instrument Transducers*, Clarendon Press, Oxford, 1963.
2. Buzdugan, Gh. and Blumenfeld, M., *Tensometria electrică rezistivă*, Editura tehnică, București, 1966.
3. * * * *Piezoelectric Accelerometers and Vibration Preamplifiers*, Brüel & Kjaer Handbook, March 1978.
4. Broch, J. T., *Mechanical Vibration and Shock Measurements*, Second Edition, Brüel & Kjaer, Naerum, 1980.
5. Holman, J. P., *Experimental Methods for Engineers*, 3rd Ed., McGraw-Hill Book Comp., New York, 1978.

5

Instrumentation for vibration measurement

5.1 General properties of measuring instruments

Considering an *instrumentation system* (Fig. 5.1), having an input $x_i(t)$ and an output $x_0(t)$, a system *frequency response function* can be established, having two parts — the *amplitude spectrum* (Fig. 5.2 a) and the *phase spectrum* (Fig. 5.2 b). These spectra describe how the input signal is altered when passing through the

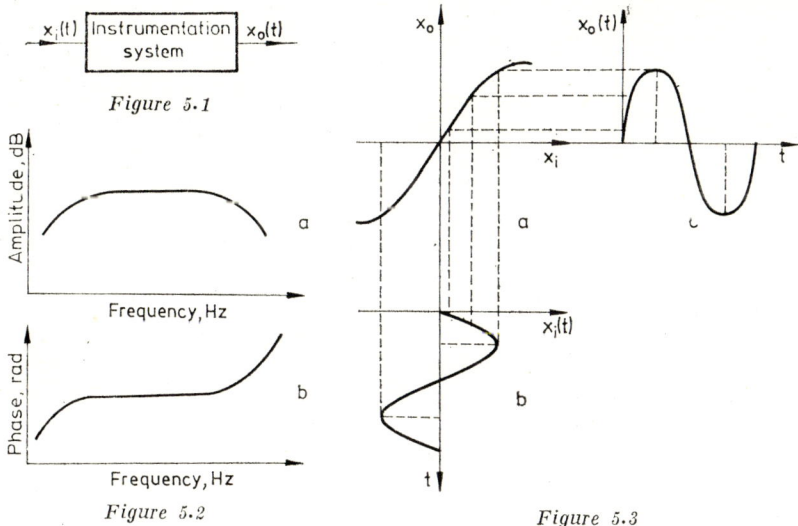

Figure 5.1

Figure 5.2

Figure 5.3

instrumentation system: the frequency components falling within the flat passband of the system remain unaltered, the others being attenuated and shifted in phase.

It is important that the system input-output characteristic be *linear* in the operating region (Fig. 5.3 a). The output versus

117

5 Instrumentation

input ratio is called *sensitivity*. Non-linear regions of the $x_i - x_0$ characteristic lead to amplitude distortions (clipping). A sine wave applied to the measurement system (Fig. 5.3 b) appears at the output as a complex periodic wave (Fig. 5.3 c). In the case presented in Figure 5.3, the *harmonic distortion* (creation of spurious harmonic components) is due to the fact that the input amplitude is too large, overpassing the linear operating region of the instrumentation system.

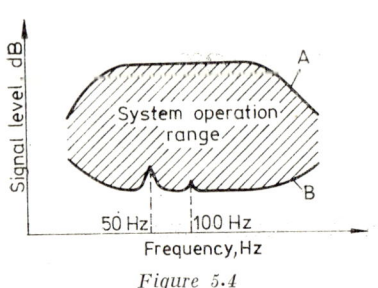

Figure 5.4

Usually, any instrument has a maximum signal amplitude, curve A in Figure 5.4, determined by the maximum permissible distortion level, and a minimum signal amplitude, curve B, determined by the inherent electrical noise (or by the sensitivity of the mechanical equipment), occasionally increased by picking-up extraneous electromagnetic fields.

The ratio, measured in dB, between the maximum amplitude (defined at some specified maximum harmonic distortion) and the minimum amplitude of the signals which can be measured is called the *system dynamic range*. Any signal whose level is outside this range is either distorted beyond the permissible limit, or buried in the hum.

Modern analog instruments have a dynamic range of 60 dB, which corresponds to a 1000 : 1 ratio of the extremal measured amplitudes. Digital equipment has greater dynamic range. For voltage functions, a 16-bit result has a theoretical dynamic range of 96-dB (6 voltage-dB per bit). Power or squared functions also have a 96-dB dynamic range if double precision results are used (32 bits times 3 power-dB per bit).

The measurement accuracy depends on other properties of the measurement system components such as: precision, resolution, sensitivity, drift, distortion, hysteresis, backlash, friction, as well as the dynamic response, i.e. the capability of the instrumentation system to follow the variations of the measurand.

Resolution is the smallest increment of the input signal which can be distinguished by the instrument. It defines the minimum value of the signal-to-noise ratio.

Instability refers to unpredictable variations of the sensitivity. In electronic instruments, it is determined by ageing of components and in mechanical devices by strain relaxation phenomena, leading to "zero" shift.

5.1 Properties of instruments

Secondary sensitivity is a measure of the dependence of output on physical variables other than the input (e.g. : temperature, humidity, transverse motions, acoustic noise, etc.). The main effect is the "zero" shift. The only effective means to deal with them is to recalibrate often the instrumentation system.

Distortions generally indicate deviations from the linear input-output relationship. By *non-linearity*, the maximum deviation of the $x_i - x_0$ curve from a straight line (expressed in terms of percent of full scale or of the input amplitude) is indicated. *Harmonic distortions* are expressed by the ratio of the magnitude of the higher harmonics to the fundamental, at instruments designed to have pure sinusoidal input signals. They appear at magnetic devices (e.g. transformers and generators) being minimized by filtering the a.c. power supplies output, as well as the input signals to multipliers used as harmonic analyzers.

Backlash defines the range of inadequate contact between driving and driven mechanical parts. Its elimination at joints is done using preloading springs, and at gears using helical teeth.

Hysteresis is a shift in the $x_i - x_0$ relationship, dependent on the point of last reversal of direction. In electrical instruments it is associated with magnetic saturation, and in mechanical devices with loading beyond the elastic limit. The effects are repeatable, so that they could be removed by careful calibration and measurements carried out in similar conditions (e.g. for displacement in the same direction). In instrumentation data sheets, the hysteresis is given by the maximum distance between the input-output curves, plotted starting from opposite full-scale points, in terms of the input signal amplitude.

Interference refers to the effect of the instrument on the measured quantity, being attenuated by proper selection and mounting of transducers on the vibrating structure, by cable screening etc.

Finally, a series of "dynamic errors", such as time lags in following the variations of the measurand, are determined by the frequency of vibration, the amount of damping in the measuring instrument and the natural frequencies of its moving parts.

A good measuring instrument should have a high and stable input impedance. This is achieved using, at the input, semiconductor circuits or special circuits with d.c. amplifiers. In order to decrease leakage currents, closed by the resistance between input and ground, the input resistance must have a high value, of the order of 10^{10} ohms.

Generally, digital equipment can be more accurate than analog equipment, because it can have greater resolution, has more repeatability, is almost insensitive to environmental factors and can handle much more data in any given time.

5 Instrumentation

5.2 Mechanical instruments for vibration measurement

This section describes three well-known mechanical instruments for the measurement of vibration, which are simple, rugged, portable and suitable for prospective measurements.

5.2.1 Tastograph

This is a quasi-static instrument, recording the relative motion between the case and the vibrating structure in a manner analogous to static measurements, hence without distortions. Hand-held or fixed on a measuring stand, the instrument schematic diagram is shown in Figure 5.5, where: 1—vibrating structure; 2—prod; 3 — spring forcing the prod against the structure surface; 4 — feeler rod; 5 — hinge joint; 6 — stylus arm; 7 — fixed pivot; 8 — recording drum; 9 — housing.

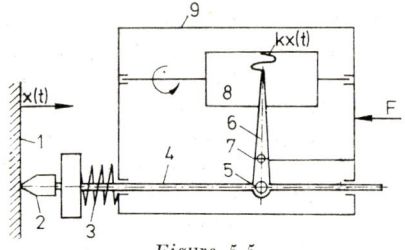

Figure 5.5

The operator holds the housing and maintains the prod in contact with the vibrating structure. Provided that the frequency of the measured vibration is lower than the natural frequency of the moving parts, the prod is able to follow the motion. The instrument is limited to measurement of vibrations having low frequency and relatively large amplitudes and to not too light structures, which can be loaded by the spring reaction.

5.2.2 Stoppani vibrograph

This is a fixed reference instrument, requiring a stationary frame of reference against which to measure the applied motion. In the schematic diagram of Figure 5.6, the following parts are shown: 1 — stylus; 2 — recording drum; 3 — metallic wire connecting the stylus arm and the vibrating structure; 4 — arm attaching clamp; 5 — drum-drive mechanism; 6 — time-base mechanism; 7 — vice for clamping the instrument to the fixed base 8. The vibrograph can also be attached to the vibrating surface, the wire 3 being fastened to a fixed base.

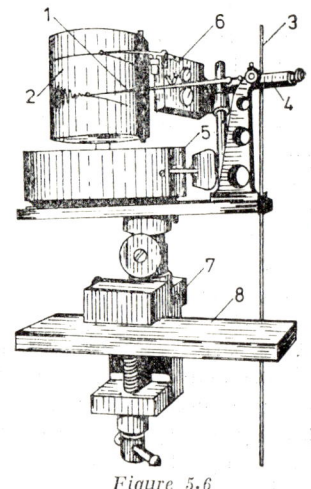

Figure 5.6

5.2 Mechanical instruments

5.2.3 Geiger vibrograph

The Geiger universal instrument is a pickup which, by simple and rapid changes, can be adapted for use as : seismic vibrograph, seismic torsiograph, tachograph, seismic accelerograph, quasistatic vibrograph, stress indicator or pressure gauge. The instrument has two main parts : the pickup and the recorder.

In the operating mode as *seismic vibrograph*, the schematic diagram is given in Figure 5.7. The seismic mass 2 oscillates around the axle 1, its proper position being adjusted by the coil spring 3. Motion of pendulum 2 is transmitted and amplified by a mechanical linkage 4 — 9 and transformed into rectilinear motion of the recording stylus.

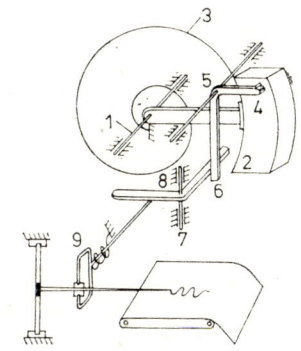

Figure 5.7

Figure 5.8 shows the instrument adapted for measurement of horizontal vibrations. One can see the seismic mass M, the housing R and the strap K for adjusting the mass position. Depending on the attached spring stiffness, the natural frequency of the instrument varies between 2 and 333 Hz, which permits its use as displacement or acceleration pickup. The displacement amplitude range is 0.01 to 15 mm.

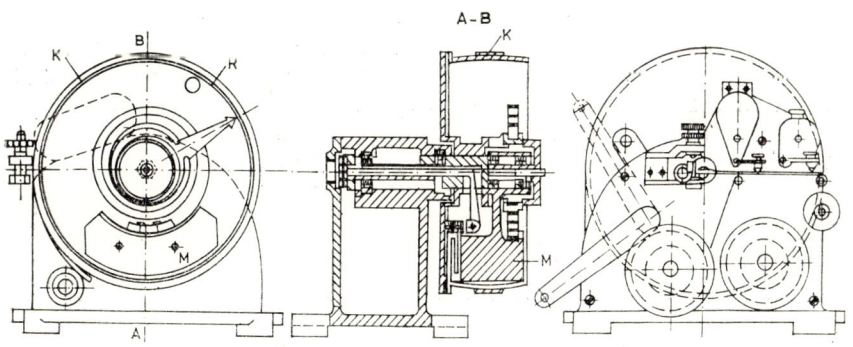

Figure 5.8

In the operating mode as torsiograph, the pickup of the Geiger instrument is shown in Figure 5.9, where : 1 — flywheel unit, playing the role of a seismic mass ; 2 — spindle ; 3 — driving disc ; 4 — coil spring between flywheel and spindle ; 5, 6, 7 — linkage

5 Instrumentation

transmitting the relative motion between drum and flywheel to a rod inside the spindle. For stationary rotation, the drum and flywheel revolve together. When torsional vibrations occur, the drum rotation is unsteady, while the seismic mass keeps rotating uniformly due to rotational inertia. Their relative motion is recorded.

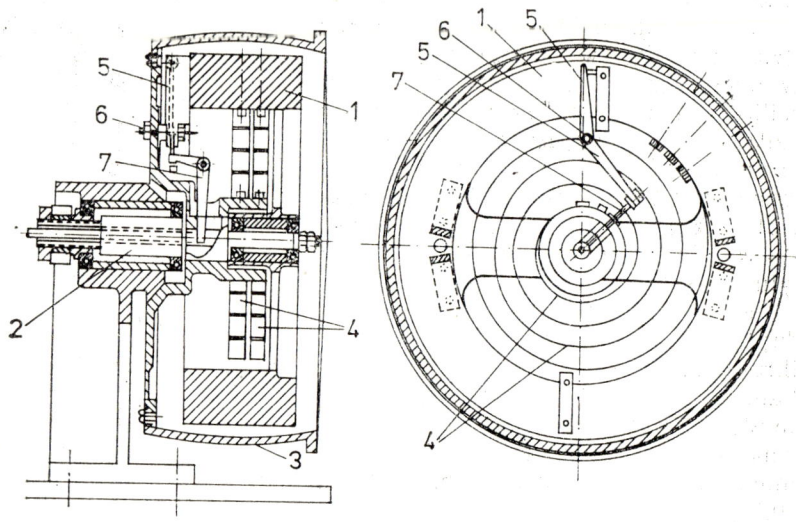

Figure 5.9

In order to operate as *quasistatic vibrograph*, the seismic mass and recording stylus are disconnected, the latter being acted through a rod, directly by the vibrating structure. When used as stress indicator, the instrument operates without seismic mass, the relative displacement between two points being transmitted to the recorder.

5.3 Conversion instruments

5.3.1 Measuring bridges

5.3.1.1. *Bridge circuits*. In instrumentation systems including passive transducers, vibration induced changes of their resistance or reactance can be measured using *bridge circuits* energized by a d.c. (for resistors only) or an a.c. voltage.

5.3 Conversion instruments

Figure 5.10 illustrates the Wheatstone bridge circuit with a single active strain gauge R_T bonded on the vibrating member or on the sensing element of the vibration pickup. Usually R_C is a dummy temperature-compensating gauge, identical with R_T, but bonded on a piece of the same material, unloaded mechanically and operating in the same environmental conditions. It eliminates output voltages due to the temperature variation of R_T. Resistors R_1 and R_2, together with the power supply and the meter connected at the output, are built in together in the so-called *measuring bridge*.

Before the measurement, after connecting the gauges R_T and R_C to the bridge, the following condition is achieved

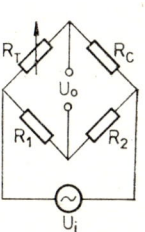

Figure 5.10

$$\frac{R_T}{R_1} = \frac{R_C}{R_2}. \tag{5.1}$$

In this case, the output voltage is set at zero $U_0 = 0$ and the bridge is said to be balanced.

When the structure under test vibrates, a change ΔR_T in the active transducer resistance occurs, inducing an output voltage

$$U_0 = U_i \frac{(R_T + \Delta R_T)R_2 - R_C R_1}{(R_T + \Delta R_T + R_C)(R_1 + R_2)}.$$

If the change ΔR_T is small, compared with its quiescent resistance R_T, so that the current through the bridge does not vary significantly, the following relationship can be established

$$U_0 = \frac{R_1 R_2}{(R_1 + R_2)^2} \frac{\Delta R_T}{R_T} U_i. \tag{5.2}$$

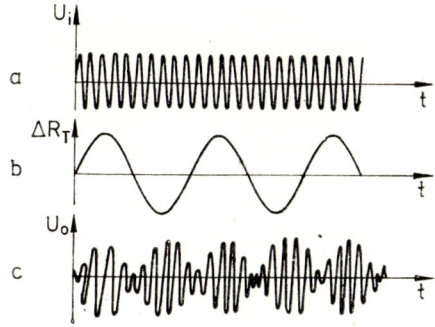

Figure 5.11

Equation (5.2) shows that the bridge output voltage is directly proportional to the product $\Delta R_T U_i$, i.e. the bridge acts as a *multiplier*.

If the input is an alternating voltage U_i (Fig. 5.11 a), then the output U_0 is also alternating and of the same frequency (Fig. 5.11 c), being amplitude-modulated by the

123

change ΔR_T of the active transducer resistance (Fig. 5.11 b). The signal U_i is called the *carrier* wave. A voltage change similar to that of ΔR_T can be obtained from the output U_0 by demodulation (detection) and filtering.

Amplitude-modulation is a multiplication of the signal bearing the information by a carrier wave of constant frequency and amplitude.

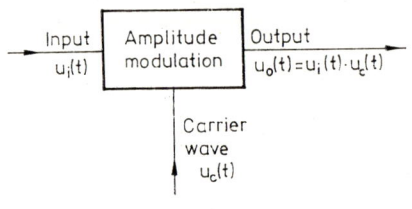

Figure 5.12

The process may be most easily explained for a sine wave $u_i(t) = U_i \sin \omega_i t$, modulating a sinusoidal carrier $u_c(t) = U_c \sin \omega_c t$ (Fig. 5.12). Usually $\omega_c \gg \omega_i$. The output signal is $u_0(t) = (U_i \sin \omega_i t)(U_c \sin \omega_c t)$. Transformation of the product of sines in a difference yields

$$u_0(t) = \frac{U_i U_c}{2} [\cos(\omega_c - \omega_i)t - \cos(\omega_c + \omega_i)t] =$$

$$= \frac{U_i U_c}{2} \sin[(\omega_c - \omega_i)t + 90°] + \frac{U_i U_c}{2} \sin[(\omega_c + \omega_i)t - 90°].$$

(5.3)

The frequency spectra of the three signals are presented in Figure 5.13. It is seen that the frequency spectrum of the output

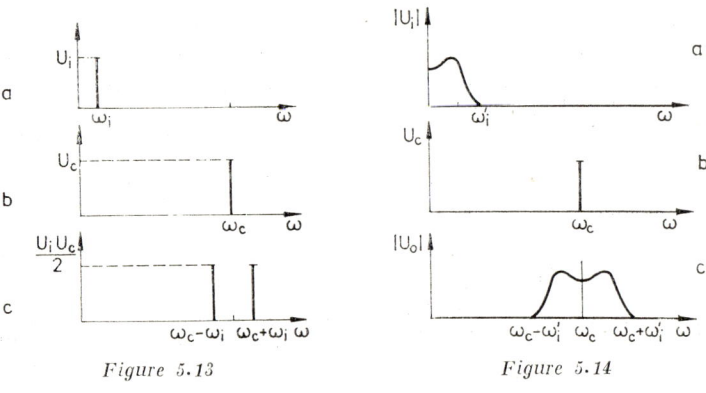

Figure 5.13   Figure 5.14

is a discrete spectrum, existing only at the frequencies $\omega_c \pm \omega_i$, called *side frequencies*.*)

When the input signal is a transient, having a continuous spectrum limited by ω'_i (Fig. 5.14 a), multiplication with the carrier wave of frequency $\omega_c > \omega'_i$ (Fig. 5.14 b) produces a shift of the spectrum up to higher frequencies, both sides of ω_c giving rise to *sidebands* (Fig. 5.14 c). The amplitude modulated signal

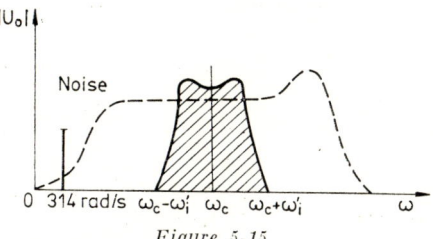

Figure 5.15

can be amplified, using an *a.c. amplifier*, whose frequency response is illustrated by the dashed line from Figure 5.15 [1]. It does not amplify constant or slowly varying voltages, so that coupled directly it is unsuitable for measuring shocks and low frequency vibrations. But connected to a strain gauge bridge, supplied with alternating voltage at frequency ω_c and whose variable resistance transducer measures a vibration having the frequency spectrum as in Figure 5.14 a, it receives an amplitude modulated input u_0. This has the frequency spectrum as in Figure 5.14 c (bounded by $\omega_c \pm \omega'_i$), which lies inside the flat portion of the response curve of the a.c. amplifier. Thus, though the amplifier does not respond to direct current, it does amplify the static component of the modulated signal.

When the wires between bridge and amplifier are subjected to a stray field at the mains frequency (50 Hz), which in certain conditions can give rise to a noise larger than the useful signal, this noise may be eliminated by designing the a.c. amplifier so that it has a negligible response at frequencies around 50 Hz (Fig. 5.15).

Generally, a.c. amplifiers are more stable than those with d.c. response, which have a strong tendency to drift and whose signal-to-noise ratio is not always adequate.

5.3.1.2 *Detection circuits.* When amplitude modulation is intentionally introduced to facilitate data processing it plays the

---

*) The case considered is that of the amplitude modulated oscillation having both side bands and the carrier suppressed.

5 Instrumentation

role of an intermediate step. The amplitude-modulated signal is not suitable for final readout and must be brought, after amplification, at the original form of the modulating signal. This process is accomplished by detection and filtering (Fig. 5.16).

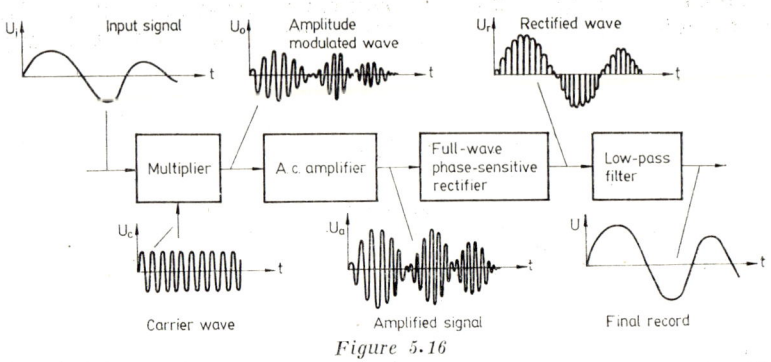

Figure 5.16

The simplest half-wave rectifier circuit with diode is illustrated in Fig. 5.17 a. When the instantaneous input voltage is positive at point M, the diode allows the current to flow through the resistor R, and the output voltage U_0 is positive at point O. When the voltage is positive at point N, the diode acts as a switch and $U_0 = 0$. If U_i is a sine wave, the circuit from Figure 5.17 a cuts the negative portions of the signal, passing only the positive portions (Fig. 5.17 b).

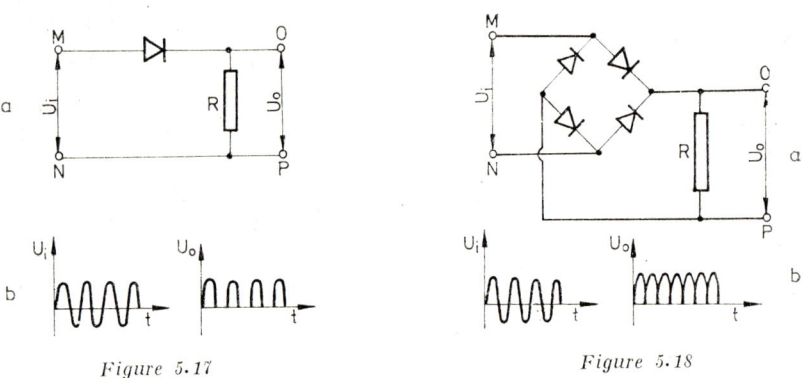

Figure 5.17   Figure 5.18

A circuit which detects both portions of the signal, called a full-wave rectifier circuit, is the *bridge rectifier* (Fig. 5.18 a). Irrespective of the polarity of point M, the output is always positive at point O, having the waveform from Figure 5.18 b.

5.3 Conversion instruments

The best reproduction of the original data is given by the full-wave phase-sensitive demodulation. Figure 5.19 illustrates the circuit arrangement for phase-sensitive demodulation using diodes and a differential transformer. Figure 5.20 presents the output voltage U_0 waveform for various core positions [1].

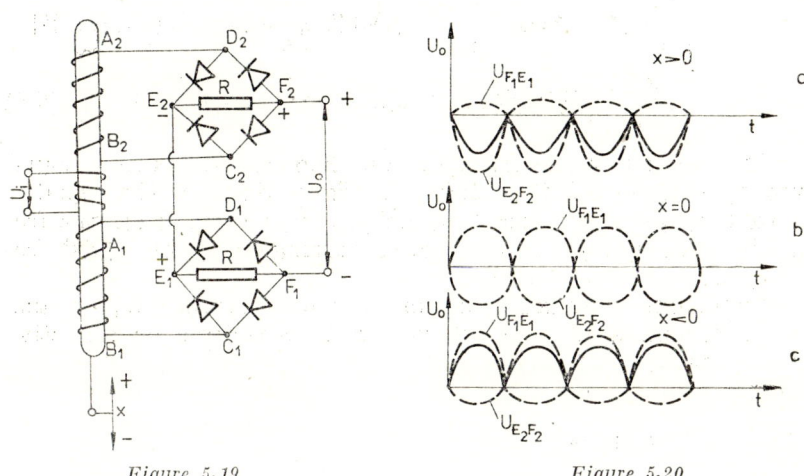

Figure 5.19   Figure 5.20

When the voltage induced in the lower secondary coil is such that A_1 is positive and B_1 is negative, the current path is $B_1A_1D_1E_1F_1C_1D_1$. When B_1 is positive and A_1 negative, the current path is $B_1C_1E_1F_1D_1A_1B_1$ so that the current through R is from E_1 to F_1 in both cases. A similar situation exists in the upper diode bridge so that when the core is in the null position (Fig. 5.20 b) $U_{F_1E_1} = |U_{E_2F_2}|$ and $U_0 = U_{F_1E_1} + U_{E_2F_2} = 0$. The core displacement from the null position induces in the secondary coils voltages of different amplitude and opposite sign so that the phase of the output voltage U_0 is dependent on the direction of the core motion, giving the phase-sensitive demodulation.

The full-wave phase-sensitive demodulation may be most easily explained for a sinusoidal signal $u_i(t) = U_i \sin \omega_i t$ and a sinusoidal carrier $u_c(t) = U_c \sin \omega_c t$. At the demodulator output, the signal is $u_0(t) = (U_i \sin \omega_i t)| U_c \sin \omega_c t|$. Because

$$|U_c \sin \omega_c t| = \frac{2}{\pi} U_c \left(1 - \frac{2}{3} \cos 2\omega_c t - \frac{2}{15} \cos 4\omega_c t + \dots \right)$$

5 Instrumentation

one obtains

$$u_0(t) = \frac{2}{\pi} U_i U_c \sin \omega_i t - \frac{4}{3\pi} U_i U_c \sin \omega_i t \cos 2\omega_c t -$$

$$- \frac{4}{15\pi} U_i U_c \sin \omega_i t \cos 4\omega_c t + \ldots$$

$$= \frac{2}{\pi} U_i U_c \sin \omega_i t - \frac{2}{3\pi} U_i U_c [\sin(2\omega_c + \omega_i)t - \sin(2\omega_c - \omega_i)t] -$$

$$- \frac{2}{15\pi} U_i U_c [\sin(4\omega_c + \omega_i)t - \sin(4\omega_c - \omega_i)t] + \ldots \quad (5.4)$$

The frequency spectrum of the demodulator output signal given by equation (5.4) is shown in Figure 5.21. If the signal is passed through a low-pass filter (Fig. 5.16), this rejects the frequencies $2\omega_c \pm \omega_i$, $4\omega_c \pm \omega_i$, etc., passing only the signal frequency ω_i.

Amplitude modulation is not limited to strain gauge transducers. It can also be employed with capacitive or inductive transducers.

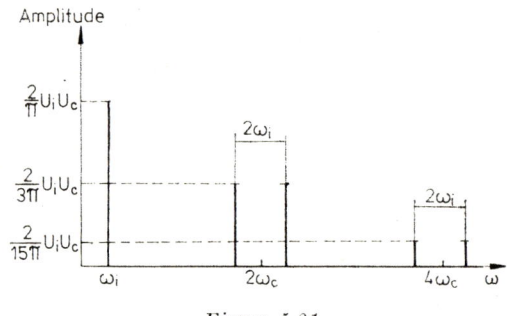

Figure 5.21

5.3.2 Frequency discriminator circuits

In frequency modulation, the data signal modifies the instantaneous frequency of a carrier wave (the carrier amplitude remains constant). The technique is used, for instance, at magnetic tape recording, having several advantages over direct recording: capability of storing signals with very low frequency components, low phase distortions and improvement of the signal-to-noise ratio (with a corresponding decrease of bandwidth).

5.3 Conversion instruments

When capacitive transducers are used, frequency-modulation is preferred to amplitude modulation. In this case, vibration of the structure under test produces a change of the transducer capacitance which controls the frequency of an oscillating circuit. Detection is performed by a frequency discriminator and the output voltage is proportional to the change of the instantaneous frequency.

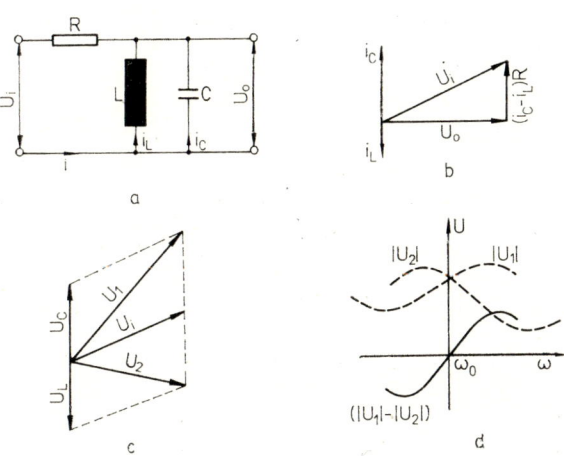

Figure 5.22

An instrument whose design is based on this principle is the wide band discriminator M 4082 manufactured by Southern Instruments Ltd. [2]. Its operation is based on the characteristics of the circuit from Figure 5.22 a.

The input voltage U_i, of frequency ω, is applied to a parallel tuned circuit LC, whose resonance frequency is $\omega_0 = \dfrac{1}{\sqrt{LC}}$. When $\omega = \omega_0$, the capacitive reactance $\dfrac{1}{\omega C}$ is equal to the inductive reactance ωL, so that the current i_C is equal to i_L but in antiphase with it. The total current $i = i_C - i_L = 0$, no current flows through R, so that $U_0 = U_i$.

At frequencies $\omega > \omega_0$, i_C increases and i_L decreases, so that a resultant component appears, in phase with i_C, which flows through R, producing an amplitude and phase difference between U_i and U_0 (Fig. 5.22 b). This phase change is converted into an amplitude change as follows.

The voltage U_i is separately added first to a voltage U_C, proportional to i_C, to produce a voltage U_1 and afterwards to a

5 Instrumentation

voltage U_L, proportional to i_L, to produce a voltage U_2 (Fig. 5.22 c). Evidently, when $\omega = \omega_0$, the amplitudes of U_1 and U_2 are equal, but for $\omega \neq \omega_0$, they vary differentially with frequency, as shown in Figure 5.22 d. If U_1 and U_2 are rectified separately and the subtraction $|U_1| - |U_2|$ is performed, the frequency-discriminator characteristic has the shape of the curve plotted with a solid line in Figure 5.22 d.

5.3.3 Amplifiers

Within the instrumentation system, amplifiers are connected between vibration transducers and data analysis or readout and recording instruments. Generally, their role is to amplify the relatively weak transducer output to a value suitable for further manipulation and recording. Additionally, in schemes using capacitive or piezoelectric transducers (with high output impedance), special types of preamplifiers are used, which, by a high input impedance, avoid decrease of both transducer sensitivity and the low-frequency operating range.

The common-mode rejection ratio is the parameter used to indicate the amplifier ability to reject spurious voltages appearing at the amplifier input due to ground loops, induced electrical hum and noise.

5.3.3.1 *Voltage amplifiers.* The basic element of a single-stage voltage amplifier is presented in Figure 5.23. The input U_i is applied (sometimes through an RC circuit) to the base of the transistor which, based on the energy received from the d.c. power supply E_B, produces a magnified signal U_0 of the same waveform as the input. The output waveform should be a precise replica of the input, without distortion.

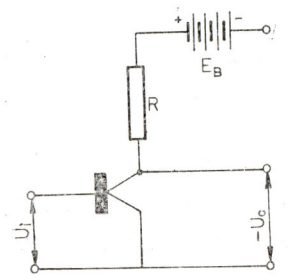

Figure 5.23

The ratio $\dfrac{U_0}{U_i} = A$ is called the voltage transfer function (gain) of the respective stage. Because the signal U_0 is 180° phase shifted with respect to U_i, this function is negative.

Amplifiers are built coupling several such stages. Coupling through resistance, impedance or transformer is used at a.c. amplifiers (so called because the coupling blocks the d.c. component of the signal). These are drift-free and have no other types

5.3 Conversion instruments

of instability which are characteristic of d.c. amplifiers but cannot amplify directly low frequency input signals. Instead, used in connection with passive transducers, in carrier-amplifier systems, they can amplify slowly varying physical quantities coming from transducers that are excited by an a.c. voltage, having the carrier frequency at least 5 to 10 times the highest signal frequency.

5.3.3.2. *Amplifiers with feedback*. Amplifier characteristics can be substantially modified using a feedback circuit (Fig. 5.24)

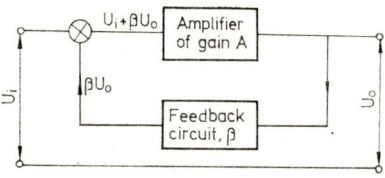

Figure 5.24

by which a fraction of the output voltage is reapplied to the input along with the original signal.

If A is the gain without feedback, the gain with feedback is

$$A_f = \frac{A}{1 - \beta A} \tag{5.5}$$

because in this case $(U_i + \beta U_0)A = U_0$ and $A_f = U_0/U_i$.

The amplifier is stable with negative feedback only, when βA is negative. When $|\beta A| \gg 1$, $A_f \cong -\dfrac{1}{\beta}$ and is practically independent of the absolute value of A, depending only on the passive elements of the feedback loop. Negative feedback extends the flat frequency response range, reduces noise and distortion produced within the amplifier.

Using operational amplifiers [1] together with feedback circuits it is possible to build up filters, integrating and differentiating circuits, charge amplifiers, etc.

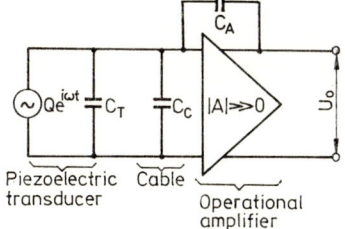

Figure 5.25

5.3.3.3. *Charge amplifiers*. These instruments are used in instrumentation systems with piezoelectric transducers (Fig. 5.25). By design, they have a large negative transfer function and a large capacitor C_A in the feedback loop.

5 Instrumentation

The following equation can be established

$$\frac{U_0}{Q} = \frac{A}{C_T + C_C - C_A(A-1)} \cong \frac{A}{-C_A A} = -\frac{1}{C_A}, \qquad (5.6)$$

where C_T is the transducer capacitance and C_C — connecting cable capacitance.

The charge amplifier output is proportional to the charge Q developed by the piezoelectric transducer and is independent of transducer and cable capacitance (provided that $C_T + C_C < C_A A$).

The charge amplifier permits use of piezoelectric transducers with cables of any length, without decreasing the sensitivity of the measuring system, for measurements at relatively low frequencies (e.g. shock measurements).

5.3.3.4. *Impedance-transforming amplifiers*. This type of "preamplifier" is used with capacitive transducers, without producing a voltage gain ($A = 0.8 - 0.9$) but only a decrease of the source impedance, hence an increase of the current delivered to subsequent instruments. It is an *impedance transformer*.

When measuring the signal of a capacitive transducer, used in a circuit with a polarizing voltage, the input impedance of the measuring instrument is a shunting resistance, permitting the leakage to ground, hence the signal distortion. The emitter follower (Fig. 5.26) is a circuit with a high input resistance which prevents this leakage, the output impedance being smaller, thus appropriate for the measuring instrument. All load is in the anodic circuit.

The emitter follower does not eliminate the connecting cable effect so that it is mounted directly on or very close to the transducer, which is favoured by its small size.

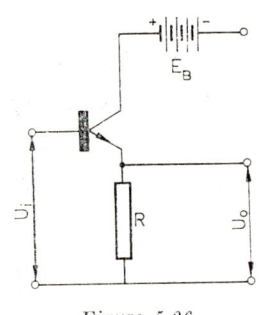
Figure 5.26

5.3.4 *Integrators. Analog low-pass and high-pass filters*

Often, within the instrumentation system, it is necessary to perform time integration or differentiation. Usually, integration is preferred because differentiation enhances unwanted noise (with high frequency components).

5.3 Conversion instruments

An *RC* integrating circuit, like that from Figure 5.27, has a frequency response function

$$\frac{U_0}{U_i} = \frac{1}{1 + i\omega RC}. \qquad (5.7)$$

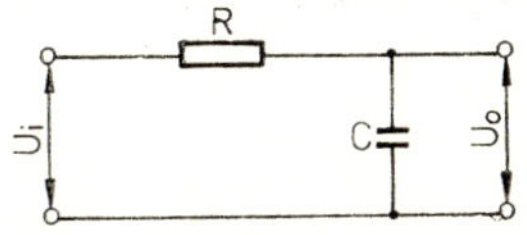

If $\omega RC \gg 1$, then $U_0 \cong \dfrac{1}{RC}\dfrac{U_i}{i\omega}$, so

Figure 5.27

that the output voltage is proportional to the time integral of the input voltage $\left(\text{if } u_i = U_i e^{i\omega t}, \text{ then } \int u_i\, dt = \int U_i e^{i\omega t}\, dt = \dfrac{U_i}{i\omega} e^{i\omega t}\right)$. The circuit acts as an *integrator*.

If $\omega RC \ll 1$, then $U_0 \cong U_i$, the output voltage equals the input voltage, the low frequency signals passing unaltered through the circuit, which acts as a *low-pass filter*. For very large values of the time constant *RC*, the circuit passes unaltered only the static component of U_i.

The frequency response characteristic of the *RC* circuit from Figure 5.27 is illustrated in Figure 5.28, together with the regions where it operates as a filter or as an integrator. There is always a limit frequency f_L below which the integration is not performed properly. This has to be considered when measuring shocks and low frequency vibrations. Operational integrators are used to extend the range in which the circuit acts as an integrator (Fig. 5.29) [3].

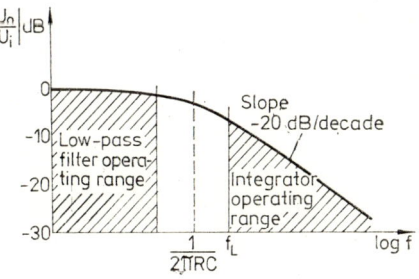

Figure 5.28

In practice, more complicated circuits are used to build up low-pass filters. A Butterworth filter has a linear amplitude characteristic and will be used when the amplitude of the signal to be filtered should remain constant. A Bessel filter has a linear phase characteristic and will be used if the shape of the signal should remain constant.

5 Instrumentation

An *RC* differentiating circuit, which can also be used as a high-pass filter, is presented in Figure 5.30. The frequency res-

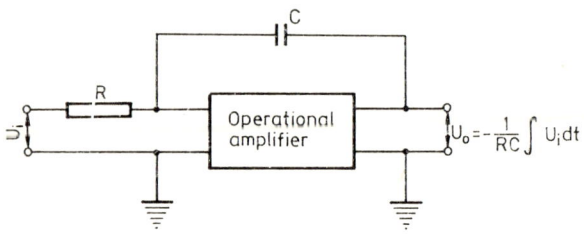

Figure 5.29

ponse function is

$$\frac{U_0}{U_i} = \frac{1}{1 + \dfrac{1}{i\omega RC}} \qquad (5.8)$$

and is plotted versus frequency in Figure 5.31 together with the operation regions of the circuit.

For $\omega RC \ll 1$, $U_0 \cong RCi\omega U_i$, hence the output voltage is proportional to the time derivative of the input voltage $\left(\text{if } u_i = U_i e^{i\omega t}, \text{ then } \dfrac{du_i}{dt} = \dfrac{d}{dt} U_i e^{i\omega t} = i\omega U_i e^{i\omega t}\right)$. For $\omega RC \gg 1$, $U_0 \cong U_i$, and the high frequency signals are passed unattenuated.

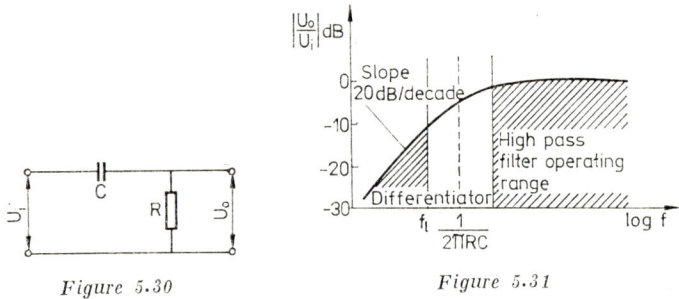

Figure 5.30   Figure 5.31

Differentiating circuits have an upper frequency limit f_l. Being very sensible to high frequency noise, they are influenced by the high frequency response of the transducer.

Generally, use of accelerometers connected to integrating circuits is preferred.

5.3 Conversion instruments

5.3.5 Analog-to-digital converters

Processing of experimental data by digital equipment requires *sampling* and *encoding* of analog signals produced by analog transducers. Signal conversion should be done as close as possible to the transducer output in order to minimize alteration of data due to noise, distortion and inaccuracies of analog instruments.

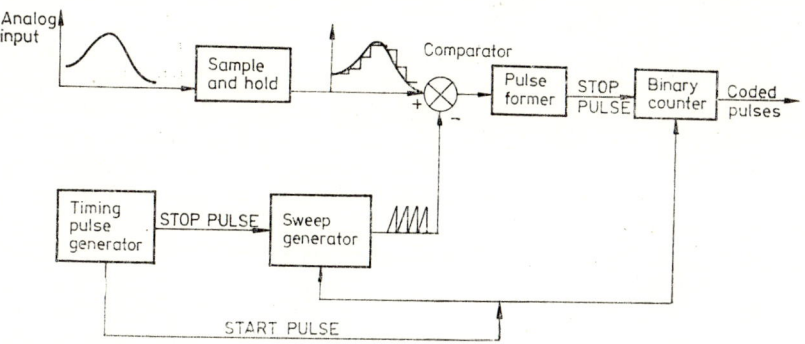

Figure 5.32

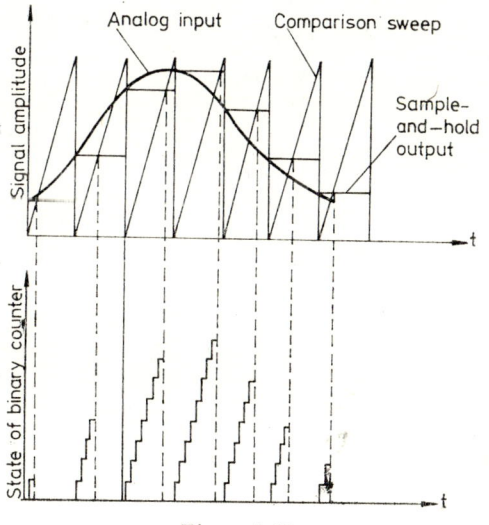

Figure 5.33

The analog-to-digital conversion, involving two basic processes — sampling and encoding — may be done according to the simplified block diagram from Figure 5.32 [1]. The analog input signal is compared with a saw-tooth sweeping signal (Fig. 5.33),

5 Instrumentation

whose peak voltage must overpass the largest expected value of the input signal.

Sampling is accomplished retaining as discrete value, during a sweep cycle, the instantaneous signal value $x(t)$ at the cycle starting, corresponding to a zero of the sweep voltage (using a sample-and-hold device).

A binary counter is started at the beginning of each sweep cycle and is stopped when the sweep voltage coincides with the sampled analog input. At the counter output, coded pulses are obtained, representing a binary number whose value depends on the time interval the counter is left on. Because this time is proportional to the sampled voltage, the analog voltage is converted to its digital equivalent. The counter is reset to zero at the end of each cycle, so that the process repeats itself and binary numbers are generated, representing the value of the analog signal at equal time increments.

Another system based on dual-slope integration is described in section 5.5.3.

In the digitizing operation, the individual samples are quantized and converted to a corresponding digital code. One of the commonest is the BCD code (Binary Coded Decimal). In this code, each decimal digit is given by its binary equivalent. For example, the decimal number 537.8 is written 0101 0011 0111.1000 in the BCD system, requiring a total of 16 bits.

The discrete amplitude levels are determined by the number of bits used in the converter to cover the amplitude range. A 10-bit converter contains a sign bit and provides $\pm 2^9$ or ± 512 discrete levels. This resolution provides sufficient accuracy for most measurements.

The total usable dynamic range can be approximated by

$$\text{Dynamic Range (dB)} = 6 \times (\text{number of bits}) + K,$$

where $K = 6-20$ dB is an additional factor created by the use of a so-called "dither" on the input analog-to-digital converter. Dither is employed to keep the one or two least significant bits of the analog-to-digital converter continuously active. When used with averaging, the dither enables the converter to more accurately define the level of very low amplitude signals, which otherwise might be lost within the large relative quantizing states of the converter.

5.3.5.1. Sampling rate. Aliasing. The main condition imposed to any sampling procedure is not to loose or deteriorate the information contained in the analog form of the signal. To this respect, requirements of the sampling theorem and selection of the

5.3 Conversion instruments

adequate number of statistically independent samples must be fulfilled in practice.

The time interval between two samples is called *sampling period* and is denoted by T_S in Figure 5.34. The reciprocal of this quantity is called *sampling frequency* (sampling rate) being denoted by f_S.

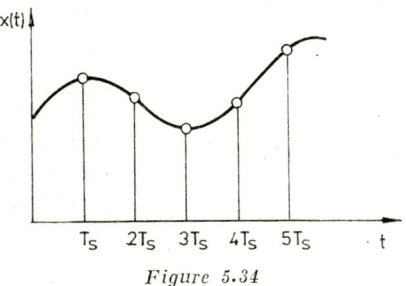

Figure 5.34

According to the *sampling theorem*, attributed to Shannon, for a band limited signal with the highest frequency B, the sampling frequency should be at least twice the highest signal frequency of interest, i.e. $f_S = 2\alpha B$, where $\alpha \geqslant 1$. Only complying with this condition, the discretized signal is completely determined and can be recovered as a continuous function of time, starting from samples. Demonstration of this theorem is given in reference [4].

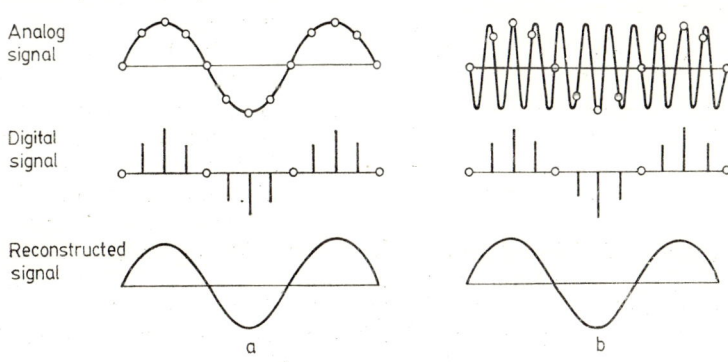

Figure 5.35

Consequences of abuse of Shannon's theorem are illustrated in Figure 5.35. For the case from Figure 5.35 a, the sampling frequency is higher than twice the sampled signal frequency (ratio 8 : 1) so that the reconstructed function is identical to the original. For the case from Figure 5.35 b, since the sampling fre-

5 Instrumentation

quency is less than twice the signal frequency (ratio 8 : 7), the reconstructed function differs considerably from the original.

Given the sampling frequency, the highest frequency of the analysed signal, called the *Nyquist frequency*, is $f_N = f_S/2$.

When the sampled signal is used for determining the power spectral density or the amplitude spectrum, incorrect choosing of the sampling frequency leads to the "folding error", phenomenon known as "aliasing". Amplitudes or r.m.s. values of frequency components higher than f_N are added to those within the frequency band $(0, f_N)$.

Sampling may be considered as a modulation process in which the signal function (carrying the information) is multiplied by a sampling function (Fig. 5.36 a) consisting of a periodic train of Dirac functions.

The Fourier spectrum of the sampling function (Fig. 5.36 b) consists of a series of lines at frequencies $0, f_S, 2f_S, \ldots$ hence a sum of sinusoids of equal amplitudes and frequencies multiples of f_S. Modulation of these sinusoids, considered as carrier waves, by a signal function having the spectrum as in Figure 5.36 c (with the highest frequency f_u), gives rise to the frequency spectrum from Figure 5.36 d, where the information spectrum is centred around each spectral line of the sampling function.

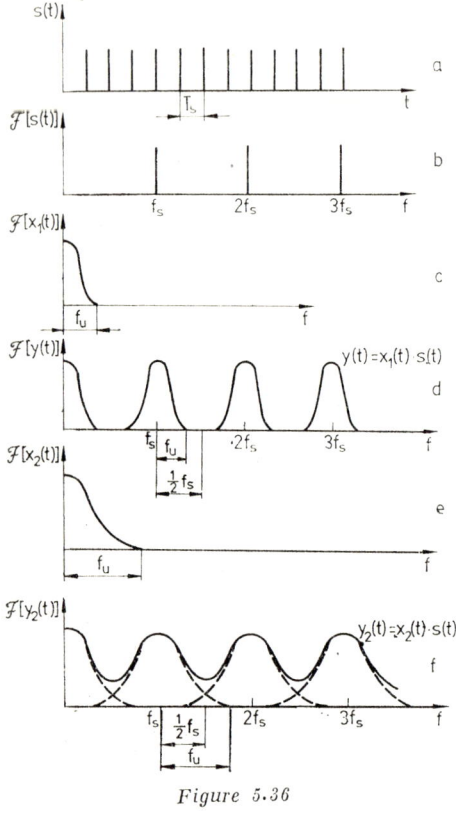

Figure 5.36

As long as $f_u \leqslant \frac{1}{2} f_S$ (oversampling), i.e. when the adjacent sidebands do not overlap (Fig. 5.36 d), no information is lost and the original function can be reconstructed by passing the modulated signal through a low-pass filter which removes the harmonics of f_S and the corresponding sidebands.

However, if the signal function has a spectrum as in Figure 5.36 e, i.e. $f_u > \frac{1}{2} f_s$ (undersampling), then the adjacent sidebands do overlap (Fig. 5.36 f), the information spectrum (the upper sideband of the zero frequency carrier) is distorted by components which originally had other frequencies. This overlapping of the frequency domain function upon itself is called *aliasing*, the frequency of the components above the half sample frequency point becoming misnamed or "aliased".

Because, quite often, input data signals have frequency components beyond the limit value f_N of the analysed frequency band, these must be attenuated by means of a lowpass filter, called "*anti-aliasing*" *filter*, located before the sampling-coding stage. Theoretically, the cutoff frequency of this filter, corresponding to a 3dB attenuation, might be chosen equal to f_N.

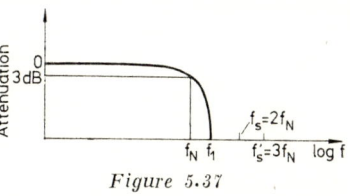

Figure 5.37

Practically realizable filters have response characteristics with finite slope in the rejection band, so that they cannot cut off completely the harmonic components above f_N (Fig. 5.37). Selecting a sampling frequency (according to Shannon's theorem) $f_S = 2f_N$, the existence of the non-attenuated components lying between f_S and f_1 ($f_1 > f_N$) gives rise to the "aliasing" phenomenon. In order to avoid it, a higher sampling frequency should be used, depending upon the rate of the filter fall-off. Usually, it is sufficient to take

$$f'_s = 3f_N.$$

Sampling at higher rates increases the requirements concerning the size of the digital memory, which leads to cost ineffective solutions.

Hence, a rule of thumb for preventing aliasing in a signal of unknown bandwidth is to analogously filter it using a cutoff frequency of $\frac{1}{2} f_u$ and sample it using a sampling frequency of minimum $2f_u$.

5.3.5.2 *Number of discrete samples.* Another problem refers to the number n of distinct samples required to completely describe a record of finite length for all t.

Consider a bandwidth limited white noise of bandwidth (0, B), recorded for the time interval (0, T). The finite bandwidth may be achieved by adequate filtering. The signal power spectral

5 Instrumentation

density is constant within the frequency interval $(-B, +B)$ and zero outside this interval. The associated autocorrelation function (Fig. 5.38) is given by

$$R_{xx}(\tau) = c\,\frac{\sin \pi 2B\tau}{\pi 2B\tau},$$

where c is a constant.

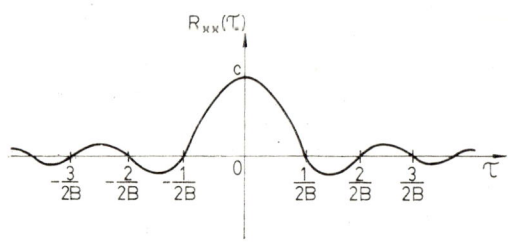

Figure 5.38

Note that for $\tau = \dfrac{k}{2B}$ ($k = 1, 2, 3, \ldots$) the autocorrelation function vanishes. Discrete values $\dfrac{1}{2B}$ apart are uncorrelated, that is statistically independent.

The total sampling time being T, the number of discrete statistically independent samples is

$$n = \frac{T}{\dfrac{1}{2B}} = 2BT.$$

There is shown [5] that the same number of discrete samples are required when sampling the fundamental intervals in the time or in the frequency domain.

The ratio $\dfrac{1}{2B}$ is the fundamental time increment, being also called the *Nyquist interval* and the fundamental frequency increment $1/T$ is called a *Nyquist cointerval*.

It should be mentioned that $n = 2BT$ represents the number of statistically independent samples, also called "the number of degrees of freedom" of the random signal, only for a Gaussian bandwidth limited white noise of bandwidth $(0, B)$.

For actual random data, the bandwidth limited white noise hypothesis will generally not be satisfied. For this reason, an "equivalent bandwidth" is defined [5], by means of which the number n is evaluated.

5.3 Conversion instruments

5.3.6 Averagers

In signal processing, two kinds of averaging are used : ensemble averaging and time averaging/integration.

Ensemble averaging is performed either in the frequency domain — on power spectra, or in the time domain — on time history records (sometimes referred to as *signal averaging*), to reduce the variance of a measurement (by reducing unwanted noise). If several time records of frequency spectra are averaged together, the noise will average to an expected value of zero and previously hidden signals will become visible.

When power averaging is applied to a mixed random process, the deterministic portion of the signal is unaffected, since its variance is zero. Only the estimate of the random portion of the spectrum will be smoothed.

In digital analysis, power averaging is directly related to the so-called "video filtering" (post detection low-pass filtering) used with conventional swept spectrum analyzers to smooth the measurement of random process spectra.

Time domain averaging is used if there exists an (trigger) independent signal, free from noise, which is synchronous with the periodic part of a mixed random process. It is used to recover periodic functions and related harmonics in applications where such periodic signals are generated (e.g. : in rotating machinery). Components which are synchronous with the trigger will average up to their mean value, while noise or nonsynchronous components will average up to zero.

Time averaging/integration is defined by the short-term average

$$\bar{g} = \frac{1}{T} \int_0^T g(t)\, \mathrm{d}t.$$

The function $g(t)$ may represent the signal $x(t)$, its square $x^2(t)$, the products $x(t)x(t-\tau)$ or $x(t)y(t-\tau)$, according to the quantity of interest.

More often, this operation is carried out after *squaring*, in order to suppress the fluctuations in the squared rectified signal from the squaring circuit, to get an expression of the mean power.

There is a difference between the averaging/integration of analog and digital signals.

5.3.6.1. *Time averaging/integration of analog signals.* Time averaging of analog signals can be performed in a number of ways [6], all implying the integration by means of operational amplifiers (Fig. 5.29) or, more often, by passive RC circuits (Fig. 5.27).

5 Instrumentation

There are four ways in which time averaging/integration is experimentally performed:

a) *Long-time averaging/integration*. In this case the averaging time T is chosen equal to the total time of observation of the signal.

b) *Step-wise averaging/integration*. The signal is integrated and averaged over a time T, then a new averaging is done over another time interval T, starting at the end of the first period. The result of averaging is given at the end of each interval T.

c) *Running averaging/integration*. The signal is integrated and averaged over the last T seconds of the record, the integrating memory continuously discarding the signal values which occurred before $(t - T)$.

d) *Weighted averaging/integration*. In analog measuring instruments, this is done using RC averaging circuits which are said to do an *exponential weighting*. Indeed, the frequency response function of the circuit from Figure 5.27 is

$$H(i\omega) = \frac{1}{1 + i\omega RC}.$$

The response to a unit impulse excitation is

$$h(\tau) = \frac{1}{2\pi} \int_{-\infty}^{\infty} H(i\omega)\, e^{i\omega\tau}\, d\omega = \frac{1}{2\pi} \int_{-\infty}^{\infty} \frac{e^{i\omega\tau}}{1 + i\omega RC}\, d\omega = \frac{1}{RC} e^{-\frac{\tau}{RC}}$$

For an input $x(t)$, the output $y(t)$ is given by Duhamel's integral

$$y(t) = \int_{-\infty}^{t} x(t) h(t - \tau)\, d\tau = \frac{1}{RC} \int_{-\infty}^{t} x(t)\, e^{-\frac{t-\tau}{RC}}\, d\tau. \tag{5.9}$$

Equation (5.9) may be written as

$$\overline{g(t)} = \frac{1}{RC} \int_{-\infty}^{t} g(t)\, e^{-\frac{t-\tau}{RC}}\, d\tau,$$

so that $\overline{g(t)}$ represents a weighted average, the weighting function being an exponential.

While running averaging gives equal weights to all signal instantaneous values, within the averaging time T, the exponential weighted averaging gives greater weight to signal values occurring at the instant of measurement.

5.3 Conversion instruments

For stationary analog signals, the result is the same irrespective of the averaging method used and in case d), on condition that $T \cong 2RC$. For non-stationary signals, the running averaging permits detection of short duration phenomena, which are very important for machine integrity. Figure 5.39 presents a compari-

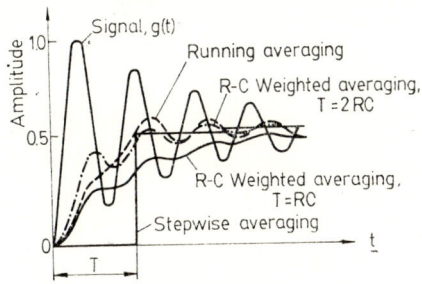

Figure 5.39

son of the results obtained using methods b), c) and d) in the case of a transient signal [6]. In the case of impulsive signals, methods c) and d) give similar results provided that $T \cong RC$. Instruments operating with exponential averaging have indicated on the front panel the equivalent averaging time (with uniform weighting) which gives the same level of fluctuations on stationary signals (equal to $2RC$).

5.3.6.2 *Time averaging/integration of digital signals.* In the case of *linear averaging* — the arithmetic mean is calculated by recurrence over a fixed time interval — the averaging time. Each sample x_k entering the averaging unit is divided by a factor N, equal to the total number of discrete samples occurring during the averaging time. The following quantity is calculated

$$m_k = m_{k-1} + \frac{x_k}{N}$$

where

$$m_{k-1} = \frac{1}{N} \sum_{j=1}^{k-1} x_j.$$

When $k = N$, the averaging time is over and m_N is the required mean value.

In the case of *exponential averaging*, a "running" mean value is calculated over a time interval which is updated with each new value x_k. The following quantity is calculated

$$m_k = m_{k-1} + \frac{x_k - m_{k-1}}{N}$$

5 Instrumentation

as a function of the weighted average of the previous sample m_{k-1} and the most recent sample value x_k, the value of k being not limited to N.

Given the difficulty to divide by an arbitrary integer N, averaging blocks of modern instruments perform division to a power of 2. The computation algorithm becomes

$$m_k = m_{k-1} + \frac{x_k - m_{k-1}}{2^n}.$$

At the beginning, n is set equal to zero, then increases in steps of one unit each time the number of samples N equals 2^n. For instance, when N becomes $64 = 2^5$, n steps from 5 to 6. The final value of n is given by the time constant of the averaging block. The averaging process continues with this value of n.

The averaging by recurrence of digital signals would resemble the weighted exponential averaging of analog signals if the RC circuit contained a capacitor whose value would increase during measurement.

5.3.7 R.M.S. detectors

R.M.S. detectors (Fig. 5.40) include a squaring stage, a time-averaging stage (described in Section 5.3.6) and a read-out instrument, calibrated according to the required quantity.

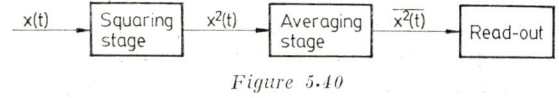

Figure 5.40

A squaring stage has an output $y(t) = x^2(t)$. This can be either an analog multiplier or a device with a parabolic characteristic. In practice, parabola is approximated by straight line segments and is obtained with diode-scaling circuits.

For measuring the r.m.s. value, a squarerooting operation is necessary, which is simply done using a non-linear meter scale.

The measurement accuracy depends on the errors introduced by each stage. The squaring stage must give the best approximation of the parabola and an integration time T must be provided for the averaging stage, at least equal to that given by the standard error equation (see Table 2.3).

For stationary signals, the meter settles after an interval equal to 3 to 4 times the time constant of the RC integrating stage, when the reading is possible. It is necessary to have sample records several times longer than the minimum integration time T and than the time constant of the RC circuit.

5.4 Instruments for signal analysis

5.4.1 Correlators

In order to obtain the correlation function (defined in Section 2.3), the following operations must be performed : 1) delaying the signal

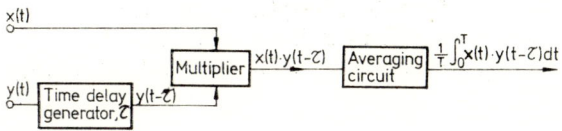

Figure 5.41

$y(t)$ by a time displacement τ, called the *lag time*; 2) multiplying at any instant the signal $x(t)$ by the value $y(t-\tau)$ that had occurred τ seconds before; 3) averaging the instantaneous product value over the sampling time T. These operations can be accomplished by analog or digital instruments called *correlators*.

The functional block diagram of a correlator is shown in Figure 5.41. A detailed description of various types of correlators may be found in Chapter 12 of reference [7].

5.4.1.1 *Stepped correlators*. Stepped correlators generate a lag time τ_k and measure the functions $\bar{R}_{xx}(\tau_k)$ or $\bar{R}_{xy}(\tau_k)$, then generate another lag time τ_{k+1} and determine $\bar{R}_{xx}(\tau_{k+1})$ or $\bar{R}_{xy}(\tau_{k+1})$, etc. The sample time history record is completely analysed for each lag time τ_k, which requires signal recording on tape, or storing in a memory. Older systems used a magnetic drum recorder with variable distance between recording head and playback head [1].

The *analysis time* of a random signal is a function of the imposed normalized standard error, as shown in Section 2.3.2. For the case of bandwidth limited Gaussian white noise, with a bandwidth B, the normalized standard error of the autocorrelation function for zero lag time ($\tau = 0$) is (see Table 2.3)

$$\varepsilon_0 = \varepsilon[\bar{R}_{xx}(0)] = \frac{1}{\sqrt{BT}}, \qquad (5.10)$$

where T is the equivalent true averaging time.

Given ε_0 and B, equation (5.10) can be solved for the averaging time T (which is approximately equal to the sample record length) yielding $T = \dfrac{1}{B\varepsilon_0^2}$.

5 Instrumentation

For instance, if $B = 20$ Hz and $\varepsilon_0 = 5\%$, then the sampling time is $T = \dfrac{1}{20 \cdot 0.05^2} = 20$ s.

When the correlation function is separately determined for k distinct values of the lag time, the analysis time will be kT.

Stepped correlators require a long operating time, because at the analysis time required for n points of the correlogram, one should add the time for rewinding the magnetic tape and changing the distance between recorder heads. Stepped correlators are of historical interest, continuous scan correlators being normally employed.

5.4.1.2 *On-line correlators.* Using automatic non-real-time correlators, the n values of the correlation function are simultaneously obtained

$$\bar{R}_{xy}(0) = \frac{1}{T} \int_0^T x(t)y(t)\, dt,$$

$$\bar{R}_{xy}(\tau) = \frac{1}{T} \int_0^T x(t)y(t-\tau)\, dt,$$

$$\bar{R}_{xy}(2\tau) = \frac{1}{T} \int_0^T x(t)y(t-2\tau)\, dt,$$

. . . .

$$\bar{R}_{xy}(n\tau) = \frac{1}{T} \int_0^T x(t)y(t-n\tau)\, dt,$$

so that it is not necessary to repass the signal through the analysis system, for each value of the lag time.

Automatic correlators use digital signals. Figure 5.42 shows schematically the sampling of signals and the multiplication sequence. Let T_S be the sampling period of the signal $x(t)$; if n is the number of lag times $\tau = T_S$ at which the correlation function has to be evaluated, the sampling period of the signal $y(t)$ is nT_S. The discrete sample $y(kT_S)$ is stored in a memory and successively multiplied with n discrete samples of the signal $x(t)$.

$$x(kT_S),\ x[(k+1)T_S],\ x[(k+2)T_S],\ \ldots,\ x[(k+n-1)T_S].$$

5.4 Instruments for signal analysis

The products are stored in memory, ordered according to the lag time duration

$x(kT_S) \cdot y(kT_S)$ 0 lag time;

$x[(k+1)T_S] \cdot y(kT_S)$ T_S lag time;

. . . .

$x[(k+n-1)T_S] \cdot y(kT_S)$ $(n-1)T_S$ lag time.

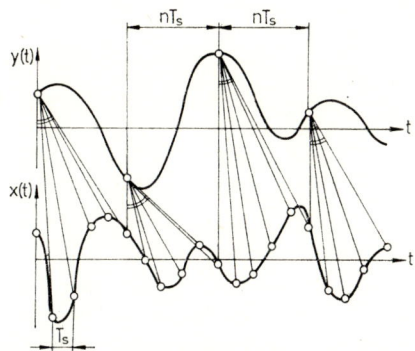

Figure 5.42

After completion of these operations, the procedure is taken again using the discrete samples $y(lT_S)$ and $x(lT_S)$, $x[(l+1)T_S]$, $x[(l+2)T_S], \ldots x[(l+n-1)T_S]$, respectively.

Supposing that each point of the correlation function plot is calculated by averaging N products, the sequence of computing the products and ordering them in memory is repeated N times. The sample record length is NnT_S. Averaging is done as the products are stored.

There is shown [5] that, for the presented procedure, the standard error for zero lag time is

$$\varepsilon_0' = \sqrt{\frac{2nT_S}{nNT_S}} = \sqrt{\frac{2}{N}} \quad (5.11)$$

hence, the estimation error diminishes when the number of averaged products increases.

For analog signals, the standard error for zero lag time is given by equation (5.10). If the signal bandwidth is $(0, B)$, then according to the sampling theorem (see Section 5.3.5.1), the sampling period is

$$T_S = \frac{1}{f_S} = \frac{1}{2\alpha B}. \quad \alpha \geqslant 1 \quad (5.12)$$

5 Instrumentation

The record length of the sampled signal is
$$T' = nNT_S. \tag{5.13}$$
Substitution of (5.12) and (5.13) into (5.11) yields
$$\varepsilon'_0 = \frac{1}{\sqrt{BT'}} \sqrt{\frac{n}{\alpha}}. \tag{5.14}$$

Comparing equations (5.14) and (5.10) it is seen that for a given value of the record length ($T = T'$), the error given by the on-line correlator is $\sqrt{\frac{n}{\alpha}}$ times greater than that obtained in the case of the stepped analog correlator. In order to get the same standard error for zero lag time ($\varepsilon'_0 = \varepsilon_0$), the record length should be $\sqrt{\frac{n}{\alpha}}$ times longer when using sampled signals than in the case of analog signals.

5.4.1.3 *Real time correlators.* Real time correlators provide automatically and simultaneously n points of the correlation function, for n different lag times, while the phenomenon is in progress. These correlators give the same amount of information on the signal as n stepped correlators which would calculate each correlation function corresponding to a given value of the lag time.

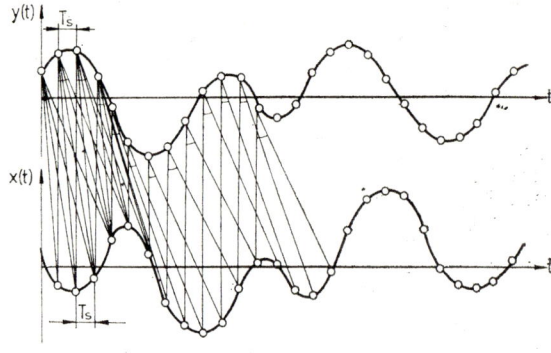

Figure 5.43

Figure 5.43 shows the sampling of the two signals for a real time correlator; the sampling period $T_S = \tau$ is the same for both input channels in the multiplier block, unlike the on-line correlator which samples in one channel at a rate of $1/T_S$ and in the other, at a rate of $\frac{1}{nT_S}$.

5.4 Instruments for signal analysis

Modern real time correlators provide auto- and cross correlation functions with incremental time delay values ranging from 0.05 μs to 2 seconds, resulting in total time delays from 20 μs to 800 seconds. These functions are simultaneously determined at 400 incremental lag points, so that a complete correlation function is displayed at one time. Examples are Solartron JM 1860/61/63 and Honeywell SAI-42-A, SAI-43-A and SAI 48.

The standard error being inversely proportional to the product BT (T-averaging time of the products $x(t)y(t - \tau)$ and B — signal bandwidth), in order to obtain a satisfactory error for band-limited signals, a prohibitive averaging time would be necessary if analog methods were used. Thus, for narrowband signals, digital integration/averaging is preferred. For wideband signals, the analog averaging time decreases and low-pass filters may be used.

With modern instruments, correlation is performed by the inverse Fourier transform of power spectra, using a digital signal processor, usually an FFT spectrum analyzer.

5.4.2 Bandpass filters

The bandpass filter is a circuit designed with the aim to approximate the ideal rectangular response illustrated in Figure 5.44 by a solid line. Frequency components of the input signal U_i between the frequencies f_l and f_u are found unaltered in the output signal U_0, those outside this band being completely rejected.

The response of physically realizable filters looks like the dashed line from Figure 5.44. It can be obtained combining the response characteristics of a low-pass (Fig. 5.27) and of a high-pass filter (Fig. 5.30), hence connecting them in tandem.

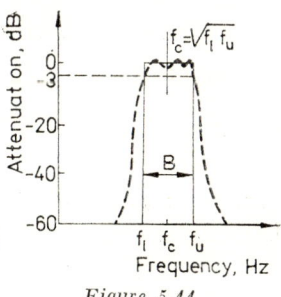

Figure 5.44

In the rejection band, the slope of such a filter characteristic is —20 dB/decade, being too small for practical applications where steeper filter skirts are required, falling off at rates up to —120 dB/decade.

Generally, one wishes to have a minimum of ripple within the passband and steeper slopes outside this band. Frequencies f_l and f_u at which the attenuation is 3 dB (the output power is half the mean passband power) are called *cutoff frequencies*. The passband at —3 dB, $B = f_u - f_l$, is called the *filter bandwidth*. At some filters, the geometrical mean $f_c = \sqrt{f_l f_u}$ of the cutoff frequencies is called *centre frequency*.

149

5 Instrumentation

Two types of electronic filters are used: *constant bandwidth filters* having $B =$ const. independent of the centre frequency f_c on which are tuned, and *constant percentage bandwidth filters* having B proportional to f_c.

5.4.2.1 Constant percentage bandwidth filters.

Examples are octave filters and third octave filters used at sound level meters and frequency analyzers.

An octave is the frequency band between any two frequencies having a ratio 2 : 1. The recommended octave bands (e.g.: by ASA Standard S1.6—1960) are listed in Table 5.1 [9]. It can be seen that

$$\frac{f_u}{f_l} = 10^{0.3} = 1.9952 \cong 2$$

so that, essentially, these are octave bands.

TABLE 5.1   *Standardized Octave Bands* [9]

f_l [Hz]	f_c [Hz]	f_u [Hz]	f_l [Hz]	f_c [Hz]	f_u [Hz]
1.41	2.0	2.82	178	250	355
2.82	4.0	5.63	355	500	709
5.63	8.0	11.2	709	1000	1410
11.2	16.0	22.4	1410	2000	2820
22.4	31.5	44.7	2820	4000	5630
44.7	63.0	89.2	5630	8000	11200
89.2	125.0	178.0	11200	16000	22400

Because

$$\frac{B}{f_c} = \frac{f_u - f_l}{\sqrt{f_u f_l}} = \frac{2f_l - f_l}{\sqrt{2f_l^2}} = \frac{f_l}{f_l\sqrt{2}} = 0.707 \qquad (5.15)$$

it follows that $B = 0.707 f_c$, so that octave filters have a 70.7 % bandwidth, being constant percentage bandwidth filters.

Practice has shown that octave bands are adequate for specifying noise levels but are too coarse for a detailed analysis of individual sources producing noise in a machine.

Better resolution is obtained using third octave filters. Their bandwidth corresponds to third octave bands, i.e. bands whose limiting frequencies have a ratio $2^{\frac{1}{3}} : 1 = 1.2599$. Third octave bands recommended by standards (ANSI S1.11—1966, IEC 225—1966, DIN 45652, ISO R 266—1975) have been designed so that $f_u/f_l = 10^{0.1}$ and the centre frequencies of two adjacent bands have a ratio of $10^{0.1}$.

5.4 Instruments for signal analysis

Because the value $f_u/f_l = 10^{0.1} = 1.2589$ is very near $2^{\frac{1}{3}} = 1.2599$, they are essentially third octave bands.

Filters of 1/10 octave (8%) and filters as narrow as 5%, 3% or 1% are used. The last mentioned can consist of two single pole Butterworth filters in series. Usually, filters having $f_c > 200$ Hz are passive LC filters and those with $f_c < 200$ Hz are active RC filters.

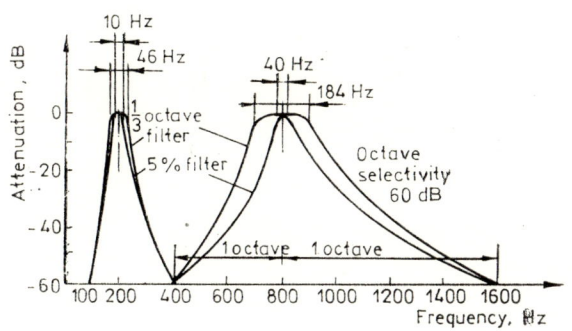

Figure 5.45

As the bandwidth B is proportional to the centre frequency f_c, the bandwidth of a constant percentage filter increases at higher frequencies, determining a decrease in the frequency resolution of the analysis. From Figure 5.45 [10] it is seen that a third octave filter has $B = 46$ Hz at $f_c = 200$ Hz and $B = 184$ Hz at $f_c = 800$ Hz.

Rejection characteristics of constant percentage filters are given in terms of *octave selectivity* which equals the attenuation one octave from the centre frequency of the filter. Thus, a typical 5% filter (manufactured by Spectral Dynamics Corp.) has an octave selectivity of 60 dB (Fig. 5.45). When tuned on 800 Hz this filter has a 40 Hz bandwidth at -3 dB and a 1200 Hz bandwidth at -60 dB [10]. The *shape factor* of this filter, defined as the ratio of the bandwidth at -60dB to the bandwidth at -3dB, is 30 :1. A filter with a high shape factor is inadequate for a detailed spectral analysis.

Constant percentage bandwidth gives uniform resolution on a logarithmic frequency scale. This scale is best suited for analysing spectra dominated by structural resonances, corresponding to constant magnification factors.

5.4.2.2. Constant bandwidth filters. These filters are found in instruments called *wave analyzers* and *heterodyne analyzers*, having bandwidths of 3, 10, 50, 100 or 1000 Hz, independent of the centre

5 Instrumentation

frequency on which are tuned. For this reason they give better resolution for high frequency measurements.

Rejection characteristics of these filters are expressed in terms of the *shape factor* defined above. Figure 5.46 [10] shows the characteristic of a filter having a shape factor of 4 : 1.

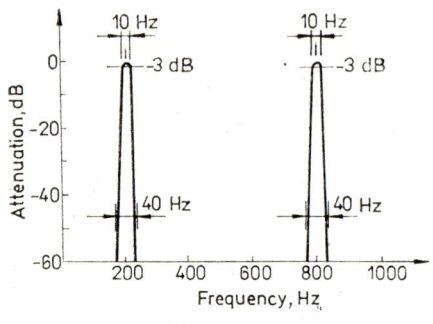

Figure 5.46

Constant bandwidth gives uniform resolution on a linear frequency scale. This scale gives equal resolution and separation of harmonically related components.

5.4.2.3. *Filter response time.* When analysing continuous deterministic signals, the sweep speed is determined by the filter response time T_R which is given approximately by

$$T_R \cong \frac{1}{B} \ [s]$$

where B is the -3dB bandwidth.

5.4.3 Non-real time spectrum analyzers

5.4.3.1 *Noise analyzers.* The first instruments used in selective filtering were *noise analyzers* having sets of octave and third octave filters. Each component filter is tuned on a different centre frequency in order to cover the entire frequency range of interest. The output signal of each filter contains only those frequency components of the input signal which are within the filter bandwidth. Filters are designed with contiguous bandwidths so that the upper cutoff frequency of one filter corresponds to the lower cutoff frequency of the next filter with adjacent bandwidth

5.4 Instruments for signal analysis

(Fig. 5.47). By successively commuting the filters to a detector, a band spectrum of the analysed signal is obtained. Examples are Brüel & Kjaer 2113 and 2114 Frequency Analyzers.

Modern Sound Level Meters are provided with octave or third octave filter sets.

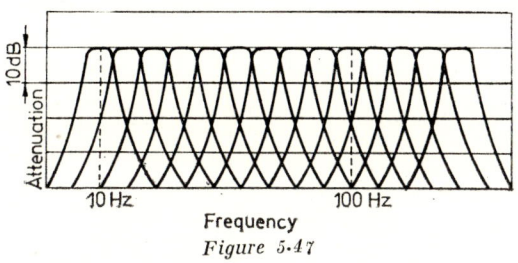

Figure 5.47

The Octave Filter Set Brüel & Kjaer 1613 contains 11 passive filters with centre frequencies from 31.5 Hz to 31.5 kHz. The Third-Octave Filter Set Brüel & Kjaer 1616 contains 34 active filters with centre frequencies from 20 Hz to 40 kHz. Other examples are General Radio 155A Octave Band-Noise Analyzer and 1564 A Sound Analyzer (1/3-octave) and the Third Octave Analyzer TOA 101 manufactured by VEB RFT Messelektronik "Otto Schön", Dresden.

5.4.3.2 *Tunable bandpass filters.* In order to obtain continuous frequency spectra (as opposed to "band spectra"), variable frequency bandpass filters can be used, covering the whole frequency range of interest. The Brüel & Kjaer 2121 Audio Frequency Analyzer contains an active, continuously variable RC-filter, having four selectable bandwidths 1%, 3%, 10% and 1/3-octave. The Brüel & Kjaer 1621 Tunable Band Pass Filter consists of a single pole Butterworth filter with switchable bandwidths of 3% and 23% (1/3 octave), covering the overall frequency range of instrument between 0.2 Hz and 20 kHz in five contiguous sub-ranges. The narrower the bandwidth, the better the frequency resolution.

5.4.3.3 *Heterodyne analyzers.* A more detailed analysis of the high frequency portion of the frequency spectrum is obtained using a single (narrow-band) constant bandwidth filter, within a continuously tuned analyzer called a *wave analyzer* or *heterodyne analyzer*, working according to the block diagram from Figure 5.48.

The analysed signal modulates the amplitude of a sine carrier wave, produced by a local oscillator. The frequency spectrum of the

5 Instrumentation

modulated signal is centred at the oscillator frequency, with the information contained in the sidebands. The carrier frequency f_p is chosen so that the filter having the centre frequency $f_c =$ = const. is tuned to a frequency within one of the sidebands. The filtered signal is then detected, leaving only the required low-frequency component.

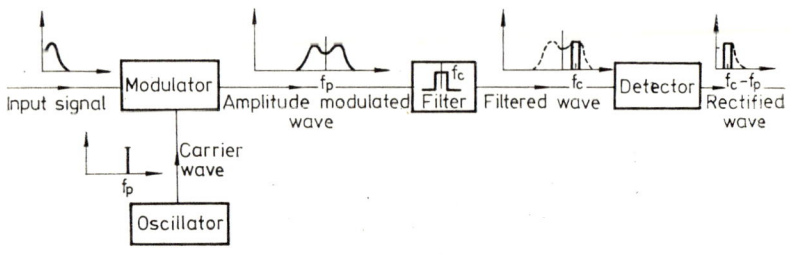

Figure 5.48

Sweeping — manually or automatically — the carrier frequency, the data spectrum is shifted ("translated") in frequency such that the filter with fixed centre frequency selects each time other frequency components of the analysed signal. Thus, the whole frequency range of interest can be explored using only one filter (tuned on a fixed frequency) by varying the carrier frequency. The filter can provide a control signal to a Level Recorder for synchronous movement of its paper (the paper advances at a rate corresponding to the sweep rate of the filter).

Examples of analyzers from this class are Brüel & Kjaer 2010, General Radio 1900 A and Hewlett Packard 302 A.

5.4.3.4 *Synchronous filters*. Tracking filters [12] are used in dynamic analyzers and the local oscillator sweeping is controlled by an external tuning signal. A simplified block diagram is shown in Figure 5.49.

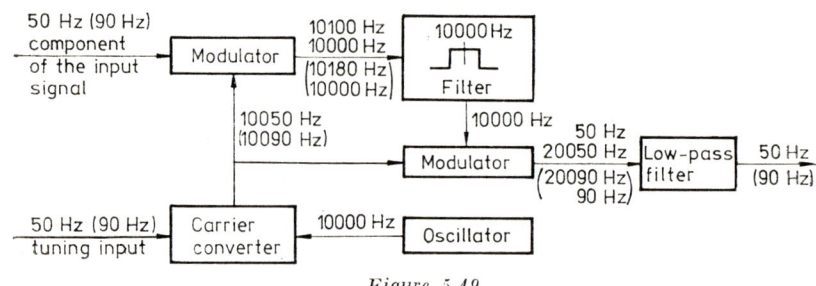

Figure 5.49

5.4 Instruments for signal analysis

Thus, for instance, for measuring the amplitude of the 50 Hz component of the data signal, the analyzer is tuned on this frequency, using an external 50 Hz sinewave at the tuning input. A 10 kHz reference oscillator signal (of frequency equal to the filter centre frequency) is added to the input tuning signal, creating a carrier of 10,050 Hz. The carrier signal is amplitude-modulated by the analysed signal. The 50 Hz component develops two sidebands of 10,100 Hz and 10,000 Hz, which are fed to the filter centred at 10 kHz. The only signal that passes through the filter is the lower sideband of 10,000 Hz. The filtered 10,000 Hz signal is then modulated by the original carrier signal of 10,050 Hz, creating two sidebands of 20,050 Hz and 50 Hz. The low-pass filter eliminates the upper sideband, leaving only the required 50 Hz component. The analysis of a frequency component of 90 Hz is illustrated by values given in parantheses.

Generally, distinction is made between true tracking filters and slave filters [13]. A true "tracking filter" can lock on the frequency of interest and follow its variations due to changes of the analysed structure parameters. A "slave filter" is externally tuned by a known parameter, proportional to the frequency of interest.

Examples are Spectral Dynamics SD 101A, Brüel & Kjaer 1623 tracking filters and Brüel & Kjaer 2020 and 2021 Slave Filters.

5.4.4 Real time spectrum analyzers

Generally, if results of an analysis can be displayed as fast as the analog data are received, then it is called a *real-time* analysis. Real time spectrum analysis is performed in the minimum possible time, being able to process the whole information contained by the analysed analog signal.

Because for a filter with bandwidth B, the response time is $1/B$, there is considered that an analysis is done in real time when data are processed so that the whole spectrum is obtained in just $1/B$ seconds. This is the shortest possible time in which the spectrum could be formed. The real time analyzer produces a spectrum having N spectral lines, where N is also called the *number of channels*.

The minimum time, required for the serial analysis of N spectrum lines, equals N times the response time $1/B$ of a filter with bandwidth B. Each frequency component is updated only once during the time N/B. At measurements of random vibrations, this time is $\frac{N}{4B\varepsilon^2}$, where ε is the standard deviation of each measurement.

In real time analysis, the spectrum level of each frequency component is updated every $1/B$ seconds $\left(\text{or each } \frac{1}{4B\varepsilon^2} \text{ seconds for random vibrations}\right)$.

In order to shorten N times the operating time of spectrum analyzers, the following methods might be used:

a) use of N parallel filters, by simultaneous measurement in the N channels, in the minimum time $1/B$;

b) signal compression in the time domain and subsequent analysis using a filter with bandwidth NB [Hz];

c) digital data processing, using high speed digital computers or FFT processors.

A significant aspect is connected to the time the signal is observed. In the case of sweeping analyzers, each of the N channels operates only $1/N$ of the time t. In a linear frequency sweep, only $100/N$ percent of each channel data is processed, the rest of $\frac{N-1}{N} \cdot 100$ percent being lost, the analyzer operating only a small fraction of the real time. In a 500 channel analyzer, $\frac{499 \cdot 100}{500}\%$ of the data would be missed and only 0.2% would be processed.

The real time frequency analysis permits a complete data analysis to be carried out, i.e. "all of the data is analysed in all of the channels for all of the time". This is very important for non-stationary signals which can be analysed directly.

5.4.4.1 *Parallel filter analyzers.* These analog instruments have a bandpass filter on each channel. The filters operate continuously and in parallel. The signal passes through all filters simultaneously. Each filter output is fed to separate analog detectors and associated circuitry whose outputs are electronically scanned and displayed on a continuously updated screen.

Analog analyzers are best suited for octave and third-octave measurements, where about 40 filters are required to cover the whole frequency range of interest. Examples are Brüel & Kjaer 3347 and GenRad 1921 analyzers.

5.4 Instruments for signal analysis

5.4.4.2 Time compression analyzers. The simplified block diagram of a time compression analyzer is shown in Figure 5.50. In the *digital time compression block*, the analog signal is converted in binary numbers and applied to a digital recirculating

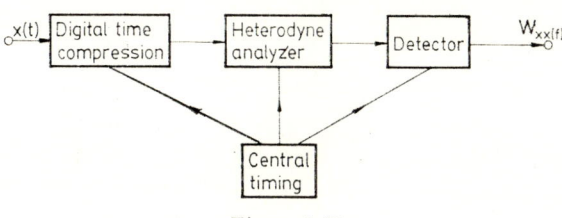

Figure 5.50

memory by which very large time compression factors are realized (up to 500,000). This way, the speeded-up high-frequency replica of the input signal can be processed in a heterodyne analyzer, using a wide filter which is stepped rapidly to specific points in the frequency spectrum, shortening considerably the analysis time.

The digital time compression block of the analyzer is presented in Figure 5.51. The analog input signal is sampled at time intervals of duration T_S (with frequency $f_S = 1/T_S$). Each sample is converted into a binary number (a word) and entered into a buffer

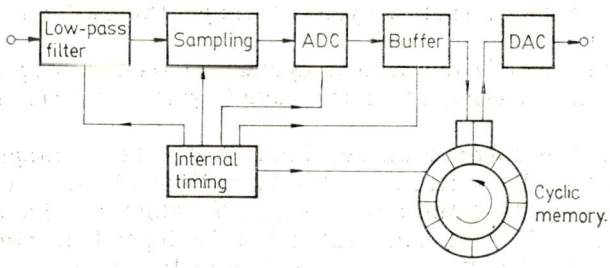

Figure 5.51

and then in a recirculating memory, capable of cycling hundreds of words at a very rapid rate. The memory cycling period T_M [s] is a characteristic of the analyzer (at some commercially available analyzers $T_M = 100$ μs). The cycling frequency is

$$f_M = \frac{1}{T_M} \quad [\text{Hz}].$$

157

5 Instrumentation

As illustrated in Figure 5.52, referring to the operation of a recirculating memory, at the beginning, the first sample is taken at a time designated by the first dot on the input signal. The digital number is loaded into the recirculating memory where it is cycled

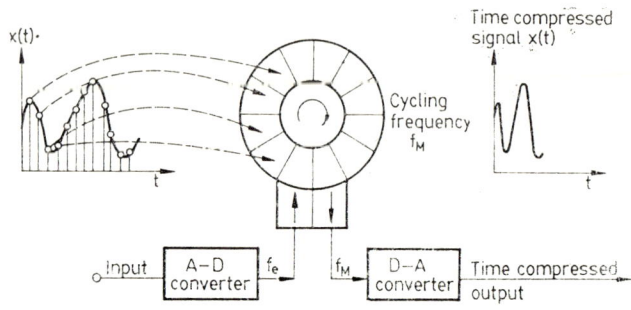

Figure 5.52

$n = \dfrac{f_M}{f_S} = \dfrac{T_S}{T_M}$ times, then the second sample is taken and the second number is loaded into the recirculating memory, directly behind the first sample. The process is continued until the memory is full of digital numbers corresponding to the segment of duration T [s] of the analysed signal.

Once the memory is full, the "oldest" sample is discarded and a new sample is taken to replace it. The updating of the memory is done at each n cycles, through the buffer controlled by the central timer.

For the analysis of transient signals, the recirculating memory can keep only a segment of the signal, without receiving new samples.

If f_u is the upper frequency limit of the chosen range and N is the number of spectral lines, the bandwidth of each "channel" is $B = f_u/N$, so that the record length T which can be analysed (in order to collect enough statistically independent samples to achieve the required number of degrees of freedom) is

$$T = \dfrac{1}{B} = \dfrac{N}{f_u}. \qquad (5.17, \text{a})$$

Since the total memory capacity is M words and at each T_S seconds a new word enters the memory, its content corresponds to an analog data segment of duration

$$T = MT_S = \dfrac{M}{f_S}. \qquad (5.17, \text{b})$$

5.4 Instruments for signal analysis

Equations (5.17) yield the digital memory capacity

$$M = \frac{f_S}{f_u} N. \quad \text{[words]} \quad (5.18)$$

Usually f_u is the cut-off frequency of the antialiasing filter. The sampling frequency should be at least $2f_u$, but due to the finite roll-off of the low-pass filter it is chosen $f_S = 3f_u$, so that in commercially available analyzers $M = 3N$.

The *compression factor* k equals the ratio of the rate at which words are read out from the memory to the rate at which they enter the memory. If M words are cycled once at each T_M seconds, i.e. f_M times in a second, the cycling rate is $M \cdot f_M$ words/second, being a constant irrespective of the analysed frequency range (M and T_M are characteristics of the analyzer).

The digital-to-analog converter reads each word at each memory cycle, at a speed of $M \cdot f_M$ words/second. The words enter the memory at a rate of one word at each T_S seconds, i.e. with a speed of $1/T_S = f_S$ words/second. The compression factor is

$$k = \frac{M f_M}{f_S} = M T_S f_M = T \cdot f_M = \frac{T}{T_M}.$$

From equation (5.18)

$$k = \frac{N f_M}{f_u}$$

so that the upper frequency limit of the reconstructed signal is $N \cdot f_M$, being always the same. This greatly simplifies the design of the heterodyne analyzer.

In order to produce N spectral lines, one can use a filter having a bandwidth $\frac{1}{N}(N \cdot f_M) = f_M$ [Hz]. As the response time of this filter is approximately $T_M = \frac{1}{f_M}$, each spectral line is obtained in T_M seconds, so that the whole spectrum is obtained in $N T_M$ seconds.

At modern analyzers, $N \cdot T_M = 500 \cdot 10^{-4} = 0.05$ seconds, so that the whole spectrum could be visualized $\frac{1}{N \cdot T_M} = \frac{1}{0.05} = 20$ times per second, which makes the operator to perceive it as being continuous. Because at each T_S seconds the memory content is updated, in the case of non-stationary signals, the spectrum appears $\frac{T_S}{N T_M}$ times, after which it changes.

5 Instrumentation

Results in the N channels appear sequentially, in $NT_M = 50$ ms. The analysis time is always the same and is independent of the data sampling time or of the data loading time. To the extent that $N \cdot T_M < \dfrac{1}{B}$, a real time analysis is obtained.

At the output from the recirculating memory, the signal is reconverted to analog form. The oscillator of the heterodyne analyzer, controlled by the central timing, generates a sine wave whose frequency is *stepped* through a range $f_2 - f_1 = \Delta f = Nf_M$ [Hz] in NT_M seconds (once at N memory cycles). This sine wave is multiplied by the time compression block and passed through a relatively broad constant frequency bandpass filter centred, for instance, on the frequency f_1. If $B = 10$ kHz, the filter response time is $\dfrac{1}{B} = 100$ μs so that it takes 100 μs to measure one channel and 50 ms to measure 500 channels. The original spectrum is obtained with the conventional heterodyne technique.

It should be underlined that, although the real time analyzer can produce an analysis in 50 ms, it cannot analyse a signal in 50 ms. The data collection time is $1/B$, being equal to the time required to fill the memory. Obviously, the memory must be first completed before the correct result can be obtained. On the other hand, the effective analysis time is represented by a sliding 50 ms window, so that a complete 500-line scan is not performed on one precise 50 ms segment of data, but on a sliding segment, due to the scanning operating principle.

Usually it is assumed that the analysed signal is stationary over a time interval of at least one memory period. Special precautions must be taken if frequencies beyond the real time rate are encountered.

Examples of commercially available time compression analyzers are Brüel & Kjaer 3348 (400 channels), Nicolet's UA-500 A "Ubiquitous" (effectively everywhere at once in the frequency domain) and 440 A "Mini-Ubiquitous", Spectral Dynamics SD 330 A "Spectrascope" (500 channels), SD 335 "Spectrascope" II, Honeywell SAI-51C and SAI-52C.

5.4.4.3 *Weighting* (Time windows). Unlike the conventional heterodyne analyzer, which usually analyses a signal whose duration can be considered to be infinite, the time compression analyzer processes a segment of finite length of the analogic signal, corresponding to the finite storage capacity of the digital memory. This segment is continuously and repetitively applied to the input of the frequency analysis block. This way, the two ends of the signal are "joined" together (Fig. 5.53 a) and a discontinuity appears at the junction.

In order to suppress this discontinuity, a time-domain weighting stage is introduced in the time compression analyzer.

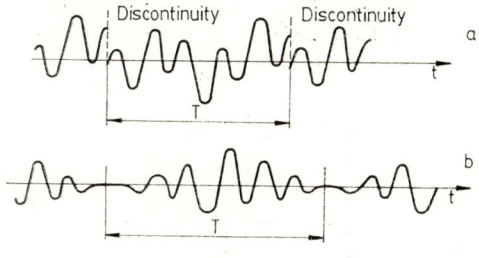

Figure 5.53

Its role is to eliminate the abrupt transition at the ends of the sample record (Fig. 5.53 b) and to emphasize the data in the centre of the sample.

The same problem arises with FFT analyzers. Due to the truncation of the time signal, the shape of the frequency spectrum is "smeared" when the data being analysed are not completely contained within the period of observation or are not exactly periodic within that time.

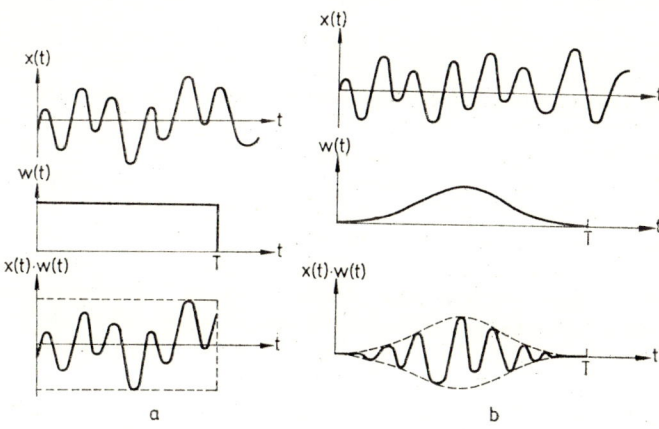

Figure 5.54

Let $x(t)$ be the signal waveform. Truncation of this waveform and selection of a sample of duration T can be done multiplying the signal $x(t)$ in the time domain by a rectangular function $w(t)$ of length T (Fig. 5.54 a). In fact, this is an amplitude-modulation of the signal prior to the frequency analysis. The weighting

5 Instrumentation

function $w(t)$ is also called "window function" or "convolution kernel". One of the most often used weighting functions is the Hanning window, whose effect on the signal is shown in Figure 5.54 b.

Multiplication in the time domain corresponds to convolution in the frequency domain. After windowing, the frequency spec-

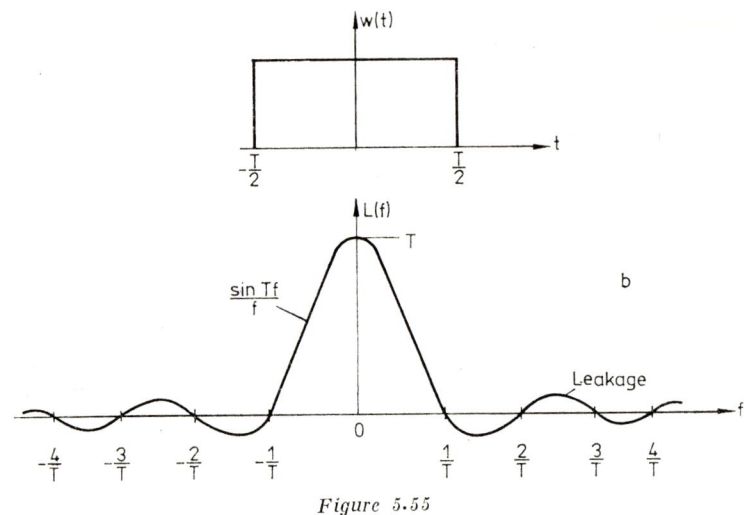

Figure 5.55

trum is the original spectrum convolved with the "line shape" of the window. The frequency line shape $L(f)$ of the rectangular time window (Fig. 5.55 a) is shown in Figure 5.55 b. It has a main lobe and several smaller side lobes. When a spectrum is convolved with such a function, the energy or power that would be associated with a particular frequency is spread over a wide range of adjacent frequencies. This phenomenon is called *leakage*. The convolution with a line shape "smears" the true frequency spectrum over a range of adjacent frequencies [15].

If the time signal is periodic, leakage can be eliminated by properly adjusting the window width to encompass an integer number of data periods. This is explained by the fact that the spectrum of periodic time signals has discrete values at frequencies which are multiples of $1/T$, where the line shape has nulls, so that no interaction is produced at these frequencies by their convolution.

When analysing transients, completely contained within the observation period, use of a weighting function is undesirable. The transient may contain more energy near the leading edge of

5.4 Instruments for signal analysis

the observation window, which is diminished by applying a weighting function which emphasizes the central part of the window. Thus, for signals that are periodic in the window interval T or for transients that completely decay within this interval, the rectangular window is best suited.

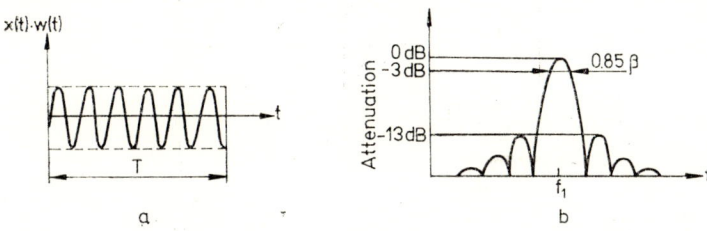

Figure 5.56

Figure 5.56 illustrates the effect of a rectangular weighting function on a sine wave of frequency f_1, in both the time domain (Fig. 5.56 a) and the frequency domain (Fig. 5.56 b). Denoting $\beta = 1/T$, where T is the weighting function length, the 3-dB width of the response is 0.85 β. The existence of high side lobes worsenes the analyzer *selectivity* (ability to resolve unequal amplitude adjacent frequency components).

For comparison, Figure 5.57 presents the effect of a Hanning function on the same sine wave. The side lobes are smaller, so that the frequency selectivity improves while the 3-dB bandwidth widens, so that the frequency *resolution* (ability to resolve equal amplitude adjacent frequency components) decreases.

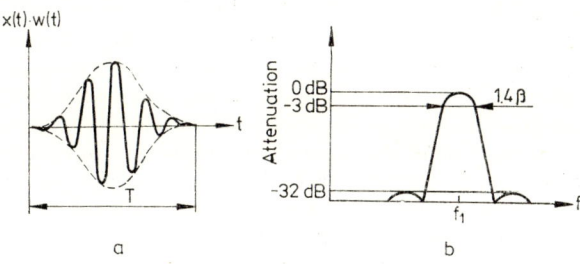

Figure 5.57

Figure 5.58 compares spectra of some common time weighting functions with that of the Gaussian function which has no sidelobes and very steep roll-off. Weighting always involves a com-

5 Instrumentation

promise between small side lobes and a narrow main lobe, i.e. between frequency resolution and selectivity, the latter being generally more important.

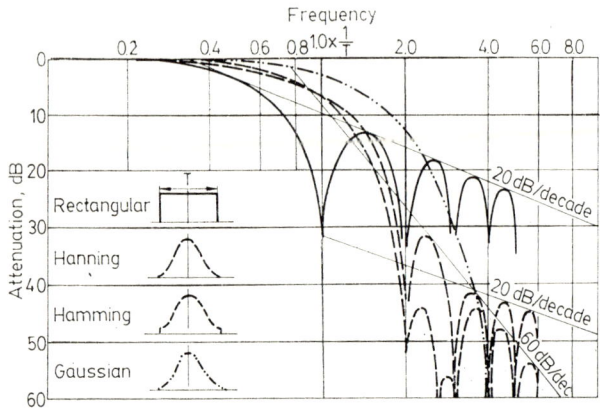

Figure 5.58

5.4.4.4 *Digital analyzers.* Low frequency spectrum analyzers are turning into digital instruments due to the implementation of efficient Fast Fourier Transform (FFT) algorithms and use of large-scale integrated (LSI) circuits.

Using an FFT algorithm, a complete frequency analysis may be performed in a fraction of the time required to do it using analog swept filters.

Direct computation of the Discrete Fourier Transform of an N-point complex valued function requires N^2 operations (an operation consists of one multiplication plus one addition). The FFT algorithms require about $N \cdot \log_2 N$ operations, where N is a power of 2. Thus, for $N = 1024$, the FFT computation is hundred times faster than standard computational procedures [15].

Digital spectrum analyzers can be classified as: a) digital filter analyzers; b) hardwired FFT analyzers (with no additional computation facilities); c) minicomputer based digital Fourier-analyzer systems; and d) Fourier transform analyzers (working in conjunction with correlators). A separate category consists of time-sharing computer based Fourier analyzer systems.

Digital filter analyzers use recursive digital filtering. In contrast to FFT, which operates on blocks of data, this is a continuous process which operates on individual samples of the input data. Digital filters are best suited for constant percentage bandwidth and logarithmic frequency scales. An example is Brüel & Kjaer 2131 Digital Frequency Analyzer used for octave and third octave analyses.

5.4 Instruments for signal analysis

FFT analyzers compute digital (sampled) forms of frequency response functions, power spectra and other frequency domain descriptors from measured (sampled) time domain signals.

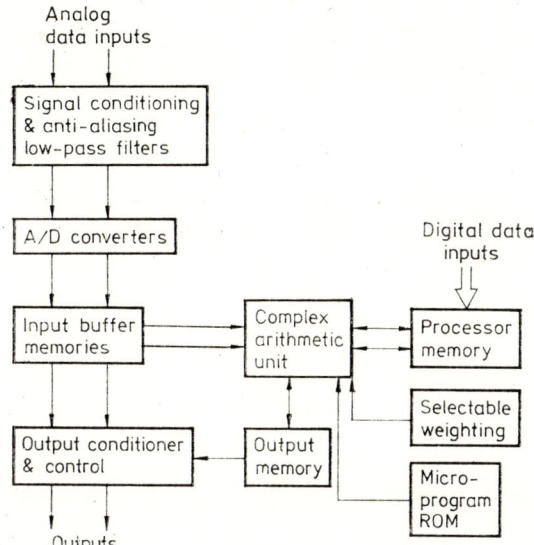

Figure 5.59

A simplified block diagram of an FFT hardwired processor is shown in Figure 5.59 [16]. Analog input signals are first conditioned through buffer amplifiers and antialiasing low-pass filters, then sampled in synchronous analog-to-digital converters and stored in input buffer memories. The contents of the input buffer memories are transferred to the processor memory at intervals determined by the selected analysis rate. Data are operated on by selectable weighting and micro-program read-only memories enabling fast processing speed. Results of the signal processing are stored in the output memory and made available through the output conditioner and control in both digital and analog forms.

FFT analyzers are designed to operate by transforming an entire block of data, each time data samples have been accumulated in the input memory.

For a single length transform ($N = 1024$), the input signal $x(t)$ is sampled N times at Δt intervals, where N is the block size and $\Delta t = T_S$ is the sampling period. This corresponds to a time record length $T = N \cdot \Delta t = N T_S = \dfrac{N}{f_S}$ where f_S is the sampling rate.

5 Instrumentation

The resulting frequency spectrum is calculated at discrete frequencies $\Delta f = 1/T$ apart, so that the time record is treated implicitly by the analyzer as one period of a periodic signal. The time record can be considered as being multiplied by a time window $w(t)$ of duration T. This results in the convolution of data spectrum with the transform of $w(t)$, what implies *leakage*. Sampling of the function $x(t)w(t)$ in the time domain, that is multiplication of $x(t)w(t)$ by a sampling function, causes replication of its spectrum at intervals $1/\Delta t$ along the frequency axis.

In order to store the results in a finite digital memory, the frequency function is sampled N times at Δf intervals, which causes replication of the time function at intervals $1/\Delta f = T$ along the time axis. The Discrete Fourier Transform (2.29) can be written

$$X(m\Delta f) = \Delta t \sum_{n=0}^{N-1} x(n \cdot \Delta t)\, e^{-\frac{i2\pi mn}{N}} \qquad (m = 0, \ldots, N-1)$$

Spectrum replication may lead to *aliasing* unless steps are taken so that the signal should contain no frequencies above half the sampling frequency.

Time function replication may lead to overlapping if frequency domain sampling is done at intervals greater than $\Delta f = 1/T$, because this implies that the corresponding time function will be replaced with a period smaller than T. But this is prevented by the fact that the time window has a duration T.

The uniform spacing Δf of the spectral lines leads to a linear frequency scale.

Let F denote the analysis (frequency) range of the processor, usually expressed as a convenient decimal value, and N_F — the number of equally spaced frequency resolution points (synthesized filters) called *spectrum lines*.

Adjacent filters are spaced by $\Delta f = \dfrac{F}{N_F}$ or, multiplying by $N/2$ both the numerator and the nominator,

$$\Delta f = \frac{\dfrac{N}{2N_F} F}{\dfrac{N}{2}} = \frac{F_{max}}{\dfrac{N}{2}},$$

where F_{max} is the upper frequency of the analysis range.

The filter bandwidth is

$$B = \frac{1}{T} = \frac{f_s}{N}.$$

5.4 Instruments for signal analysis

Usually, the filter bandwidth equals the filter spacing

$$B = \frac{f_S}{N} = \Delta f = \frac{F}{N_F} = \frac{F_{max}}{\frac{N}{2}}.$$

The sampling rate is $f_S = \frac{N}{N_F} F = 2F_{max}$, so that F_{max} corresponds to the Nyquist frequency.

The memory period is

$$T = \frac{N}{f_S}.$$

As an example, for $N = 1024$ words, $N_F = 500$ lines,

$$f_S = \frac{1024}{500} F = 2.048F, \quad T = \frac{500}{F}, \quad \text{and} \quad B = \frac{F}{500} = \frac{1.024F}{512}.$$

The frequency resolution is $\Delta f = \frac{F_{max}}{N/2}$. Because F_{max} is fixed by aliasing considerations and by the range of interest in a given experiment, the only way to improve frequency resolution seems to be the increase of the block size N. But this number is limited by the memory size and by the simultaneous increase of the processing time. The solution is use of "zoom transform" techniques, which provide greatly increased resolution without increasing N [11].

Picket-fence effect. The non-rectangular shape and finite spacing of the individual synthesized filters causes the so-called "picket-fence" effect. The response characteristic of a bank of such contiguous filters (Fig. 5.60) [17] evokes the image of a

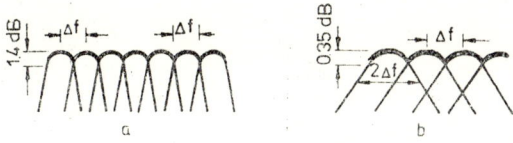

Figure 5.60

picket fence. If the frequency of a sine wave is exactly at the centre of any filter, it will have an amplitude which can be defined as 0 dB. As the frequency is slightly changed, the amplitude decreases to the minimum corresponding to the trough between two adjacent filters. The error is dependent on the filter selectivity,

5 Instrumentation

which is determined by the window function used. Thus, using a Hanning window, the maximum error will be 1.4 dB, while for a rectangular window it is 4.2 dB. Doubling the width of each filter $B = 2\Delta f$ and maintaining their spacing Δf, the maximum amplitude error can be reduced to 0.35 dB (with Hanning window), with the expense of reducing the nominal resolution to $N_F/2$ lines (Fig. 5.60 b).

The picket-fence effect occurs at the analysis of periodic signals. It will not occur at the analysis of broad band noise, whose spectrum is almost constant over the bandwidth of a synthesized filter.

Wrap-around error. Another error introduced by the discrete nature of the DFT is the "wrap-around error" encountered at measurements of correlation functions and auto- or cross-spectra, involving the product of two frequency functions.

Multiplication of two functions in the frequency domain causes convolution of their inverse transforms in the time domain. Sampling in the frequency domain implies replication in the time domain so that, for discrete transforms, the convolution process works with periodic functions. The convolution with one

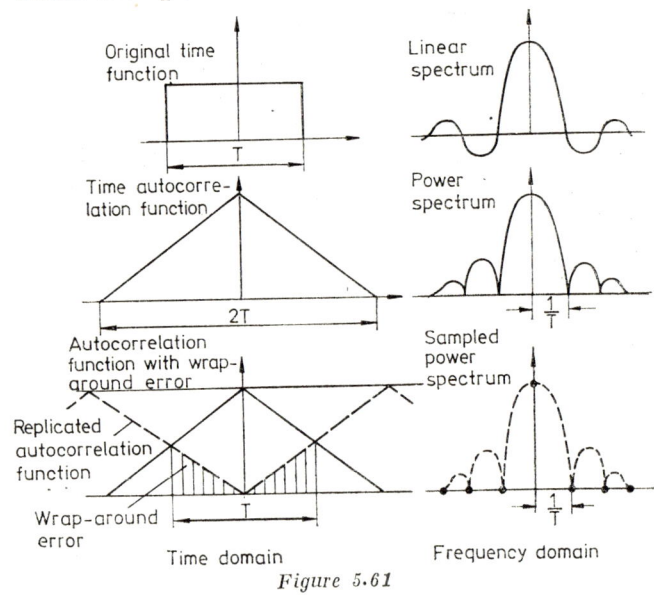

Figure 5.61

period of a function overlaps that of adjacent periods. This is analogous to aliasing in the frequency domain and is caused by undersampling the product of two functions in the frequency domain.

Figure 5.61 [15] shows that the autocorrelation function of a rectangle of length T is a triangle of base $2T$. If the corresponding

5.4 Instruments for signal analysis

power spectrum is sampled at intervals of $1/T$, instead of $1/2T$, the composite time function is replicated at intervals of T and the sum of these overlapped triangles is a constant. The "tails" of the expected true triangular shape might be visualized as being "wrapped-around" a cylinder of circumference equal to the time function length. Added to the remaining portion of the triangle, these tails produce a rectangle. This overlapping of replicated (periodic) time functions is called *wrap-around error*.

In many applications, the results are averaged to obtain signal-to-noise enhancement or to improve the statistical confidence in the results. Usually, the analysis of a given sample of data is performed with two statistical degrees of freedom, $2BT = 2\frac{F}{N_F}\frac{N_F}{F}=2$. By averaging the spectral ensembles, the statistical degree of freedom figure is increased by the number of independent samples averaged.

There is a large choice of FFT signal analyzers. They include GenRad 2512 Spectrum Analyzer, Brüel & Kjaer 2031 and 2033 Narrow Band Spectrum Analyzers, EMR Schlumberger 1510 Spectrameter, Hewlett Packard 3582A Spectrum Analyzer, Nicolet 444 A and 446 B "Mini-Ubiquitous" FFT Computing Spectrum Analyzers, Nicolet 660 A—2 D Dual-Channel FFT Analyzer, Rockland FFT 512/S Series II & III Narrow Band Octave and Third-Octave Real Time Spectrum Analyzer, Spectral Dynamics SD 345, SD 350, and SD 360 Digital Signal Processor and SD 375 Dynamic Analyzer II. Using IEEE 488 or RS 232 interfaces, these instruments can be connected to calculators like Hewlett Packard 9800 Series, Tektronix 4051, DEC PDP-11 or Data General NOVA to build up powerful signal processing and modal analysis computerized systems. Other examples are UNIGON FFT 1024A Fast Fourier Analyzer, Princeton Applied Research 4512 Real Time Spectrum Analyzer, Nicolet "Omniferous" 400B FFT Analyzer, Wavetek Rockland Scientific 5810A Mini-Analyzer and Brüel & Kjaer 2032 and 2034 Dual Channel Signal Analyzers.

A detailed description of the dual channel FFT analysis is given in [18].

Mini-computer based systems have increased flexibility to provide analysis, synthesis and manipulation of time-varying data. The data handling capabilities and speed of the mini-computer allow the storage, display and manipulation of large amounts of data. They are operated by conversational languages so that computer programming is not required. Preprogrammed functions include direct and inverse FFT, auto- and cross spectrum with selectable time windows, transfer and coherence functions, impulse response, auto- and cross correlation, amplitude histograms etc.

5 Instrumentation

Increases in processing speed are being accomplished using FPE microprogrammed Fourier processing elements and Fast Fourier Processors. Thus, the Time Data T/D 90A Processor needs 12 ms for the direct Fourier transform of a block of $N = 1024$ real words, compared to 348 ms required by the standard TDA3 Time Series Analyzer or 117 ms taken by the basic TDA51 Time Series Analyzer.

Hewlett Packard 5451C Fourier Analyzer contains a modified 2113 MX—E computer as the system processor. The HP 5451B Fourier Analyzer has a HP 2100 S computer. Nicolet 6601 B Modal Analysis System has a Data General micro NOVA computer. Time/Data Time Series Analyzers are based on PDP-11/05 or PDP-11/35 mini-computers.

Modern processing digital oscilloscopes can be added as examples. Norland Instruments 2001 A "Waveform and Data Analysis System" contains an Intel 8080 microprocessor and Tektronix 7704A oscilloscope has been completed with a P7001 Processor and a PDP-11/05 minicomputer.

In recent years, dedicated, turn-key, function-based instruments have been developed for *modal analysis* studies. Examples are GenRad 2507, 2508 Structural Analysis Systems and 2510 Micro Modal Analyzer, Hewlett Packard 5423A Structural Dynamics Analyzer and Hewlett Packard 5451B Fourier Analyzer plus Option 402. They perform identification of natural frequencies and damping, measurement of modal vectors and extraction of modal coefficients using either a least square circle fit procedure or a peak picking method [19]. GenRad 2515 Computer-Aided Test System has multichannel data acquisition capabilities to speed data collection.

Other mini-computer based systems are Spectral Dynamics SD 2001 DM Modal Analyzer, Nicolet OF-400 B FFT Analyzer plus Option MA and Solartron 1191.

Stand alone instruments using the FFT are becoming simpler and less expensive when compared to mini-computer systems [20].

Using a *terminal based analysis system* it is possible to have a low cost instrumentation system and be able to use advanced software necessary for modelling. Such a hybrid system can be used with commercial time-sharing services on in-house computer facilities. An example is Zonic's DMS/TSA (Data Memory System/ Time Sharing Analysis) hardware-based testing system which is linked via telephone lines to a computer containing sophisticated dynamic analysis programs [21].

Fourier Transform Analyzers are fully digital instruments which perform a Fourier analysis of any function computed by a correlator. An example is Honeywell SAI-470 FTA used in conjunction with SAI-42A, SAI-43A or SAI-48 Correlation and Probability Analyzers.

5.4 Instruments for signal analysis

5.4.5 Shock Spectrum Analyzers

The shock spectrum of an impulse (or transient) can be measured using an analyzer which utilizes a set of resonant RLC circuits (having different resonance frequencies and variable damping), each connected to a peak-holding detector. The detectors retain the maximum output of each circuit when a transient is applied to the input. By spacing the resonance frequencies of the oscillators over the frequency range of interest, a pattern of energy distribution versus frequency can be recorded, the resulting curve defining the shock spectrum of the particular impulse.

Modern shock spectrum analyzers, like Spectral Dynamics SD 320, are hybrid instruments, using analog single-degree-of-freedom filtering and two digital memories. The first memory is used to recirculate input signals with variable speed-up ratios to synthesize the various filter bandwidths. The outputs of the synthesized s.d.o.f. filters, in terms of primary, residual or maximax response, are then sampled and stored in the second memory. This enables display of one analysis while a second transient is being loaded into the input memory.

5.4.6 Amplitude distribution analyzers

For a sample record $x(t)$ of a stationary random signal, the probability density function $p(x)$ may be estimated by

$$p(x) = \frac{\sum_{i=1}^{n} (\Delta t_i)}{T \Delta x} \qquad (5.19)$$

where $\sum_{i=1}^{n} (\Delta t_i)$ is the time spent by the signal $x(t)$ in the interval $(x, x + \Delta x)$ and T is the sample record length (see Fig. 2.5).

The analog amplitude probability density analyzer samples the time history record and classifies the instantaneous values according to their amplitude. The content of each register corresponds to the number of samples having a given amplitude.

A functional block diagram for a sequential amplitude probability density analyzer is shown in Figure 5.62 [5]. It requires reintroduction of the signal in the analysis circuit for each point of the probability density plot.

The variable d.c. signal generator is used to produce a bias voltage which sets the lower limit value of the window width. The bias voltage x is subtracted from the input signal and the

5 Instrumentation

difference $[x(t) - x]$ is fed into the narrow amplitude window of width $(0, \Delta x)$. Each time the level of the signal $[x(t) - x]$ enters the window $(0, \Delta x)$, i.e. $x < x(t) < x + \Delta x$, the AND gate opens and passes the clock generated pulses. When the signal level leaves

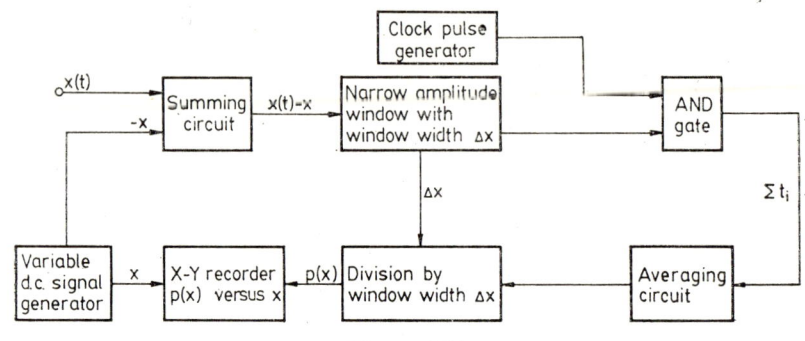

Figure 5.62

he window, the AND gate closes. This way, the time $\sum (\Delta t_i)$ spent by the signal within the window is measured. Then, two divisions are performed: by T and by Δx. The first is done by averaging over the total sampling time, the second by a proper scale calibration.

Analysis of the amplitude probability density can also be performed in parallel, obtaining simultaneously all the points of the probability density plot.

Amplitude distribution analyzers are instruments typically used to obtain measurements over relatively long time spans, especially for noise environments. Examples are Brüel & Kjaer 4426, BBN 614, GenRad 1945 and Metrosonics db-601 and 611.

5.4.7 Resolved component indicators

Vector components of complex frequency response functions are measured using wattmetric instruments [22]. A wattmeter responds to the average of the product of two applied signals. If one of these signals is $x(t) = \dfrac{a_0}{2} + \sum\limits_{n=1}^{\infty}(a_n \cos n\omega t + b_n \sin n\omega t)$ and the other is $e_0 \cos \omega t$, then the quantity indicated by the wattmeter is proportional to the Fourier coefficient $a_1 = \dfrac{2}{T}\displaystyle\int_0^T x(t) \cos \omega t \, dt$,

5.4 Instruments for signal analysis

where $T = \dfrac{2\pi}{\omega}$. Consequently

$$[x(t)\, e_0 \cos \omega t]_{\text{average}} = \dfrac{e_0}{2}\, a_1.$$

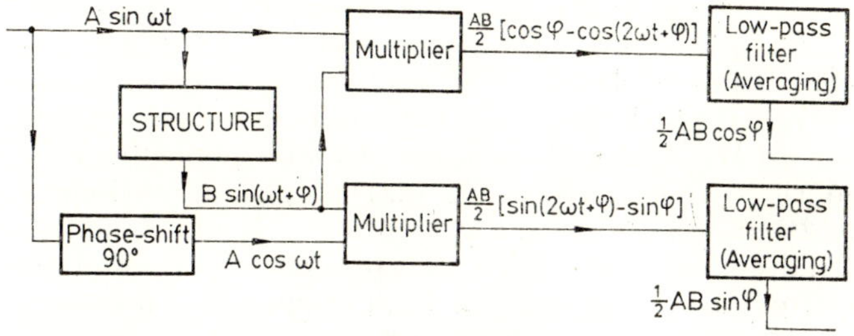

Figure 5.63

Actually, a wattmeter indicates a product averaged over some effective "averaging time", whose duration depends upon the time constant of the averaging instrument. If the time constant is too small, the output fluctuates and makes difficult the reading of the mean value. If the time constant is too large, the time required to reach steady-state conditions may be prohibitive. Usually, a compromise is realized, choosing the time constant 3 to 6 times greater than the period of the signal being analysed.

Figure 5.63 illustrates the application of the wattmeter principle for measuring the vector components of the frequency response for a linear system. Operations indicated by the block diagram can be performed in analog or digital instruments. The upper filter–averaging block calculates $\dfrac{1}{nT}\displaystyle\int_0^{nT} AB \sin \omega t \sin(\omega t + \varphi)\,\mathrm{d}t,$

while the lower block calculates $\dfrac{1}{nT}\displaystyle\int_0^{nT} AB \cos \omega t \sin(\omega t + \varphi)\,\mathrm{d}t,$

where $T = \dfrac{2\pi}{\omega}$ and n is the number of integration cycles. The scheme corresponds to the orthogonal correlation process.

Spectral Dynamics' Co/Quad SD 109A, Solartron's 1170 Series and DA 1310 Frequency Response Analyzers as well as Schenck's Vibrovid III are based on this principle. Strain gauge transducers can be used as multiplying devices when the bridge circuit is energized by a sinusoidal reference signal.

5 *Instrumentation*

5.5 Display and recording instruments

5.5.1 Stroboscopes

The stroboscope produces the intermittent illumination of a target glued on the (vibrating and rotating) body, at a frequency which could be controlled and measured. When the frequency of the stroboscopic lamp equals the frequency of the vibrating body, the target is illuminated each time at the same position, hence it appears motionless. The low frequency limit of the observed motion is 15 Hz, being determined by the persistence of vision.

Modern stroboscopes enable analysis of non-stationary phenomena to be done, the flashing frequency being synchronized to an external signal produced by a photoelectric tachometer probe or a control sensor. Analysis of high frequency vibrations is also possible, "slowing" the motion by an adequate adjustment of the flashing rate.

Examples are Philips PR 9111 and PR 9112 stroboscopes, GenRad 1540 Strobolume, 1531, 1538, 1542 — 1544 Strobotac, 1539 Stroboslave and Brüel & Kjaer 4911 Stroboscope.

5.5.2 Analog meters

The most common analog read-out instruments are the moving-coil d.c. meters. (Fig. 5.64) based on the action of a magnetic field on a d.c. current passing through a coil of wire. The pointer 1 is joined together with the moving coil 2, which can rotate in the field of a permanent magnet 3, about pivot 4, under the action of the electromagnetic torque, against the restoring torque produced by the spiral spring 5. The pointer moves through an angle proportional to the applied current. The quantity indicated on the scale 6 depends on the detector which precedes the meter.

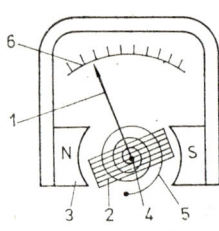

Figure 5.64

Another common instrument is the so-called "electronic voltmeter" which has the advantage that its input impedance is very high, so that it does not significantly loads the circuit being measured. Examples are Brüel & Kjaer 2425, 2426 and 2427, the last having a digital display.

5.5.3 Digital meters and printers

Digital multimeters and voltmeters accept analog voltage inputs and display the voltage reading in decimal digits. They consist of an analog-to-digital converter and some means of visual display.

5.5 Display and recording instruments

Modern digital meters use dual-slope integration (Fig. 5.65 a) for converting the voltage input to time output intervals via counter and clock [23].

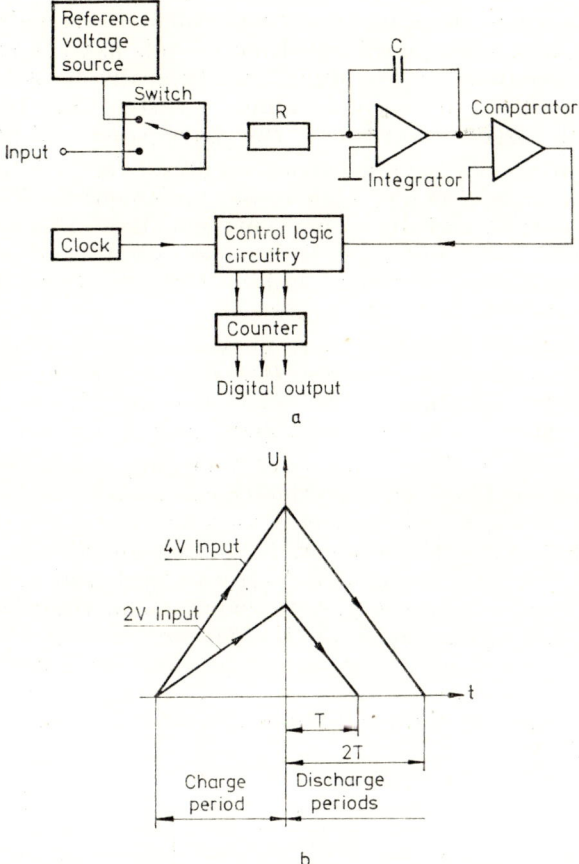

Figure 5.65

The input signal is switched into the input of the integrator via the charging resistor R. During this period the capacitor C is charged and, at the end of the integration period, determined by the control-logic circuitry, the capacitor charge is proportional to the input voltage. At this moment, the switch connects the integrator (an operational amplifier) to the reference voltage source, which is always of opposite polarity to the incoming signal. The capacitor is discharged linearly by the reference voltage, the discharge time being proportional to the input voltage level.

5 Instrumentation

At the beginning of the discharge period, a counter is started by the output from the oscillator clock and stopped when the capacitor is fully discharged. The counter reading is thus a direct measure of the input voltage level. Figure 5.65 b shows the charge-discharge slope diagrams for two input signals having a ratio 2 : 1 and whose discharge time periods have the same ratio. Readings accurate to five digits can typically be achieved at a rate of about 15 per second, which is faster than the speed at which data loggers can transfer the chart paper and print.

Digital meters have increased accuracy and reliability, sensitivities down to 100 nV, high resolution (0.001% of full scale), electrical output and digital presentation. Examples are Schlumberger 4445, 7040 and 7054 Multimeters, A210 and A230 Series Digital Voltmeters, as well as Philips PM 2421 (with HF input), PM 2441 and PM 2442 DC Voltmeters.

Data logs are listings of current readings from all (or selected) channels of a measuring system. *Data logging systems* are especially useful when simultaneous measurements at a large number of remote points have to be done rapidly and accurately.

A simplified block diagram of a logging system is presented in Figure 5.66. The digital voltmeter can measure only one signal at a time. Each input channel is switched in turn to the digital voltmeter input by the scanner. Both are controlled by the logger which receives measurement data from the digital voltmeter and converts them to an appropriate code for the recorder. Because the data are in digital form, the logger can manipulate and process the reading without loss of accuracy or sensitivity. Data logs can be presented visually on a CRT terminal or printed on a teletype writer or high speed line printer if a permanent record is desired.

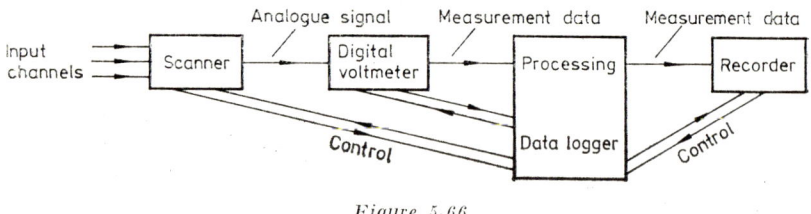

Figure 5.66

A data logger system can detect when results are outside defined limits and initiate corrective commands or alarm indications. It can pass all readings to a computer. A typical example is Schlumberger's Compact-33 Data Logger and 3366 Computer Logger.

5.5 Display and recording instruments

5.5.4 Analog strip-chart recorders

If a pen is attached at the pointer end of the instrument from Figure 5.67 and a paper is moved past the pen at constant speed, a *vibrogram* is obtained, describing the time history of the observed phenomenon. In practice, the moving parts of strip chart recorders are more sturdy. Because pen motors require more power than the magnetoelectric instruments, suitable amplifiers are necessary which, in addition to gain, produce a widening of the flat frequency response range. Usually, direct-coupled amplifiers are used with constant response down to zero frequency.

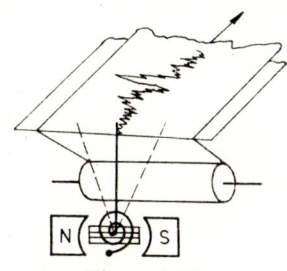

Figure 5.67

The moving parts of pen motors form a single degree of freedom system. Its natural frequency determines an upper frequency limit of the recorded signals (usually 50 — 60 Hz). Similar to seismic instruments, the improvement of the low frequency response is obtained with moderate amounts of damping ($\zeta = 0.6 - 0.7$), using either a shorted-turn coil or the currents induced in the driving coil. In order to prevent bending resonances, the pen is made as light and sturdy as possible.

Improved designs use an electrically heated wire stylus recording on wax-coated paper. Strip-chart recorders have paper-drive motors with several speeds (from 0.25 mm/min to 100 mm/s) specially designed to use the calibrated chart paper as a time base for the record.

Examples are Philips PM 8202 (single-line), PM 8222 (double-line) and PM 8235 (multi-point) recorders, GenRad 1985 DC Recorder, Schlumberger SERVOTRACE and SERVORAC based on the SEFRAM range of recorders, Hellige He-1. Recent multi-point recorders use d.c. servo-potentiometers (see Section 5.5.5), as for example Speedomax M Mark III and Speedomax 165 and 250 Series, manufactured by Leeds & Northrup.

Modern recorders use tungsten carbide tips (in ink recorders) or ceramic tips (in thermal recorders), having extremely low wear characteristics. Examples are Hewlett Packard 7402A and 7404A Ink Recorders and 7414A, 7418A and 7702B Thermal Recorders.

5.5.5 X — Y recorders

The $X - Y$ plotter is a moving pen recorder with two inputs, driving the pen in two mutually perpendicular directions.

5 Instrumentation

A simplified diagram of the servomechanism moving the pen (in one direction) is presented in Figure 5.68. The input signal is applied to a resistor with a sliding contact. The potentiometer output, which is a function of the cursor position, is compared to a fixed reference voltage. The difference of the two voltages

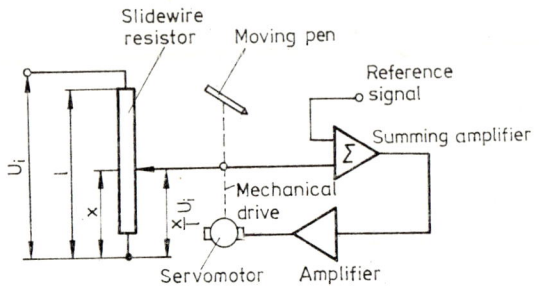

Figure 5.68

is amplified to drive a motor which moves the sliding contact and the attached pen to the position of zero voltage difference. The pen displacement is inversely proportional to the input signal voltage, tending to maintain the potentiometer output voltage constant.

The instrument contains two independent systems of this type. The carriage 1 (Fig. 5.69) moves in the X-direction; the cursor 2 and the attached pen move up and down in the Y-direction along the carriage. The maximum operating frequency is very low (5 — 10 Hz) so that these instruments cannot be used for direct plotting of high frequency time histories.

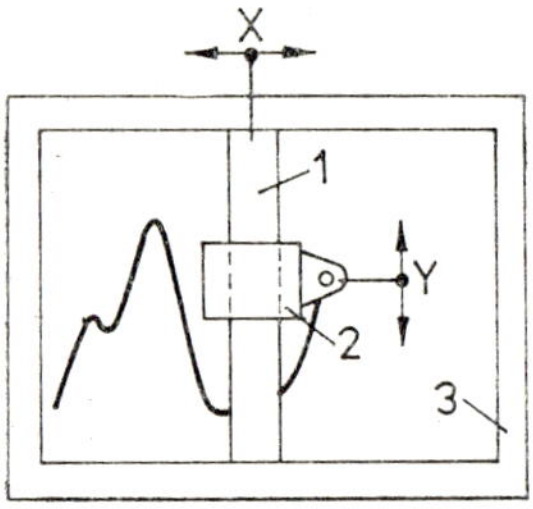

Depending on the d.c. signals applied at the two inputs, the instrument is used for recording frequency spectra, frequency response curves and even vibrograms. Examples are Hewlett Packard 7044A to 7047A recorders, 7225A Graphics Plotter, Schlumberger TGM and TRP, Philips PM 8120 (single function) and PM 8125 (multirange) recorders, IRD Mechanalysis 1080 Vibration Baseline Recorder and 1082 Vibration Baseline/Time Plotter. Typical features are: sensitivities from 20 µV/cm to 5 V/cm, slowing speeds of 76 cm/s and peak accelerations of 76 m/s² in the Y axis.

Figure 5.69

5.5 Display and recording instruments

5.5.6 Graphic level recorders

Graphic level recorders of the self-balancing potentiometer type are used especially for plotting noise and vibration frequency spectra. When operated in conjunction with frequency analyzers, the paper is driven by the frequency sweeping.

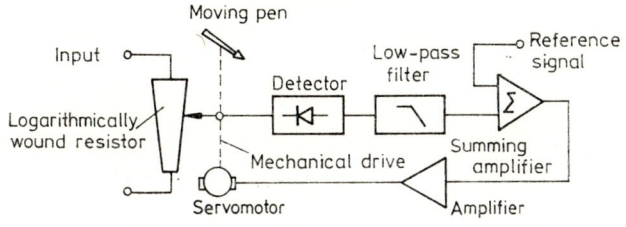

Figure 5.70

The pen moving mechanism (Fig. 5.70) is based on the scheme from Figure 5.68 where an additional detector and a low-pass filter are introduced, to permit the application of dynamic waveforms directly to its input [8].

By design, the slidewire resistance varies logarithmically with its length, yielding a logarithmic scale which can be graded in dB, resulting in an instrument which plots levels. Examples are GenRad 1521 and 1523, Brüel & Kjaer 2306 and 2307.

5.5.7 Magnetic oscillographs

Magnetic oscillographs, sometimes referred to as string oscillographs, record on light-sensitive paper the time-history of a phenomenon by means of a light beam reflected from a small mirror attached to a coil-type or a loop-type galvanometer.

Construction of a bifilar galvanometer is shown in Figure 5.71, where: 1 — permanent magnet; 2 — mirror; 3 — metallic string or ribbon of the loop; 4 — insulating bridges; 5 — loop spool; 6 — tensioning spring. When current flows through the loop, the electromagnetic forces move the ribbons in opposite directions, rotating the mirror. A light beam from an electric lamp is reflected by the mirror to a light-sensitive paper driven at constant speed by a motor.

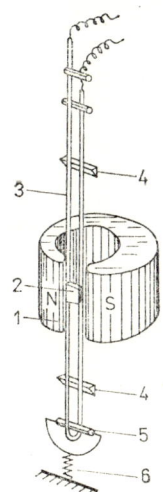

Figure 5.71

Substitution of the stylus used by direct-writing recorders by the massless light beam reduces the torque required to rotate

5 Instrumentation

the galvanometer, which may be made small and light, with natural frequencies of the order of 10,000 Hz. The loop-mirror assembly is sealed in a housing which may be filled with damping fluid. When the damping ratio is adjusted to 0.65 the frequency response is practically flat in the 0 — 3000 Hz range.

A low-frequency galvanometer is properly damped for a given value of the external resistance. Resistive coupling networks are interposed between transducer and galvanometer to obtain optimum damping. High-frequency low-sensitivity galvanometers require fluid damping, the external resistance having small influence on the total damping.

Magnetic oscillographs have the advantages (over pen recorders) of better frequency range and simpler multi-channel recording.

Modern oscillographic recorders use ultraviolet recording lamps (mercury vapour or tungsten filament) and photosensitive direct print paper which is developed by simple exposure to normal room illumination.

Examples are Schlumberger UV Optical Recorders OM 4501 (8 channels) and OH 4525 (12 — 24 channels), Honeywell 1508C and the Visicorder.

5.5.8 Cathode-ray oscilloscopes

The popularity of cathode-ray oscilloscopes in visual presentation and recording of dynamic phenomena is due mainly to the use of the cathode ray tube (CRT) electron beam as pointer, whose negligible inertia does not impose mechanical limits on the high frequency response. It is possible to have several traces on one screen, but usually single and dual beam oscilloscopes are used.

Using the CRT oscilloscope as an $X - Y$ recorder, one can made frequency, phase shift and time delay measurements, comparing the stationary patterns from the screen to the Lissajous patterns.

Recording of the time history of the studied phenomenon can be made applying the signal to the vertical deflection plates, turning off the sweep and passing a moving film continuously past the CRT screen. New storage-oscilloscopes can be used for a variety of applications. These include: capturing part of a waveform or a non-repetitive event for detailed analysis; observing changes in a signal; comparing two or more signals; observing a slow-moving signal; "babysitting", that is unattended recording of signal excursions over selected time frame or occurring while waiting to capture a transient; "glitch-catching", that is

5.5 Display and recording instruments

capturing fast variations of slower signals; and varying brightness of random signals relative to repetitive signals.

Both analog and digital storage oscilloscopes have been developed, but while the former capture and retain signals within the CRT itself, the latter store data representing waveforms in a digital memory. Examples are Tektronix 7000-Series laboratory oscilloscopes which have different analog storage modes (e.g. phosphor target bistable, bright bistable, variable persistance and transfer storage), several digital plug-ins and sampling capabilities.

Simpler portable instruments like Tektronix 464, 465, 466 and 475 have a built-in digital multimeter and a numerical time interval readout. The more complex digital processing oscilloscopes as Tektronix 7000 Series and Norland 3001 are computer aided, comprehensive, powerful and flexible data acquisition and analysis systems. Conventional oscilloscopes contain only acquisition and display units. Digital processing oscilloscopes also contain a processor for linking the scope to a minicomputer [24]. Oscilloscopes with improved characteristics are used for recording on ultraviolet light sensitive paper in Honeywell's 1858 CRT-Visicorder.

5.5.9 Magnetic tape recorders

Use of tape recorders is determined by some peculiar beneficial features: a) data storing in analogic form for subsequent processing; b) reproduction of the record at any time and a great number of times without tape damage; c) contraction or expansion of time, which permits frequency translation of the data spectrum, by recording at one speed and reproducing at another speed.

Instrumentation tape recorders usually operate in one of two modes: direct recording and frequency modulation recording.

Common tape recorders using *direct recording* have bad low-frequency response characteristics. The recording head magnetizes the tape to an intensity proportional to the signal amplitude. The playback head picks-up a signal proportional to the rate of change of magnetic flux and therefore proportional to the time derivative of the original signal. An integrating amplifier is required to reproduce the original signal. Zero and low-frequency signals are not picked-up at a level sufficiently greater than stray electrical noise signals on the tape.

In the direct recording mode the data waveform is mixed with a high frequency pre-magnetising (bias) waveform and recorded on the magnetic tape. The bias current allows recording of data waveform within the linear region of the magnetization curve of the magnetic tape. Depending on the tape speed, spurious noise limits the lower frequency of the recorded signal to 100 — 300 Hz, for an accuracy of 5 % [25].

In frequency modulation recording, the waveform is first used to frequency-modulate a high-frequency carrier, and this FM modulated signal is then recorded on magnetic tape. This practically eliminates the effect of the instantaneous magnetic characteristics of the oxide coating, allows the lowest frequencies (starting from d.c.) to be recorded, giving better signal-to-noise ratios, less phase distortion, hence improved accuracy than direct recording [26].

Multichannel tape recorders are available with up to 14 tracks. Start-stop tape recorders are used for storing digital data at rates of 20 kbytes/s. Examples of instrumentation tape recorders are Hewlett Packard 3964A (4-channel) and 3968A (8-channel), Schlumberger MT 5530 (14-channel), MP 5522 (8-channel), MP 5421, MP 5419 and ML 2600 (high performance laboratory recorder), Brüel & Kjaer 7003 to 7006 Magnetic Recorders. Most features of tape recorders are standardized (e.g. : IRIG 106—71).

Clean erasing of entire reels of tape can be done using Hewlett Packard 13064A Tape Degausser.

5.5.10 Digital recorders

Digital recorders, also called *waveform digitizers* are generally designed to capture short duration events and to reproduce them when desired on a CRO or pen recorder or output them to a digital device. They can hold single-shot events, track signals, capture regions of interest or provide visual monitoring of slow or irregular waveforms. Most instruments consist of a sampling circuit, an analog-to-digital converter, a memory and a digital-to-analog converter with associated command and control circuitry.

Modern instruments have three different recording modes : a) In the *pre-trigger mode*, new samples are continuously taken and fed to the memory as the oldest samples are discarded. When a trigger occurs, the sampling continues for a period specified by the delay control, to produce a record containing a controlled amount of pre- and post-trigger signal information. b) The *delayed sweep mode* allows sweep initiation to be delayed following detection of a trigger and placed at the desired part of the signal. c) Slowly moving waveforms are observed by automatically recirculating the memory between samples.

Examples are Datalab 900 and 2000 Series Transient Recorders, Brüel & Kjaer 7502 Digital Event Recorder and Difa TR 1010 Transient Recorder. Tektronix R 7912 Transient Digitizer, in order to avoid analog-to-digital conversion errors, digitizes the stored image of the event rather than the event itself.

5.6 Computerized vibration analysis systems

To cope with the high (real time) analysis speed required in most applications, minicomputers have been incorporated into the measurement systems used for vibration excitation, data acquisition, data reduction and data analysis. Usually, these are performed by providing dedicated instruments for each individual

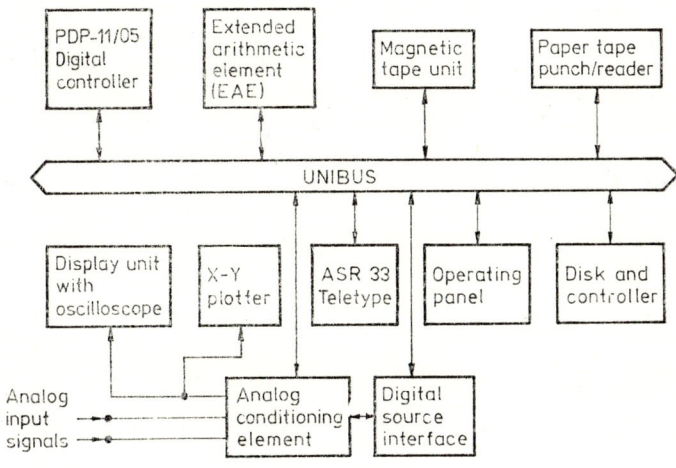

Figure 5.72

task. This approach limits the overall capability of a laboratory facility, so that systems with general purpose capabilities and flexible in operation have been designed.

A schematic block diagram of the TDA 1 Time Series Analyzer system plus some add-on options is shown in Figure 5.72. The central processor may be a DEC PDP-11 minicomputer with 16K or 32K of MOS memory. It uses the advanced PDP-11 UNIBUS architecture which is a bidirectional data path to which all major system components are connected, being easily added, removed or changed. Operator interface is effected through a *teletype* or a *graphics terminal* and *hard copy unit* for a permanent record of the computer output. *Cartridge disks* (or *flexible disks*) can be used for program loading and data storage. GenRad TDA 20 and TDA 25 systems have floppy disk-based operation, which avoids fragile, slow and inconvenient paper tape. The capacity of single floppy disk memories of desktop instruments is of 250,000 words (16-bit).

The computer can accept several channels of conditioned analog data, multiplexed from sample-and-hold analog-to-digital

5 Instrumentation

converters. Data generated in the computer can be output using digital-to-analog converters.

A full range of peripherals can be used as a paper tape punch reader, magnetic tape units, operating panels, FFT processors and output devices as X—Y plotters, line printers and large-screen oscilloscopes [27, 28].

In order to get an idea on the operating speed characteristics of some items of a computerized system, the following features are worth mentioning. The PDP 11/05 computer with 8K memory has a 1.2 μs cycle time. A Tektronix 4911 Tape Punch/Reader reads into the computer at up to 200 characters per second or punches new tapes at up to 75 characters per second. A Tektronix 4610 Hard Copy Unit needs 18 seconds for the first copy and 10 seconds for each subsequent copy. A Tektronix 4010 Graphics Computer Terminal has a direct view storage CRT whose screen holds up to 34 lines of alphanumeric data at one time (a maximum of 2590 characters) and generates 1200 characters per second.

References for Chapter 5

1. Doebelin, E. O., *Measurement Systems : Application and Design*, McGraw-Hill Book Comp., New York, 1966.
2. Willis, J., New frequency discriminator circuit, *Electronics Letters*, **1**, 194—195 (Sept. 1965).
3. Barr, W. H., Integrators, *Electromechanical Design*, p. 57 (Oct. 1961).
4. Cartianu, Gh., *Analiza și sinteza circuitelor electrice*, Editura didactică și pedagogică, București, 1972.
5. Bendat, J. S., Piersol, A. G., *Measurement and Analysis of Random Data*, John Wiley & Sons Inc., New York, 1966.
6. Wahrmann, G. G., Broch, J. T., On the averaging time of RMS measurements, *Brüel & Kjaer Technical Review*, *2*, 3—21 (1975) and *3*, 3—35 (1975).
7. Skudrzyk, E., *The Foundations of Acoustics*, Springer Verlag, Wien, New York, 1971.
8. Keast, D. N., *Measurements in Mechanical Dynamics*, McGraw-Hill Book Comp., New York, 1967.
9. Campbell, J. S., The basic terminology of noise, *Machine Design*, **34**, *9*, 113—118 (1962).
10. Fieldhouse, K. N., Techniques for identifying sources of noise and vibration, *S/V, Sound and Vibration*, *4*, *12*, 14—18 (Dec. 1970).
11. Randall, R. B., *Application of Brüel & Kjaer Equipment to Frequency Analysis*, Brüel & Kjaer, Sept. 1977.
12. Keller, A. C., The tracking filter, *Electronic Instruments Digest*, *4*, *6*, 10—15 (1968).

References for Chapter 5

13. Broch, J. T., *Mechanical Vibration and Shock Measurements*, Brüel & Kjaer, Oct. 1980.
14. ₓ*ₓ *Real Time Signal Processing in the Frequency Domain*, Nicolet, Monograph No. *3*, March 1973.
15. Richardson, M. H., Fundamentals of the Discrete Fourier Transform, *S/V, Sound and Vibration*, **12**, *3*, 40−46 (March 1978).
16. ₓ*ₓ *SD 360 Digital Signal Processor*, Spectral Dynamics Corp.
17. Flink, J., Amplitude accuracy limitations of real-time single channel spectrum analyzers, *S/V, Sound and Vibration*, **12**, *3*, 52−54 (March 1978).
18. Herlufsen, H., Dual Channel FFT Analysis (Part I), *Brüel & Kjaer Technical Review*, *1* (1984).
19. Enochson, L., Desktop instruments for modal analysis, *Machine Design*, **53**, *10*, 81−86 (May 7, 1981).
20. Yates, W., Update! Low frequency, real time FFT spectrum analyzers, *Electronic Products*, **25**, *1*, Feb. 7, 1983.
21. Russell, R. H. and Deel, J. C., Modal analysis : Trouble-shooting to product design, *S/V, Sound and Vibration*, *11*, 22−38 (Nov. 1977).
22. Olsen, U., Wattmetrische Messgeräte zur Schwingungsuntersuchung von Kraftfahrzeuge, *Automobiltechnische Zeitschrift*, **71**, *3* (1969).
23. Quinn, G. C., Recording Instruments (Part II), *Power*, Jan. (1978).
24. Moriyasu, H., Hamilton, B., Navarro, L., and Eshelman, W., Digital processing interface brings computer power to oscilloscope, *Electronics*, March 1973.
25. ₓ*ₓ *Magnetic Tape Recorders*, Schlumberger Catalogue, 1975.
26. ₓ*ₓ *Tape Recording Principles*, Brüel & Kjaer Lecture No. 3011.
27. Ibáñez, P. and Spencer R. B., Experience with a field computerized vibration analysis system, *SAE Paper* No. 791074 (Dec. 1979).
28. ₓ*ₓ *SYAME 02, Système d'analyse et de modélisation expérimentale des comportements vibratoires des structures mécaniques*, Metravib, France.

6

Vibration Exciters

This chapter describes several types of vibrators (shakers) which are currently used either in reliability testing, fatigue testing and instrument calibration or for determining the dynamic characteristics of structures, machinery, foundations, etc. They may be parts of vibration testing machines or may be included in vibrating mechanisms for conveyers or process installations. Piezoelectric, magnetostrictive and pneumatic vibrators are not presented herein.

6.1 Mechanical vibration exciters

6.1.1 Reciprocating vibration exciters

Figure 6.1 shows three types of mechanical vibrators whose table, driven by a rotating eccentric or cam, has a rectilinear motion. Under ideal conditions, i.e. infinite rigidity and no loose bearings,

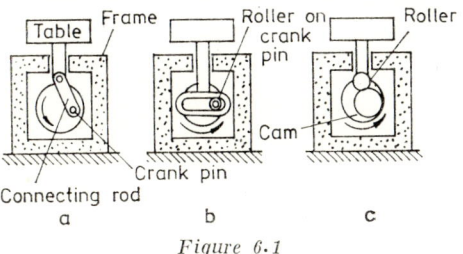

Figure 6.1

the driving rod has a constant amplitude displacement, independent of the operating speed and loading force (hence the name "kinematic vibrators") [1].

In comparison with the "connecting rod — crank" mechanism (Fig. 6.1 a) that one with roller on crank pin (Fig. 6.1 b) has the

6 Vibration exciters

advantage of generating harmonic motion. Generally, for adjusting the crank offset it is necessary to stop the machine; there are also vibrators by which the adjustment can be made during operation. The simplest reciprocating mechanical vibrator is driven by a constant-speed motor via a speed changer. More elaborate vibration machines employ variable speed motors with electronic speed control.

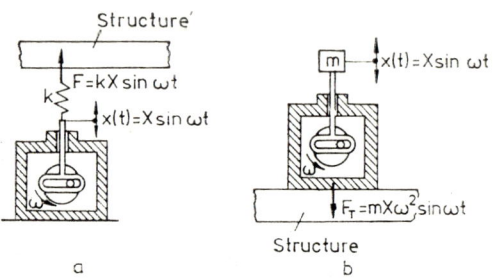

Figure 6.2

Figure 6.2 shows the two modes of operation of a reciprocating vibration exciter: a) direct-drive (Fig. 6.2 a) and b) inertial (reaction) drive (Fig. 6.2 b).

In the case of direct-drive through an elastic element (Fig. 6.2 a), the structure is driven by a constant amplitude harmonic force on condition that the vibrator has a backup fixture.

In the case of inertial drive (Fig. 6.2 b), the vibrator frame is attached to the tested structure, transmitting to it a frequency-dependent force which equals the inertia force of the mass m mounted at the driving rod top.

In both cases, change of the force amplitude is achieved by adjusting the crank offset (hence the rod stroke). In practice, the elastic deformation of load-carrying members and the inherent clearances limit the use of reciprocating vibrators at frequencies below 30 Hz and average rated loads under 700 N.

6.1.2 Rotating unbalance vibration exciters

For relatively large rated loads (between 400 and 20,000 N), rotating unbalance reaction-type vibrators are used. A rectilinear motion vibrator can be designed using two identical rotating unbalances, turning in opposite directions and symmetrically located with respect to the median vertical plane (Fig. 6.3). The horizontal components $mr\omega^2 \sin \omega t$ of the unbalance forces cancel

6.1 Mechanical vibrators

each other, being taken over by the vibrator frame. The vertical components add, giving the resultant

$$F(t) = 2m\,r\omega^2 \cos \omega t \tag{6.1}$$

which is transmitted through bearings to the vibrator frame and then directly to the tested structure.

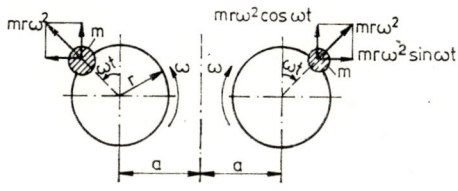

Figure 6.3

In order to have larger possibilities for adjusting the amplitude of unbalanced forces (at $\omega = $ const.), each mass unbalance is made from two (or more) identical masses which can be placed at different fixed locations on the periphery of a disc (Fig. 6.4). However, adjustment of the eccentric radius r and the angle α is done stepwise, the masses being attached by screws, passing through holes made in the two discs, so that a stepless adjustment of the generated force amplitude is impossible.

Rotating unbalance vibrators are driven by variable-speed electric motors: for high power, by d.c. motors and for low power, by collector monophase a.c. motors [2].

This type of vibrators have the following drawbacks: difficult adjustment of speed and force amplitude, distortion of the generated force waveform (due to clearances in bearings and gears) and limited operating range (up to 60 Hz).

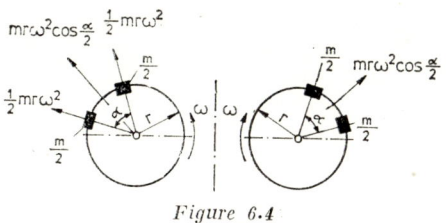

Figure 6.4

When the two counterrotating masses are arranged as in Figure 6.5, passing at a given time through points A_1 and A_2, respectively, the horizontal force components are added, developing a harmonic force of the form (6.1). The vertical components

6 Vibration exciters

have no resultant force but produce a harmonic rocking couple, 90° phase-shifted behind the horizontal force and given by

$$M(t) = 2\ mra\ \omega^2 \sin \omega t. \tag{6.2}$$

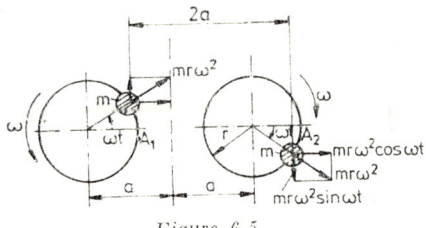

Figure 6.5

A more flexible design has the vibrator with four eccentric masses arranged one at each end of two parallel shafts (Fig. 6.6). With the arrangement of weights from Figure 6.6 a, a vertical harmonic force is produced. With the arrangement from Figure 6.6 b a torsional couple about a vertical axis is obtained. With the arrangement from Figure 6.6 c a rocking couple about a horizontal axis is developed [3].

Vibrators of this type are manufactured by Losenhausen in Düsseldorf. The vibrator type 8020/100 generates forces up to 800 N, at frequencies of 5 — 25 Hz or 20 — 100 Hz, while the type 2000 4/20 generates forces up to 20,000 N at frequencies of 2 — 8 Hz or 5 — 20 Hz [4]. Similar vibrators are types V1 and VMME-250 developed at the Strength of Materials Laboratory of the Polytechnic Institute of Bucharest (see Table 6.1). The latter is driven by two 7 kW d.c. motors supplied via converter unit, the two shafts being synchronized by a geared coupling.

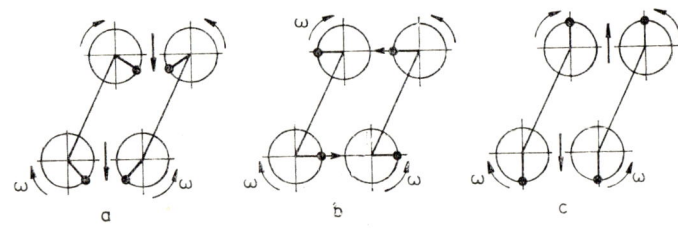

Figure 6.6

Performance characteristics of other rotating-mass vibrators are given in reference [1]. Table 6.2 lists performance data and specifications of vibrators developed by ANCO Engrs.Inc. at Santa Monica, Ca., U.S.A. [5].

6.1 Mechanical vibrators

On the same principle was based the vibrator developed by DEGEBO (Deutsche Forschungsgesellschaft für Bodenmechanik, Berlin) during the period 1928—1936 at the first experiments on the dynamic response of model footings resting on soil [6]. The four eccentric mass vibrator was arranged as in Figure 6.7, with four shafts driven by bevel gears.

TABLE 6.1 *Performance Characteristics of Rotating Mass Vibrators Developed at the Polytechnic Institute of Bucharest*

Designation	Unbalance, kgm	Frequency range, Hz	Force, N	Power, kW
V1	0.18	1—20	40—11,000	3
VM1	0.0064—0.0544	10—50	25—5370	1
VM2	0.0405—0.737	5—25	40—18,185	3
VMME-250	7.597	5—50	1170—231,700	2×7

Very compact mechanical vibrators with two eccentric masses and bevel gears are of types VM1 and VM2, developed at the Strength of Materials Laboratory of the Polytechnic Institute of Bucharest (see Table 6.1) [4]. On Figure 6.8 there can be seen: the driving motor 1, on whose shaft is mounted the bevel pinion 2, that drives gears 3 and 4, on which are attached the eccentric masses 5 and 6. These masses are inversely proportional to their

TABLE 6.2 *Performance Characteristics of Rotating-Mass ANCO-Eng. Vibrators [5]*

Designation	Maximum force, N	Minimum frequency for maximum force, Hz	Upper frequency limit, Hz	Unbalance, kgm	Mass of vibrator, kg	Power, kW
MK-11	40,000	100	100	0.1	10	1.5
MK-12	40,000	40	100	0.5	40	1.5
MK-13	40,000	6	20	30	150	1.5
MK-14	40,000	1.5	10	370	400	1.5
MK-15	1,000,000	2.5	30	4,000	8,000	75.0

6 Vibration exciters

eccentricity, generating equal amplitude centrifugal forces. The taped form of the two masses is calculated so that the two forces be in the same plane. The vibrator generates forces between 60 and 5000 N, at frequencies between 10 and 50 Hz.

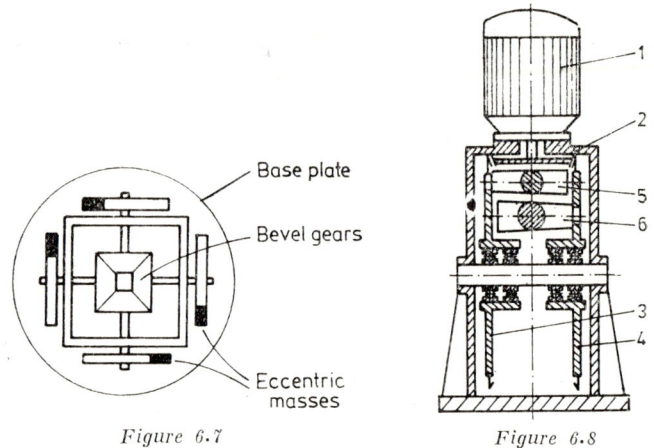

Figure 6.7   Figure 6.8

6.2 Electromagnetic vibration exciters

Electromagnetic vibrators are used in applications where contactless excitation is recommended and where, besides sinusoidal forces, static preloading forces are required. Their practical application is limited by two disadvantages: the difficulty of measuring the generated force and the non-linearity of the "magnetic flux density — force" characteristic. Recent designs using Hall-effect transducers and feedback loops, eliminating these undesirable features, have been used at the dynamic analysis of machine tools [7].

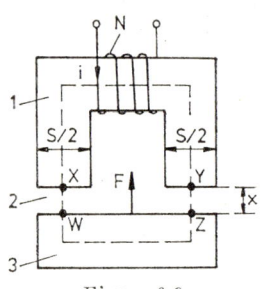

Figure 6.9

6.2.1 Force generated by an electromagnet

Figure 6.9 illustrates a magnetic circuit consisting of the iron core 1, with a single winding of N turns, the air gap 2 and the movable part 3 which can be the excited structure itself or an armature connected to it through a force transducer.

192

6.2 Electromagnetic vibrators

The energy of the magnetic field is expressed as

$$W_m = \frac{1}{2} L i^2, \tag{6.3}$$

where i is the input magnetizing current and L is the inductance. The latter is given by

$$L = \frac{N^2}{\dfrac{2l_f}{\mu S} + \dfrac{4x}{\mu_0 S_0}} = \frac{N^2 \mu_0 S_0}{\dfrac{2l_f S_0}{\mu_r S} + 4x}, \tag{6.4}$$

where $l_f = XY + ZW$ is the length of iron section, $2x = YZ + WX$ is the air gap length, S_0 — air gap area, S — iron cross section area, μ and μ_0 — permeabilities of iron and air, respectively.

The attractive force between armature and core is

$$F = \frac{dW_m}{dx} = \frac{1}{2} i^2 \frac{dL}{dx} = \frac{1}{2} i^2 \left(-\frac{4L^2}{N^2 \mu_0 S_0} \right) = -\frac{B^2 S_0}{2\mu_0} \tag{6.5}$$

where

$$B = \frac{2Li}{N S_0} \tag{6.6}$$

is the magnetic flux density in the air gap.

An alternating current $i = I_1 \cos \omega t$ produces a magnetic flux density $B = B_1 \cos \omega t$, which, according to (6.5), generates a force

$$F(t) = -\frac{B_1^2 S_0}{2\mu_0} \cos^2 \omega t = F_0' + F' \cos 2\omega t. \tag{6.7}$$

It consists of a steady force F_0' and an alternating force having twice the fundamental frequency of the magnetizing current (second harmonic).

To generate a force of frequency ω, a second winding may be added, to which a d.c. bias current may be applied, so that the magnetizing current becomes

$$i = I_0 + I_1 \cos \omega t. \tag{6.8}$$

Alternatively, a bias magnetic field can be used, so that in both cases the flux density has the form

$$B = B_0 + B_1 \cos \omega t \tag{6.9}$$

6 Vibration exciters

generating a force

$$F = -\frac{(B_0 + B_1\cos\omega t)^2 S_0}{2\mu_0} = F_0 + F_1\cos\omega t + F_2\cos 2\omega t. \quad (6.10)$$

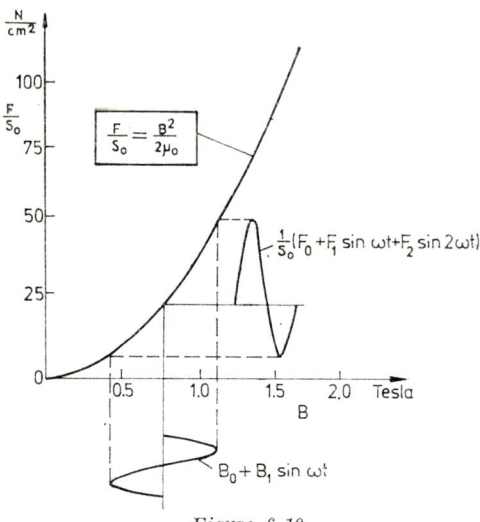

Figure 6.10

As can be seen from Figure 6.10, equation (6.10) is a direct result of the non-linear relationship (6.5). The force varies as the second power of the flux density.

A simplified analysis can be done for the case when the armature 3 (Fig. 6.9) has a translatory sinusoidal motion $x_1\cos\omega t$. In this case $YZ = WX = x_0 + x_1\cos\omega t$.

Neglecting the iron reluctance with respect to that of the air, one can write

$$L \cong \frac{N^2\mu_0 S_0}{4}\frac{1}{x} = \frac{K}{x}, \quad \frac{dL}{dx} = -\frac{K}{x^2}.$$

$$F = -\frac{1}{2}i^2\frac{K}{x^2} = -\frac{K}{2}\frac{(I_0 + I_1\cos\omega t)^2}{(x_0 + x_1\cos\omega t)^2} =$$

$$= -\frac{KI_0^2}{2x_0^2}\left(1 + \frac{I_1}{I_0}\cos\omega t\right)^2\left(1 + \frac{x_1}{x_0}\cos\omega t\right)^{-2}.$$

If $I_0 \gg I_1$ and $x_0 \gg x_1$, considering only the first two terms in the binomial series expansion

$$F \cong -\frac{K}{2}\frac{I_0^2}{x_0^2}\left(1 + 2\frac{I_1}{I_0}\cos\omega t\right)\left(1 - 2\frac{x_1}{x_0}\cos\omega t\right)$$

6.2 Electromagnetic vibrators

or, neglecting the term in $\cos^2 \omega t$,

$$F \cong - \frac{KI_0^2}{2x_0^2} - \frac{KI_0 I_1}{x_0^2} \cos \omega t + \frac{KI_0^2}{x_0^3} x_1 \cos \omega t. \qquad (6.11)$$

In equation (6.11), the first term from the right-hand side represents the steady force, the second, the alternating force, directly proportional to I_1 and of the same frequency. The last term, proportional to the amplitude x_1 of armature vibration, has the form of an elastic force, indicating the existence of a *negative stiffness* which adds to the stiffness of the excited structure. The force has also higher harmonic components that do not appear in equation (6.11) being eliminated by simplifications.

The steady force component can be minimized by arranging the armature in a plane of symmetry of the magnetic circuit (using two symmetrically arranged magnets), i.e. realizing a self-balancing configuration. This way, the alternating force can be doubled, but at the same time the supplementary stiffness is doubled.

6.2.2 Measurement of electromagnet force

The force generated by an electromagnetic vibrator can be determined either using a force transducer or measuring the magnetic flux density B in the air gap. Use of the force transducer requires a thorough analysis of the dynamic response of the "vibrator — force transducer — structure" system in order to make adequate corrections. Measurement of B may be done : 1) based on the value of the magnetizing current i; 2) using coils placed on pole pieces in the vicinity of the air gap (the fluxmeter dynamometer); and 3) using a Hall-effect transducer [7].

a) *Measurement of magnetizing current.* Magnetic flux density computation based on equation (6.6) and measurement of the magnetizing current i is a rather inaccurate method. The value of B depends on the inductance L, which is a function of the air gap length $2x$ and of the iron equivalent reluctance determined taking into account hysteresis and eddy current losses.

b) *Measurement using a fluxmeter dynamometer.* A coil placed on a pole in the vicinity of the air gap is used to measure the alternating flux density in the magnetic circuit. The electromotive force (e.m.f.) induced in an N turn winding is

$$u = -nS \frac{dB}{dt} = n S \omega B_1 \sin \omega t, \qquad (6.12)$$

6 Vibration exciters

where B is given by equation (6.9). Integration with respect to time yields

$$\int u \, dt = -n \, S \, B_1 \cos \omega t, \qquad (6.13)$$

where the integration constant is zero, because $\int u \, dt = 0$ for $B_1 = 0$.

According to equation (6.5), the force generated by vibrator is

$$F = -\frac{(B_0 + B_1 \cos \omega t)^2 S_0}{2\mu_0} = -\frac{S}{2\mu_0}\left(B_0 - \frac{\int u \, dt}{nS}\right)^2$$

or

$$F = -\frac{S_0 B_0^2}{2\mu_0} + \frac{B_0 S_0}{\mu_0 nS}\int u \, dt - \frac{S_0}{2\mu_0 n^2 S^2}\left(\int u \, dt\right)^2 \qquad (6.14)$$

which is similar to equation (6.10).

The first term from the right-hand side of equation (6.14), representing the steady force, cannot be measured using the fluxmeter coil. It can be determined from static calibration curves of the dependence $F_0 = F_0(I_0, x)$ plotted using a dynamometer [8]. The second term represents the desired alternating force, which is proportional to the integral of the induced e.m.f. It can be also determined from calibration plots giving the dependence

$$\frac{|F_1 \cos \omega t|}{\left|\int u \, dt\right|} = \frac{B_0 S_0}{\mu_0 nS} = f(I_0, x).$$

The third term produces the distortion of the force waveform generated by the vibrator, being a function of the ratio B_1/B_0. Other sources of distortion are the hysteresis, eddy currents and the imperfection of the amplifier included in the circuit of the magnetizing winding. A supplementary source of errors is the fact that the fluxmeter coil senses also the stray flux which does not participate in the force generation.

c) *Measurement with a Hall transducer.* Analog Hall-effect transducers provide a d.c. output voltage u_H proportional to the magnetic flux density B, perpendicular to its surface

$$u_H = K_H B. \qquad (6.15)$$

6.2 Electromagnetic vibrators

Squaring equation (6.15) and taking into account equation (6.5), one obtains

$$u_H^2 = \frac{2\mu_0 K_H^2}{S_0} F = K_1 F. \qquad (6.16)$$

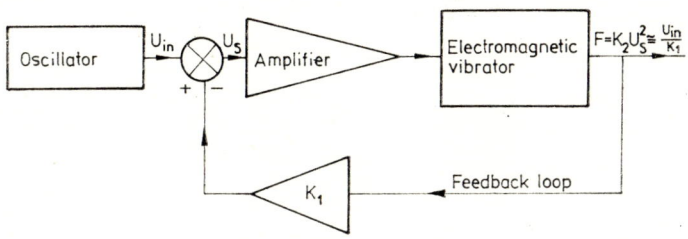

Figure 6.11

Equation (6.16) shows that measuring the voltage u_H, the total force F can be determined, including the steady component and all harmonics, the result being independent of the air gap.

An oscillator and a power amplifier are used in the vibrator circuit (Fig. 6.11). The "force — magnetic flux density" characteristic (6.5) shows that the force is directly proportional to the square of the oscillator signal. For the linearization of this characteristic, a high gain feedback loop can be used.

Modern Hall-effect transducers combine digital signal conditioning circuits on the Hall-sensor chip. Typical circuits consist of amplifier, hysteresis flip-flop, offset voltage adjustment and comparator.

6.2.3 General performance characteristics

According to equation (6.5), for a maximum value $B = 2$ Teslas of the magnetic flux density, the ratio of the maximum force generated by the vibrator to the air gap section area is

$$\left(\frac{F}{S_0}\right)_{max} = \frac{1}{2}\frac{B^2}{\mu_0} = \frac{1}{2}\frac{4}{4\cdot\pi\cdot 10^{-7}} = 1.59\cdot 10^6 \frac{N}{m^2} = 159 \frac{N}{cm^2}. \qquad (6.17)$$

The maximum available dynamic force depends on the steady force and the frequency. For a vibrator with Hall-effect transducer and feed-back loop [7], having a total mass of 3.4 kg, this dependence is graphically shown in Figure 6.12.

6 Vibration exciters

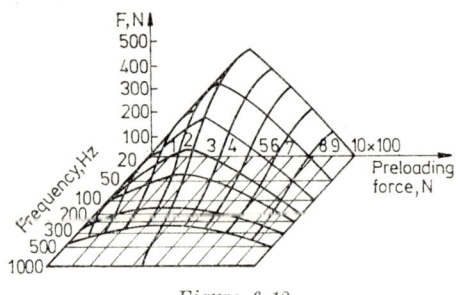

Figure 6.12

6.3 Electrodynamic vibration exciters

6.3.1 Principle of operation

The electrodynamic vibrator *) is based on the principle of the ordinary moving coil radio loudspeaker.

A section through a very simple vibrator is shown in Figure 6.13. The drive coil 1, with the windings wound on and bonded to an antimagnetic former 2, is supplied with current from an electronic oscillator via a power amplifier. It can move axially in the radial magnetic field existing in the annular air gap between the top ring of the magnetic pot 3 and the centre pole 4, due to

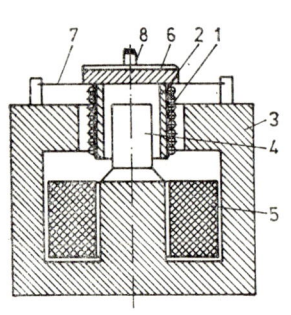

Figure 6.13

the d.c. field coil 5. The top table 6 is used for mounting small test articles. The drive coil and table assembly, referred to as the *armature*, is supported and centred in the air gap by the suspension springs 7, allowing motion in the axial direction, but minimizing any sideways or rocking motion of the armature. In other designs, the piece 6 has reduced size and is provided with a threaded rod 8, which permits the attachment of small vibrators to the tested structures.

The force generated by the interaction between the current flowing through the drive coil and the magnetic field from the air gap (causing the coil to move), is

$$F = B l i = \Gamma i, \text{ [Newtons]} \tag{6.18}$$

---
*) Sometimes referred to as *electromagnetic* vibrators (e.g.: by Derritron).

198

6.3 Electrodynamic vibrators

where B is the magnetic flux density, in Teslas; l — total length of the coil, in meters; i — current through the coil, in Amperes.

The magnetic field may be set up either by an electromagnet (for large forces) or by a permanent magnet (for small forces).

Unlike the electromagnetic vibrator, in order to obtain an alternating force, the current i may be varied instead of the flux density B. This way, a force is obtained which is directly proportional to the current through the drive coil and is practically independent of the magnetic circuit reluctance. If the coil is moving in a homogeneous magnetic field, the force component proportional to the displacement does not appear, hence no supplementary stiffness is introduced.

When the current flowing through the coil varies harmonically with the time, the force generated is also harmonic. Its value is limited by cooling, by materials (at high excitation currents, the heat dissipation is high) and by the mechanical strength of moving parts. The peak value of the alternating force produced by vibrator is referred to as the *thrust*.

When the drive coil is energized by direct current, the vibrator generates a steady force. The electrical input signal can have any waveform, the forces generated by an electrodynamic vibrator being of very different nature: periodic, transient, random, impacts, fact which extends considerably its field of application.

The coil displacement in the magnetic field generates an induced voltage, referred to as the *back electromotive force* $e = -Blv$. This voltage reduces the excitation current and decreases the thrust, if the input voltage is kept constant. In order to produce constant amplitude forces, electrodynamic vibrators have the input signal derived from a specially designed power amplifier, supplying a *constant* (amplitude) *current*. Due to reversibility of the phenomenon, a moving coil velocity transducer may be also used as a force-producing transducer, i.e. a vibrator.

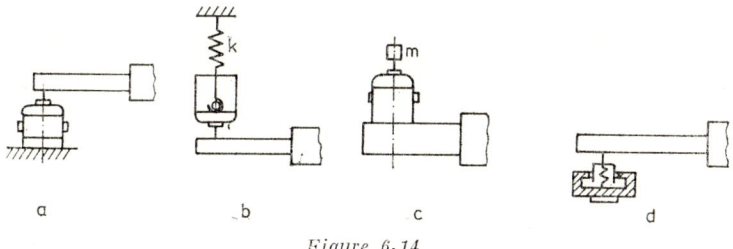

Figure 6.14

Electrodynamic vibrators are used with both direct drive (Figs 6.14 a and b) and inertial drive (Figs 6.14 c and d) and corresponding different configurations [9, 10].

6 Vibration exciters

At direct-drive, the armature is attached to the tested object or structure and the vibrator body is either mounted on a very rigid base (Fig. 6.14 a) or suspended on very soft springs (Fig. 6.14 b). Generally, the arrangement from Figure 6.14 a is used in vibration or qualification testing, when a "large vibrator" is used, on whose table top a "small component" is mounted, e.g. an electronic instrument, a scale model, etc. The arrangement from Figure 6.14 b corresponds to tests for experimental modal analysis or mobility measurements, where a "small vibrator" is used to excite a "large structure", e.g. aircraft, pipelines, metallic structures, etc. This is due to the requirement of matching the mechanical impedances of the tested structure and of the vibrator, which has to be taken into account when selecting a vibrator for a given test.

The inertial drive can be performed in two ways. For exciting relatively "large structures" — like foundation blocks, floors, columns, etc., conventional vibrators (built for direct drive) can be used, mounted as in Figure 6.14 c. The vibrator body is rigidly attached to the structure and a supplementary mass is added to the table. The force acting on the structure equals the inertia force corresponding to the armature acceleration.

Sometimes the vibrator is mounted upside-down, connecting the armature to the tested structure and introducing a supplementary spring (or coil membrane) to support the vibrator own weight. Based on this principle, vibrators have been specially designed for inertial drive (Fig. 6.14 d), having the size (and therefore the mass of magnetic circuit) sensibly reduced. They are used especially at flight testing of airplanes.

In the aforementioned cases, the electrodynamic vibrator operates in contact with the tested structure. It is possible to attach only the moving coil to the tested structure, so that a contactless excitation can be applied, eliminating the coil elastic suspension. In this case a larger annular gap is provided. At the same time, perfect alignment of stationary magnet and moving coil is required, as well as measurement at a point of the structure with rectilinear translational motion, in order to avoid rubbing.

6.3.2 Interaction between vibrator and tested structure

Consider a simple single-degree-of-freedom system excited by an electrodynamic vibrator (Fig. 6.15). The exciter acts as a voltage generator when the coil moves in the magnetic field. Denoting R_B — coil resistance, L — coil inductance, $\Gamma\dot{x}$ — back e.m.f.

6.3 Electrodynamic vibrators

induced in the coil, u — source voltage, R_A — source resistance, the electrical circuit equation is

$$L\frac{di}{dt} + (R_B + R_A)i + \Gamma\dot{x} = u. \qquad (6.19)$$

The equation of motion for the mass m is

$$m\ddot{x} + c\dot{x} + kx = \Gamma i, \qquad (6.20)$$

where the effects of coil inertia and its suspension are neglected.

The steady-state solutions of these equations can be written

$$x = Xe^{i\omega t}, \quad u = Ue^{i\omega t}, \quad i = Ie^{i\omega t} \qquad (6.21)$$

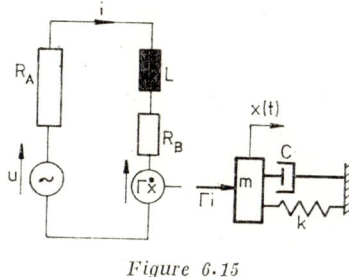

Figure 6.15

so that equations (6.19) and (6.20) become

$$(R + i\omega L)I + i\omega \Gamma X = U,$$
$$\Gamma I - (k - m\omega^2 + i\omega c)X = 0, \qquad (6.22)$$

where $R = R_B + R_A$.

By solving equations (6.22) one obtains X and I.
The exciting force $f = Fe^{i\omega t}$ has the complex amplitude

$$F = \Gamma I = \frac{\Gamma U}{R + i\omega L + i\omega \dfrac{\Gamma^2}{k - m\omega^2 + i\omega c}}. \qquad (6.23)$$

If a *constant* amplitude *voltage* source is used, the exciting force is frequency dependent, its amplitude having a minimum at the resonance frequency of the tested system.

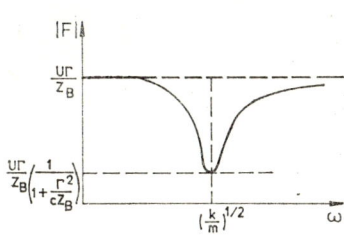

Figure 6.16

If the coil impedance, at rest, is predominantly resistive, for $L = 0$, equation (6.23) becomes

$$F = \frac{U}{\dfrac{R}{\Gamma} + \dfrac{\Gamma}{c + i\left(\omega m - \dfrac{k}{\omega}\right)}}. \qquad (6.24)$$

The modulus of this expression is plotted against frequency in Figure 6.16. To minimize the trough it is necessary that Γ

6 Vibration exciters

be small (low force per unit current) and R be large, both implying inefficiency [11].

Using correction circuits or shaping the input voltage it is possible to compensate the troughs produced by structural resonances. The simplest solution is to use a *constant* amplitude *current* source, so that, according to the first equation (6.23), the exciting force be independent of frequency.

Disconnecting the coil ($u = 0$) and considering that the steady-state vibrations are entertained by a force $f_1 = F_1 e^{i\omega t}$, equations (6.22) become

$$(R + i\omega L)I + i\omega \Gamma X = 0,$$

$$-\Gamma I + (k - m\omega^2 + i\omega c)X = F_1,$$

wherefrom, by elimination of I, one obtains

$$\left[-m\omega^2 + i\omega \left(c + \frac{\Gamma^2 R}{R^2 + \omega^2 L^2} \right) + \left(k + \frac{\Gamma^2 L}{\frac{R^2}{\omega^2} + L^2} \right) \right] X = F_1. \quad (6.25)$$

It follows that, due to the interaction between the electrical circuit and the mechanical system, supplementary damping and stiffness are introduced in the system. The apparent viscous damping decreases with increasing frequency and the apparent stiffness increases with increasing frequency [12]. This can have serious consequences in free decay experiments.

6.3.3 Frequency response

For constant amplitude input voltage to the drive coil, the variation of table acceleration is shown in Figure 6.17 [1] as a function of frequency. The damped peak at low frequencies corres-

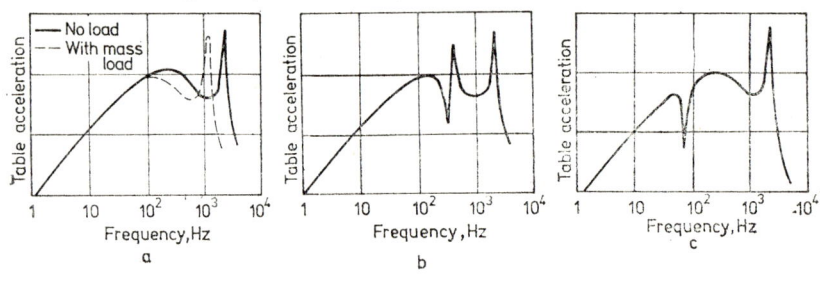

Figure 6.17

ponds to the series resonance between the drive coil inductance and the drive coil — table assembly mass, referred to as *electrical resonance*. The decrease which follows is determined by the coil

6.3 Electrodynamic vibrators

inductance. The peak at high frequencies is produced by the axial mechanical resonance of the coil supporting structure, referred to as the *axial resonance*.

Figure 6.17 a shows the table acceleration without load (solid line) and with a nonresonant load (dashed line), which acts as a rigid mass. Addition of a mass load has the effect of decreasing the frequency of the axial resonance.

Resonant loads, having both mass and stiffness, give a characteristic "peak-trough" response as in Figure 6.17 b, where the trough (antiresonance) frequency is higher than the electrical resonance frequency. The trough indicates a pronounced decrease of the table acceleration at the frequency where the attached structure acts as a dynamic absorber. The peak indicates an increase in the table acceleration due to a "reflected resonance" of the tested structure.

If the antiresonance frequency is lower than the electrical resonance frequency, the low frequency peak is highly damped and does not exceed the asymptotic response (Fig. 6.17 c).

Generally, at the excitation of actual structures, exhibiting several resonances, the frequency response characteristic of the vibrator is more complex than that from Figure 6.17, having a peak-trough at each resonance.

6.3.4 Measurement of the force applied to the structure

Generally, the force applied by the electrodynamic vibrator to the excited structure differs from the force (6.18) generated by the drive coil.

Figure 6.18 a shows schematically a direct-drive vibrator attached to a single-degree-of-freedom system. The equivalent lumped parameter system is presented in Figure 6.18 b, where: m_1 is the armature mass, m_2 — vibrator body mass, m_s — structure equivalent mass, k_1 — coil suspension stiffness, c_1 — coil suspension damping coefficient, k_2 — vibrator suspension stiffness, c_2 — vibrator suspension damping coefficient, k_s and c_s — stiffness and damping of the structure, $f = \Gamma i$ — the force generated by the coil.

If the force f_s applied to the mass m_s is pointed out (Fig. 6.18 c), the equations of motion for the three masses are the following

$$m_2\ddot{x}_2 + (c_1 + c_2)\dot{x}_2 + (k_1 + k_2)x_2 = c_1\dot{x}_1 + k_1x_1 - \Gamma i,$$
$$m_1\ddot{x}_1 + c_1\dot{x}_1 + k_1x_1 = c_1\dot{x}_2 + k_1x_2 + \Gamma i - f_s, \qquad (6.26)$$
$$m_s\ddot{x}_1 + c_s\dot{x}_1 + k_sx_1 = f_s.$$

6 Vibration exciters

The second equation (6.26) can also be written

$$\Gamma i - f_s = m_1 \ddot{x}_1 + c_1(\dot{x}_1 - \dot{x}_2) + k_1(x_1 - x_2). \tag{6.27}$$

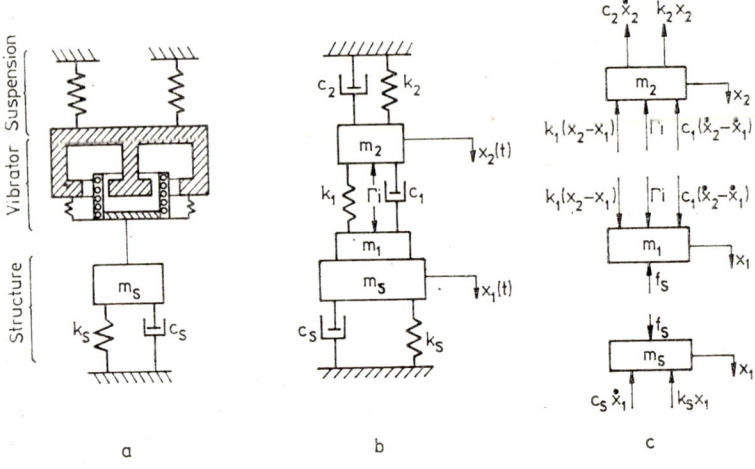

Figure 6.18

It shows that the difference between the force generated by the coil Γi and the force actually applied to the structure f_s depends on the moving parts inertia and the elastic and damping characteristics of the coil suspension. At the testing of small scale models, the masses m_1 and m_s can be comparable; at the excitation of lightly damped structures, the coefficient c_1 can be much greater than c_s. These have to be taken into account when an accurate measurement of the force and of the structural damping is required. In this case, the exciting force cannot be evaluated by simply measuring the current passing through the drive coil and using equation (6.18); a force transducer is used, mounted between the structure and the vibrator table (or threaded rod).

Generally, the suspension k_2 is very soft, and the mass m_2 is much larger than $(m_s + m_1)$, so that the resonance of the mass m_2 on its springs appears at low frequencies. As for very large vibrator units, there is considered that $x_2 = 0$ and calculation is performed based on the simplified model from Figure 6.19.

Figure 6.20 a depicts an inertial drive vibrator attached to a resonant structure modelled by a single degree of freedom system. At reaction-type vibrators, m_1 is the moving coil assembly mass, the spring k_1 having a relatively high stiffness in order to support the vibrator mass m_2. At inverted direct-drive vibrators

204

6.3 Electrodynamic vibrators

used for inertial drive, m_2 is the mass of the armature plus the mass attached for increasing the inertia force, and m_1 is the mass of the vibrator body which is rigidly attached to the structure.

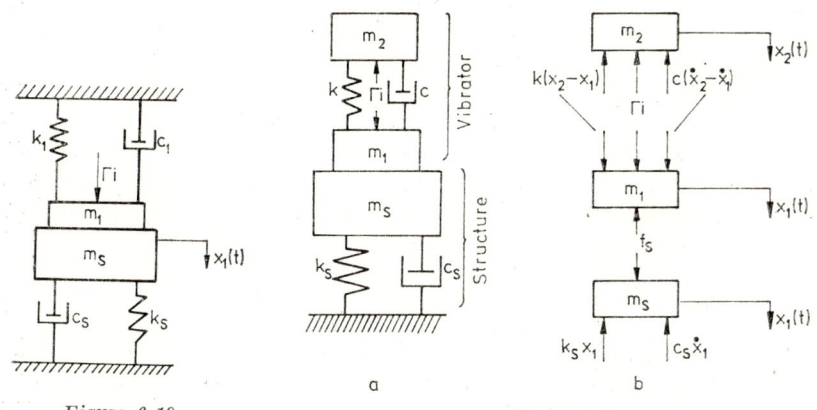

Figure 6.19   Figure 6.20

If the force f_s applied to the structure is pointed out (Fig. 6.20 b), the equations of motion for the two masses are

$$m_2 \ddot{x}_2 + c(\dot{x}_2 - \dot{x}_1) + k(x_2 - x_1) = -\Gamma i,$$
$$m_1 \ddot{x}_1 + c(\dot{x}_1 - \dot{x}_2) + k(x_1 - x_2) + f_s = \Gamma i,$$

wherefrom one obtains

$$f_s = -m_1 \ddot{x}_1 - m_2 \ddot{x}_2.$$

Usually $x_2 > x_1$ and $m_2 \gg m_1$ so that $f_s \cong -m_2 \ddot{x}_2$ and can be evaluated by measuring the acceleration of the mass m_2.

The damping c determines a phase lag between current and force. Coil suspensions, obtained by inserting a disc of foam nylon between the moving element and the magnet core (like those used at some loudspeakers and vibrators for transducer calibration, e.g. Brüel & Kjaer 4290), must be avoided where measurements are carried out taking as reference the amplitude and phase of the current passing through the drive coil.

6.3.5 Features of a vibrator used in structural testing

An electrodynamic vibrator used for vibration studies should meet the following requirements: a) the table motion or the excitation force waveform should not be distorted; b) motion of the excited structure should not produce a reaction on the excitation force

6 Vibration exciters

produced by vibrator; c) the vibrator should not alter the dynamic characteristics of the excited system by adding mass, stiffness or damping; d) the force amplitude and frequency should have independent control; e) when several vibrators are simultaneously used, the phase of their forces should be invariable.

The efficiency of the conversion of electrical energy into mechanical energy is a secondary problem. The operation being intermittent, the cost is proportional to the required accuracy.

Generally it is impossible to design a vibrator to meet all these requirements. Some of them can be fulfilled assuming that : a) the steady magnetic field is much stronger than the alternating magnetic field generated by the drive coil; b) the source impedance from the coil electrical circuit is very large, to ensure that the two magnetic fluxes in the air gap are independent of each other; c) the moving coil has small (relative) displacements with respect to the air gap length; d) magnetic saturation is avoided; e) the d.c. current in the drive coil is avoided; f) eddy currents are reduced to a minimum; g) the stray field around the main magnetic circuit is minimized using degaussing coils or magnetic shields, especially near the table; h) the axial resonance frequency of the coil and of its fixture is as high as possible. Requirements concerning weight, overall dimensions, handling and long time fidelity are added to these [13].

6.3.6 Application

Practical use of electrodynamic vibrators is limited (by their performance characteristics) to testing with relatively small forces (under 30,000 N), small displacements (up to 25 mm) and relatively high frequencies (5 Hz to 20 kHz). They have a low resistance to side loads and to overtravel, require large driving power supplies and have a reduced resistance to overloading. Auxiliary support is required for large masses. Maximum loads are of the order of 800 kg.

The ratio $\dfrac{\text{Excitation force}}{\text{Moving coil weight}} = \dfrac{\dfrac{i}{S}BV}{V\gamma} = \dfrac{i}{S}\dfrac{B}{\gamma}$, where γ is the specific weight of the coil wire material, has values up to 150, while for electromagnetic vibrators the corresponding ratio can be maximum 20.

At small vibrators, the constant $\Gamma = \dfrac{F}{i} = 10\ldots20$ [N/A].

6.4 Hydraulic vibrators

The use of the electronic oscillator for generating the input signal enables a high degree of accuracy and stability of the exciting frequency to be maintained. Successful matching with the power amplifier over a wide frequency range is obtained with moving coils having a low impedance with a flat frequency response characteristic. Various types of transients and shocks can also be easily simulated by a proper choice of the signal generated by the oscillator.

Besides vibrators for motion in translation, there are vibrators producing oscillatory motion in rotation. They are of much more limited use.

6.4 Hydraulic vibration exciters

6.4.1 Construction

The hydraulic vibrator is a device which uses the energy of a high pressure flow of fluid to generate a reciprocating motion of the piston of a servo system. The basic elements of an electrohydraulic vibration-generation system are shown in the block diagram shown in Figure 6.21.

The main features are the hydraulic amplification of the input signal which produces the displacement of the actuator (and of the tested structure) and the control of the servo valve by a (force or displacement) transducer, which feeds back a signal to the servo amplifier, in opposition to the input. With a suitable device

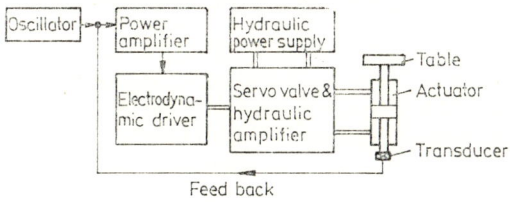

Figure 6.21

included in the feedback path, the system can be made to yield constant amplitude displacement, velocity, acceleration or strain. Performances are limited by the maximum acceleration attainable at the servo valve, by the fluid flow which can be obtained at high frequencies, by the effective area and the stroke of the actuator, by the mass of the specimen, etc.

207

6 Vibration exciters

6.4.2 Frequency response

Manufacturers present the frequency response of hydraulic vibrators by performance graphs like that shown in Figure 6.22.

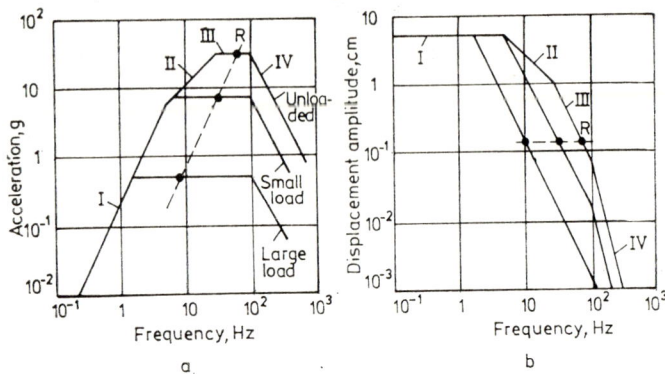

Figure 6.22

At very low frequencies, a limit is imposed by the stroke of the actuator (region I). At higher frequencies, the oil pump flow, the area of the actuator and the servo valve, and the loss of pressure in the valve limit the maximum velocity (region II). The mass of the tested object limits the acceleration which can be maintained constant (region III) until the required flow exceeds that available from the valve. Above this limit, the compressibility of the hydraulic oil reduces the effective thrust and acceleration varies inversely as the square of frequency (region IV). Point R locates the oil-column resonance in the servo-actuator. To maintain constant acceleration within region III, below this frequency the flow requirement is inversely proportional to frequency, while above it, the flow must be directly proportional to the frequency. For large load weights, region II can miss from the response line.

Generally, the actuator stroke should be kept to a minimum in order to reduce the volume of the oil enclosed in the actuator. The stroke length is determined by the performance required at very low frequencies, where high accelerations call for long strokes.

6.4.3 General performance characteristics

Hydraulic vibrators have no lower frequency limit. The normal frequency range extends up to 150 Hz, but special short-stroke actuators are manufactured for operation at frequencies up to 500 Hz. Typical actuators have stroke lengths up to ± 50 mm, but

6.4 Hydraulic vibrators

strokes up to ±150 mm are readily available at low frequencies. Force capability is high, thrusts ranging from 2000 to 450,000 N. The weight of hydraulic vibrators is in general low relative to the generated force, however the complete hydraulic system is rather bulky. The waveform of the table motion is not so good as that of the electrodynamic vibrators, due to distortion introduced by the valve itself and the friction in the actuator. Performance characteristics of some typical hydraulic vibrators used in structural resonance testing are given in Table 6.3 [5].

TABLE 6.3 *Performance Characteristics of Typical Hydraulic Vibrators*

Manufacturer	Type	Maximum force, N	Min. freq. for max. force, Hz	Upper frequency limit at full force, Hz	Mass of vibrator, kg	Power, kW
Sandia Lab.	Linear inertial mass	56,000	2	50	8,800	64
Zonic ES-302-1	Linear inertial mass	9,000	20	200	318	8
Zonic 1306	Actuator against strong wall	90,000	0	10	35	25
Boeing	Linear inertial mass	300,000	—	—	30,000	300

Improved performances have Zonic XCITE[R] Systems capable of providing both static preload and dynamic forces. They contain small sized exciter heads (5 kg), high frequency response servovalves (0—1000 Hz) and dual-loop servo controllers. Strokes of 25 (or 50) mm are obtained together with forces of 4500 N. Inertial mass exciters for dynamic forces of 4500 or 9000 N are also built providing excellent excitation in the ranges required for rotating and moving machinery.

References for Chapter 6

1. Harris, C. M. and Crede, Ch. (Red.), *Shock and Vibration Handbook*, 2nd Ed., McGraw-Hill Comp., 1976.
2. Buzdugan Gh., *Dinamica fundațiilor de mașini*, Editura Academiei, București, 1968.

6 Vibration exciters

3. Richart, F. E., Jr., Hall, J. R., Jr., and Woods, R. D., *Vibrations of Soils and Foundations*, Prentice Hall Inc., Englewood Cliffs, N. J., 1970.
4. Buzdugan, Gh., *Dynamique des fondations de machines*, Éditions Eyrolles, Paris, 1972.
5. Ibáñez, P., Review of analytical and experimental techniques for improving structural dynamic models, *Welding Research Council Bulletin*, 249 (June 1979).
6. Lorenz, H., *Grundbau Dynamik*, Springer Verlag, Berlin, Göttingen, Heidelberg, 1960.
7. de Ro, M., Mesure et linéarisation de la force de l'excitateur magnétique au moyen de générateurs de Hall, *Annals of the C.I.R.P.*, 17, 401—408 (1969).
8. Lombard, J., Mirski, F., Mesure de l'effort dynamique developpé par un excitateur électromagnétique, *Annals of the C.I.R.P.*, 17, 503—510 (1969).
9. Grootenhuis, P., Gearing, J. W., Design of electromagnetic vibration generators, Proc. Conf. "Machines for Materials and Environmental Testing", *Inst. Mech. Engrs. and S.E.E.*, Part 3 (1965).
10. de Vries, G., Quelques points particuliers de la technique d'excitation en vol par vibreurs harmoniques, *La Recherche Aéronautique*, 68, 47—53 (1959).
11. White, R. G., Vibration testing machines, Lecture Notes, Bucharest, 1972.
12. de Vries, G., Les excitateurs électriques, *La Recherche Aéronautique*, 41, 35—45 (1954).
13. de Vries, G., Utilisation de l'excitateur électrodynamique en tant qu'instrument de mesure de précision, *La Recherche Aéronautique*, 78, 47—55 (1960).
14. * * * *Hydraulic Vibration Systems*, Fairey Surveys Ltd.
15. Magrab, E. B. (Ed.), *Vibration Testing — Instrumentation and Data Analysis*, A.S.M.E. Monograph, A.M.D. vol. 12, 1975

7

Instrument set-ups and techniques for vibration measurement

7.1 Selection of equipment

Selection of the adequate components of an instrumentation system depends on several factors which must be weighted at the planning stage of any measurement program. The following are among the most significant : a) measurement location and direction, available space for transducer and equipment installation; b) frequency range and amplitude range; e) required accuracy for the data; d) environmental conditions (temperature, humidity, magnetic fields, radiation, noise); e) form of the final record which is desirable; f) number of identic or different physical quantities to be measured simultaneously; g) type of analysis : time domain or frequency domain; h) duration of test program; i) cost considerations; j) personnel skill and experience.

Transducer selection is determined by the (presumptive or predetermined) amplitude of the measurand, the operating frequency range, the relative mass with respect to the effective mass of the measured structure, the environment and the limitations set up by other instruments. As has been shown in section 1.4, low-frequency large-amplitude vibrations are measured using *displacement pickups,* while shocks and high frequency vibrations are usually measured using *accelerometers.*

Piezoelectric *accelerometers* have come into general use relatively recently, due to the small size and light weight, lack of parts in relative motion, wide frequency range and possibility to obtain, by integration, the velocity or the displacement of vibration. They are best suited for measurements at high frequencies, to detect blade and gear defects or to help at early warning on ball bearing deterioration.

Velocity pickups are sometimes preferred because a whole range of standards and codes specify limits for the severity of

7 Instrument set-ups and techniques

vibration, expressed in terms of the root-mean-square value of the velocity of vibration, which is directly proportional to the energy producing it. However, they often are large in size, which raises problems at installation, and contain delicate moving parts prone to friction and wear. Velocity pickups are suitable for rugged industrial application, being relatively insensitive to dirt, dust, water and attachment configurations. They are often used as contact pickups to record bearing vibrations.

Displacement-sensing pickups have a relatively narrow frequency range, their size requiring large mounting surface areas, which can increase the local stiffness of flexible structures. They are recommended when measurements are to be made on relatively heavy structures, exhibiting large amplitude vibrations.

Non-contacting transducers are used especially to measure shaft vibrations of machines having heavy casings and relatively lightweight rotors. Mounted in pairs 90° apart in a journal bearing, they are used for recording shaft orbits and detecting shaft unbalance, misalignment or whipping. Eddy-current transducers are the most widely used, since they are not affected by oil, gas or other non-conductive materials. The peak value or the peak-to-peak value of the vibration *displacement* is measured in radial direction, but the axial position of the shaft can also be monitored.

Piezoelectric accelerometers are used with preamplifiers called voltage or charge amplifiers. When a *voltage amplifier* is used, the circuit total capacitance is very much decreased, so that the signal transmitted to subsequent instruments has the highest possible voltage. When a *charge amplifier* is used, the accelerometer is loaded with a high shunt capacitance so that the output signal becomes independent of small changes in cable capacitance (due to changes in cable length between accelerometer and preamplifier).

The form of the final record which is desirable determines a good deal of the instrumentation system. Recording on paper or film offers a simple and convenient form for determining trends, being yet a "final record", which cannot be analysed by instruments, but must be read by a human operator before using it in subsequent computations.

Electrical recording, either on magnetic tape or in memory devices developed for digital instruments, offers substantial advantages. Data with slow rates of change in time can be speeded up by time compression techniques. This permits the frequency translation of spectra, hence use of spectrum analyzers with better resolution and higher analysis speed. Small sections of data which are particularly interesting (e.g., large amplitude shocks or transients) can be removed from the large amount of data and can be analysed in detail. Similarly, periodic functions

7.2 Measurement of signal waveform

can be created out of any segment of data (using tape loops or event recorders) which can be analysed later on by using simpler techniques.

Instruments used for frequency analysis are more complex, the higher is the frequency resolution and the shorter is the analysis time. Thus, if octave or third-octave filters are sufficient for establishing the noise level in a room or a workshop, then narrow band constant bandwidth filters are required for identifying sources of noise and vibration by frequency analysis. Real time analyzers with digital memories and heterodyne filtering are employed for the analysis of non-stationary signals.

Digital instruments are coming into current use due to obvious advantages : a) high accuracy, good reliability and high measurement speed ; b) simultaneous processing of large amounts of data ; c) capabilities to be used in minicomputer controlled schemes or in time-sharing systems, using large digital computers.

To a large extent, instrumentation selection is conditioned by the understanding and clear definition of the dynamic problem to be studied, of the conditions under which the test will be performed and of the particular technique adopted.

The same results can be obtained in several ways, using different set-ups, the final choice being determined by cost, installation and maintenance problems.

7.2 Basic set-ups for signal waveform measurement

In this section, some "classical" vibration measurement techniques are mentioned, utilizing a minimum of equipment, with the purpose of determining the overall vibration level. This type of measurement is referred to as a "linear measurement", in the sense that no filter is used for signal analysis, hence all harmonics of the input signal are processed alike, evidently within the linear operating range of the instrumentation system.

Analysis of these measurement data is conclusive provided that the signals have a fairly simple waveform, i.e. harmonics within a narrow frequency range. It is considered that if a signal contains the fundamental and more than five harmonics, then the visual analysis of its time history record becomes difficult (if not impossible) giving inaccurate information on the phenomenon, so that signal filtering is required.

For prospective measurements, performed for determining the order of magnitude of vibrations, as well as for some field measurements, where a high accuracy or a frequency analysis is not required, the equipment consists of several basic items, often portable and battery-operated.

When only the vibration amplitude has to be measured, a set-up like that shown in Figure 7.1 is sufficient, where the transducer sensitivity must be known and where the voltmeter

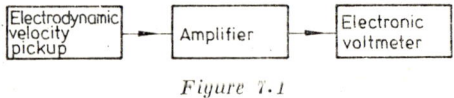

Figure 7.1

readout indicates, according to the scale calibration : the maximum positive or negative elongation, the peak amplitude, the peak-to--peak amplitude, the mean square or the root-mean-square value.

This arrangement forms the basis of some portable vibrometers, used either for measuring the vibration level or for the first part of an experimental study aiming to locate regions and/or components exhibiting large vibration amplitudes.

The pickup is hand-held. The case, containing the amplifier, integrating circuits and the direct reading meter (with a scale calibrated in displacement, velocity or acceleration units), is held by a belt neck-carried by the operator. The order of magnitude of the vibration is thus determined, making possible to plot "modal maps" — either as lines of constant vibration amplitude or as modal shapes, where nodal lines are also drawn. Portability, ruggedness and ease of operation are the main advantages obtained at the expense of less accuracy.

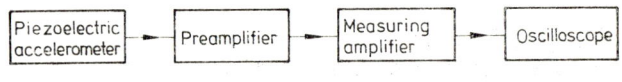

Figure 7.2

The set-up shown in Figure 7.2 is used to the same purpose. When a direct reading meter is not available, the vibration amplitude can be measured by the vertical deflection of the spot of a cathode ray oscilloscope relative to the millimeter coordinate scale of the screen. A prior calibration is needed in this case.

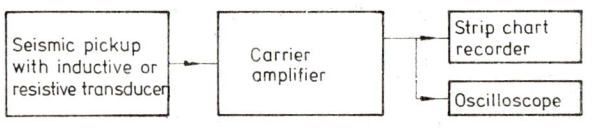

Figure 7.3

More information can be obtained using the set-up presented in Figure 7.3, by examining the time history record (vibrogram) of the vibration, either from a chart tracing or from the c.r.o. recordings made with a camera.

When a displacement sensing pickup with strain gauges is used, the "zero line" A is first plotted (Fig. 7.4), driving the chart paper when the pickup is dismounted and the bridge is balanced. The line B is then traced, corresponding to a known static displacement d of the pickup seismic mass, produced in the dismounted pickup by means of a screw with a calibrated thread whose axial displacement is given by the product of the thread pitch to the number of turns. Finally, the signal waveform C is recorded.

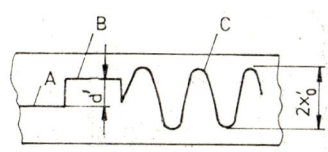

Figure 7.4

If d' is the distance between lines A and B, and $2x_0'$ is the peak-to-peak displacement measured on the chart paper, then the peak amplitude of the actual displacement is given by

$$x_0 = \frac{x_0'}{d'} d. \qquad (7.1)$$

In the case of periodic vibrations, the frequency can be measured from chart tracings using the simultaneously recorded timing marks. Usually, the chart paper has millimeter line markings and the paper drive velocity (past the stylus tip) is known. The number N_1 of millimeters corresponding to the chart paper displacement in one second is therefore known and the operator has to determine the number N_2 of millimeters corresponding to n periods of vibration (Fig. 7.5). The frequency of vibration is given by

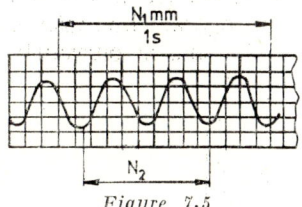

Figure 7.5

$$f = n \frac{N_1}{N_2} \; [\text{Hz}], \qquad (7.2)$$

Sometimes, during field measurements, small variations of the supplying voltage might produce variations of the chart paper speed. In these cases it is recommended either to use equation (7.2) for large values of n, determining an average frequency, or to use a time-marking trace on the same record.

For recording non-stationary signals, e.g. vibrations generated by a variable speed rotating machine, it is useful to trace the time marking using the output from an inductive or eddy-current non-contacting transducer. Each time a prominent part revolving with the machine shaft passes close in front of the transducer

7 Instrument set-ups and techniques

face, the output voltage increases, giving a pulse in the time record (Fig. 7.6).

Time history records may also be obtained using the cathode ray oscilloscope. Moving-film recording with the sweep turned off is generally used with shutterless cameras placed in front of the oscilloscope screen.

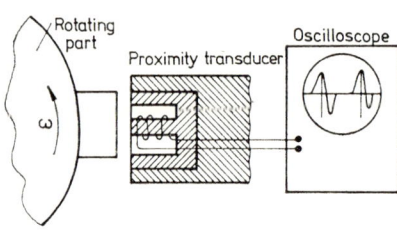

Figure 7.6

Using the oscilloscope timebase, the sweep frequency can be adjusted until stationary patterns of vibration waveforms are obtained, then, using the millimeter coordinate scale of the screen, the signal amplitude can be measured (usually the peak-to-peak amplitude). Using a dual-beam oscilloscope, both amplitude and frequency measurements can be performed.

A reference signal from an oscillator may be used, adjusting its amplitude and frequency until the stationary patterns of the two signal waveforms coincide with each other. For more accurate measurements of the frequency of sinusoidal signals, use of digital multimeters or frequency meters is recommended.

The frequency of the analysed signal, displayed along the vertical axis of a cathode-ray oscilloscope, can be determined by comparing it to a second signal, whose frequency is variable and known, displayed along the horizontal axis, using the set-up shown in the block diagram presented in Figure 7.7. The frequency of the reference signal is adjusted until a stationary Lissajous pattern is obtained on the screen, when the frequencies of the two signals are related to each other as a ratio of integers. The method better applies to sinusoidal stationary signals and is currently used for the production control of turbine blades.

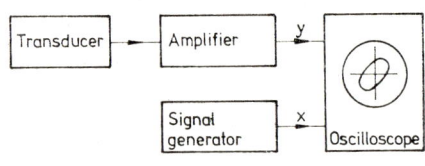

Figure 7.7

The orbit analysis of a journal in an oil lubricated bearing is based on the same principle. The orbit is obtained by displaying, on the X- and Y-axis, respectively, the signals from two non-contacting pickups mounted in the bearing, 90° apart, preferably on the vertical and horizontal direction.

7.3 Random vibration analysis

By means of a cathode-ray oscilloscope, the phase difference between two signals of the same frequency can be measured, using the elliptical-pattern measuring technique. Connecting the two signals to the X and Y axes of the scope (Fig. 7.8 a), an ellipti-

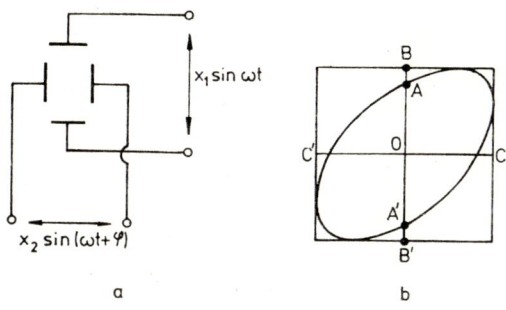

Figure 7.8

cal pattern is obtained (Fig. 7.8 b). The absolute value of the phase angle φ between the two voltages is given by

$$\varphi = \sin^{-1} \frac{AA'}{BB'}, \qquad (7.3)$$

where AA' is the distance between points of Y-axis crossing and BB' is the distance between the horizontal tangents to the ellipse. The sign of the angle φ can be determined introducing an additional phase difference and observing the change of the elliptical pattern.

The phase shift introduced by amplifiers on the two channels should be checked before measurement. Connecting the same signal to both axes, a straight line must be seen on the screen, indicating zero phase difference between channels. Other methods are used for more accurate measurements, utilizing either phase-meters or a phase comparator in conjunction with a calibrated phase-shift network [1].

7.3 Procedures for analysing random vibration records

In the case of random signals, the shape of time-history records, obtained from experiments carried out under identical testing conditions, differs from a record to the other. This requires determination of some statistical properties, obtained by processing a single record or a collection of sample records.

7 Instrument set-ups and techniques

The collected time-history records should have sufficient length so that the statistical properties be representative for the phenomenon under investigation.

Data processing procedures differ according to the number of sample records analysed.

7.3.1 Analysis of a single record

A general procedure for analysing the statistical properties of a single sample record is presented in Figure 7.9 [2].

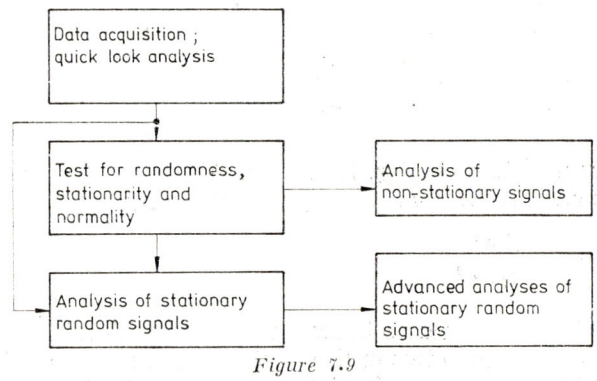

Figure 7.9

Data acquisition must be followed by a *quick look* at the time history data (e.g., on the screen of a cathode ray oscilloscope) to detect relevant properties of the studied phenomenon and to see whether or not the recorded data is random, stationary and/or normal. Sometimes, the presence of a periodicity, a lack of stationarity and a significant deviation from normality can be detected by simple observation of the time history data, but more usually, in order to validate the results of an analysis, this requires inspection of the statistical functions.

For three usual signals : sine wave, narrow band random noise and sine wave plus random noise, typical time histories together with theoretical forms for their probability density functions, autocorrelation functions and power spectral density functions are shown in Table 7.1.

The presence of a sinusoidal component in otherwise random data may be detected on either of the three graphs from the table : a) the probability density function has a well defined form, being a mixture of the bell-shaped Gaussian characteristic of a random signal and of the dish-shaped function of a sine wave; b) the

218

7.3 *Random vibration analysis*

TABLE 7.1 *Statistical Functions for Different Signals*

Signal	Time history	Probability density function	Autocorrelation function	Power spectral density function
Sine wave	$x(t)$	$p(x)$	$R_{xx}(\tau)$	$S_{xx}(\omega)$
Narrow band random noise	$x(t)$	$p(x)$	$R_{xx}(\tau)$	$S_{xx}(\omega)$
Sine wave plus random noise	$x(t)$	$p(x)$	$R_{xx}(\tau)$	$S_{xx}(\omega)$

7 Instrument set-ups and techniques

autocorrelation function is periodic, with the same period as the sine wave; c) the power spectrum has a sharp peak at the frequency of the sine wave.

The most efficient method of detecting the presence of sinusoids in random data is the analysis of an autocorrelogram. Use of power spectra requires a frequency analysis with very narrowband filters, in order to detect the harmonic component.

Practical tests for stationarity are based on the assumption that the mean value and the mean square value computed by time averaging over each of a sequence of short time intervals from a single sample record will not vary significantly from one time interval to the next. Some time after the beginning of the measurement, these values tend to a stationary value, otherwise the data are non-stationary.

Tests for normality are described in books on statistical mathematics. The most convenient is Pearson's chi-square goodness-of-fit test [3].

Analysis of stationary random signals implies measurement of the following data properties, defined in Chapter 2:
— mean (average) value $\bar{x}$ and mean square value $\overline{x^2}$;
— probability density function $p(x)$ of instantaneous values;
— autocorrelation function $\bar{R}_{xx}(\tau)$;
— power spectral density function $S_{xx}(\omega)$ or $W_{xx}(f)$;
— amplitude density spectrum $|X(i\omega)|$, calculated as the modulus of the Fourier transform of the signal $x(t)$.

Measurement of all statistical properties is seldom performed because part of the information can be obtained in many different ways.

The interpretation of experimental data may require *specialized types of analysis of stationary random signals* such as: extreme value analysis, peak value distribution analysis, zero crossing or threshold crossing analysis.

As for the analysis of *non-stationary* signals, specialized instrumentation and experimental techniques have been developed for each different type of non-stationarity.

7.3.2 Analysis of a collection of records

A procedure for analysing a collection of sample records is schematically presented in Figure 7.10 [2].

As in the case of single record analysis, *data collection* is accompanied by a *quick look analysis*; at the same time, the *analysis of individual records* is performed as shown in Section 7.3.1. The next step is the *test for equivalence* carried out on several or all of the sample records and the pooling of equivalent data.

7.4 Frequency analysis

The analysis of pairs of sample records consists in determining the following functions: the joint probability density $p(x, y)$; the cross-correlation function $\bar{R}_{xy}(\tau)$; the cross-spectral density $S_{xy}(i\omega)$ or $W_{xy}(if)$.

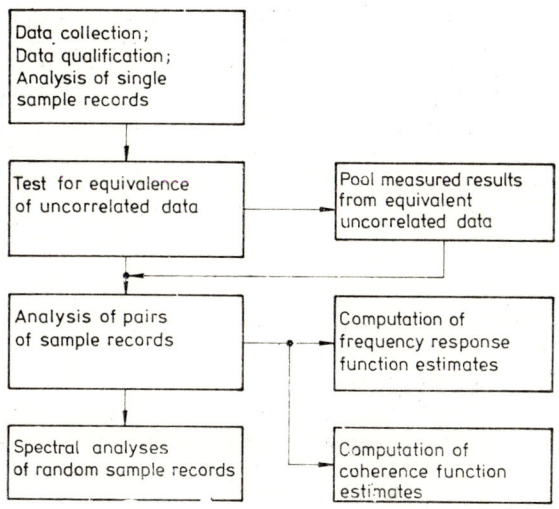

Figure 7.10

For linear vibrating systems it is possible to determine the *frequency response function* (2.73) based on power and cross-spectra measurements of input and output signals. The *coherence function* can also be calculated from power- and cross-spectra of two signals.

Special analyses are also required for the interpretation of random data. They might include measurement of joint peak value probability density and conditional probability density.

7.4 Frequency analysis

Frequency (spectral) analysis is a procedure for determining the frequency distribution of the power (or energy) of a signal. It is performed separating the components of distinct frequency from a complex signal and indicating their magnitude (or determining the power spectral density distribution in given frequency bands). This separation is achieved using either a set of band filters centred

7 Instrument set-ups and techniques

on various frequencies or a single filter, stepped or swept over the frequency range of interest.

Frequency analysis can be performed with analog equipment consisting of vibration pickups (or microphones), preamplifiers, measuring amplifiers, band filters, means to switch the filters or scan the spectrum (see Section 5.4.3), readout devices (meters, dials, c.r.t. displays) and other equipment, including level recorders, eventually a magnetic tape recorder. Modern spectrum analyzers are connected to transducers via signal conditioning instruments (usually charge amplifiers) and have output sockets for $X-Y$ recorders when hard copies of the spectra are required. Frequency analysis using the computer has replaced the analog and hybrid methods because of the versatility of the FFT algorithm and dedicated hardware.

7.4.1 Frequency analysis of stationary signals

7.4.1.1 *Selective filtering*. The classical technique for the analysis of stationary signals is *selective filtering* because the time during which the signal is at the operator's disposal is generally sufficient to carry out a non-real time analysis, scanning the spectrum serially, band by band.

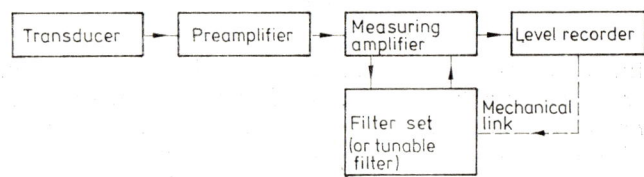

Figure 7.11

The simplest experimental set-up using analog instruments is shown in Figure 7.11. Usually, the filters are connected to the measuring amplifier together with a detector consisting of a multiplying (squaring) stage and an averaging (integrating) stage (Fig. 7.12).

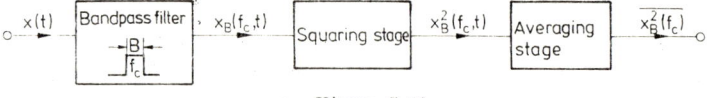

Figure 7.12

When a(n) (octave or third octave) filter set is used, the input waveform is applied to a number of *filters in parallel* and the

7.4 Frequency analysis

output of any filter is manually selectable by a switch. The output of the selected band is then detected and displayed and the level of the output power in each frequency band is written in a notebook and used for subsequent plotting of a *band spectrum*. Often the rate of stepping through the filters is controlled by and synchronized with the paper speed of a level recorder, used to plot the resulting spectrum.

For narrowband analysis, a single continuously tunable filter is commonly used, obtaining a *continuous frequency spectrum*. The filter can be tuned either manually or automatically, being synchronized with the level recorder.

7.4.1.2 *Time compression analysis.* Results of frequency analyses performed with filters having different bandwidths are compared in Figure 7.13. There can be seen that the smaller the filter bandwidth, the higher the frequency resolution. However, use of narrow band filters leads to a considerable increase of the analysis time, required by the increase of the filter response time.

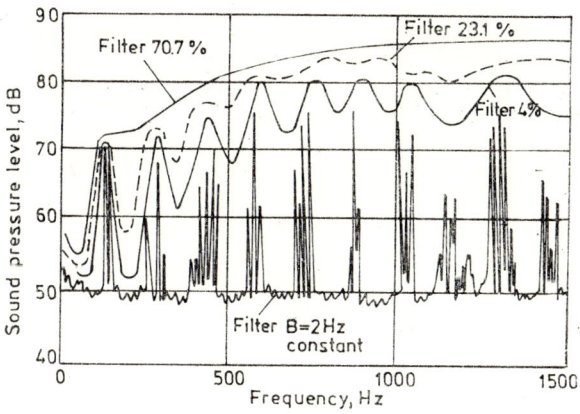

Figure 7.13

It has been determined empirically [5] that the time T required for the output of a filter (with a shape factor of 4) to reach 63% of the input level is $T = 1/B$, where B is the -3 dB bandwidth. For a filter to reach 99% of the input level, $4T$ is required. The filter can "move" one bandwidth B every time it has responded, so that the scanning rate is

$$v_s = \frac{B}{4T} = \frac{B^2}{4} \cdot [\text{Hz/s}]$$

7 Instrument set-ups and techniques

Scanning faster than this rate will result in errors, the filter having not enough time to respond to the input signal.

The time required to obtain a complete spectrum, using a B, Hz, wide filter and analysing from f_1, Hz, to f_2, Hz, is:

$$t = \frac{f_2 - f_1}{v_s} = \frac{4(f_2 - f_1)}{B^2}. \quad [\text{s}] \tag{7.4}$$

In order to reduce the analysis time, the signal can be speeded up, compressing it in the time domain, hence expanding it in the frequency domain.

Analog time compression can be done using the equipment shown in the block diagram presented in Figure 7.14. The data signal (Fig. 7.15 a) is recorded on magnetic tape at a speed v_r and played-back at a faster speed v_p.

Figure 7.14

If $v_p = k \cdot v_r$ ($k > 1$), the data time interval is compressed k times (Fig. 7.15 c) so that all the frequency components of the initial spectrum (Fig. 7.15 b) are translated by k times.

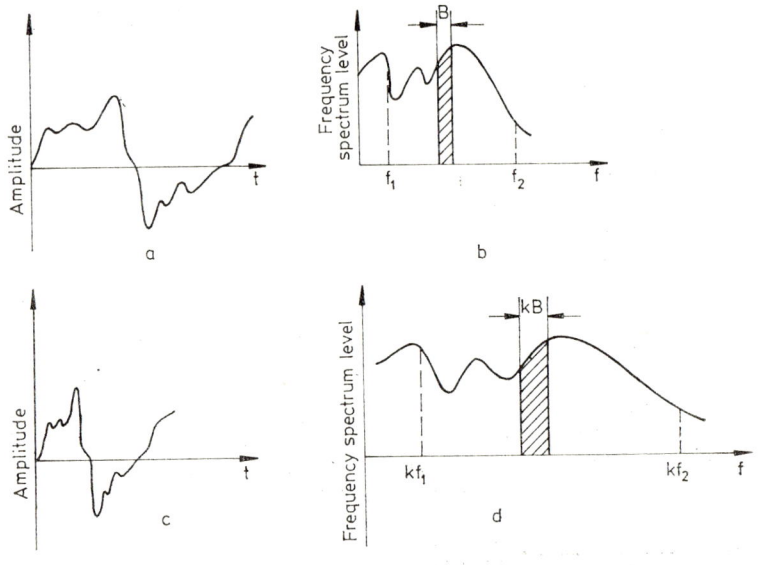

Figure 7.15

7.4 Frequency analysis

The analysed frequency range is expanded k times, lying now between kf_1, Hz, and kf_2, Hz (Fig. 7.15 d). If the same frequency resolution has to be maintained during spectrum analysis, then a filter k times as wide can be used, of bandwidth kB, and a sweep rate k^2 times faster may be employed, $\dfrac{k^2 B^2}{4}$, which leads to a k times reduction of the total analysis time t_k, which becomes

$$t_k = 4\,\frac{k(f_2 - f_1)}{k^2 B^2} = \frac{t}{k}. \quad [\text{s}] \tag{7.5}$$

Analog time compression cannot be performed on-line and certain tape recorders have speed limitations.

Digital techniques are preferred to accomplish *time compression*, being implemented on digital event recorders and on real time spectrum analyzers in which compression factors as high as $k = 500$ can be obtained.

The analog data are first sampled and speeded up in a recirculating memory in digital form. The memory output is then converted back to analog form to give the time compressed signal which is subsequently analysed using a narrow band swept filter of high centre frequency (heterodyne analysis).

7.4.1.3 *Digital analysis.* Digital frequency analysis is implemented on digital filters and FFT analyzers. Modern FFT analyzers allow a full range of data manipulations between and within the time and the frequency domains. Most real-time FFT spectrum analyzers perform only the transformation from the time domain to the frequency domain, but modern instruments allow transformation in both directions as well as data manipulations in the frequency domain.

FFT spectrum analyzers permit calculation of: Fourier transform, inverse Fourier transform, auto spectra, cross spectra, which can be used for computing correlation functions, convolution, frequency response functions, cross spectra and coherence functions. Arithmetic operations may be carried out on complex data arrays in the frequency domain such as addition and subtraction, multiplication and division, multiplication by weighting functions, as well as change of coordinates from rectangular to polar or the reverse. Capabilities for probability analysis are also included in some models.

Both hardwired and minicomputer based systems are available. The latter give the accuracy and flexibility of digital processing while allowing the operator to control the analysis with an easy-to-use control panel, without requiring knowledge of computer programming.

7 Instrument set-ups and techniques

7.4.1.4 Analysis of stationary random signals. The analysis of stationary random signals can be performed for determining the *power spectral density* defined by equation (2.25):

$$W_{xx}(f_c) = \lim_{B \to 0} \lim_{T \to \infty} \frac{1}{BT} \int_{-T/2}^{T/2} x_B^2(f_c, t) \, dt = \lim_{B \to 0} \frac{\overline{x_B^2(f_c)}}{B}. \tag{7.6}$$

It can be obtained by dividing the mean square value $\overline{x_B^2}$, measured in a frequency band centred on the frequency f_c, by the bandwidth B.

When constant bandwidth ($B = $ const.) filters are used, the frequency spectrum obtained with the equipment shown in Figure 7.12 is proportional to the power spectral density. When constant percentage bandwidth filters are used, $B = \alpha f_c$, where α is a proportionality factor, so that the actual spectrum is no more proportional to the spectral density and a correction has to be applied (either graphically or by special filters).

It follows that, if the measured vibrations were periodic (deterministic), the experimentally obtained spectrum would indicate the mean square (effective) value regardless of the filter bandwidth, while at the measurement of random vibrations, the obtained spectrum gives the power spectral density which does depend on the filter bandwidth. The narrower is the filter bandwidth, the more accurately is measured the function $W_{xx}(f_c)$.

The principle of operation of an analog cross-spectrum analyzer is shown in Figure 7.16. On condition that constant bandwidth filters are used, the co- and quad-spectral density functions are given by

$$C_{xy}(f_c) = \lim_{B \to 0} \frac{\overline{x_B(f_c) \cdot y_B(f_c)}}{B}, \quad Q_{xy}(f_c) = \lim_{B \to 0} \frac{\overline{x_B^{90°}(f_c) \cdot y_B(f_c)}}{B}, \tag{7.7}$$

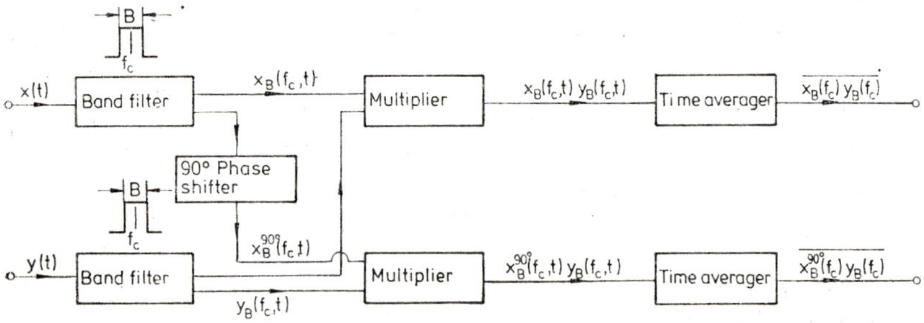

Figure 7.16

7.4 Frequency analysis

The spectrum of a random signal is in itself a random variable so that *averaging* is required to obtain good estimates of "mean" spectral characteristics. The statistical fluctuations in the data decrease as the number of statistically independent spectrum samples which are averaged increases.

The *analysis time* for a random signal, necessary to determine the value of the power spectral density with a given standard error, can be obtained using the information given in Section 2.3.

Starting from the expression of the normalized standard error for the power spectral density $\varepsilon = \dfrac{1}{\sqrt{BT}}$, where B is the filter bandwidth (and BT is the number of statistically independent samples collected in T seconds), the duration of each measurement can be obtained as

$$T' = \frac{1}{B\varepsilon^2}$$

which is longer than the time $1/B$ required for deterministic signals because of averaging. If the analysed signal has frequencies between f_1, Hz, and f_2, Hz, in order to obtain $N = \dfrac{f_2 - f_1}{B}$ points on the spectrum, the total analysis time will be

$$t = NT' = \frac{f_2 - f_1}{B^2 \varepsilon^2}. \qquad (7.8)$$

7.4.2 Frequency analysis of shocks

Generally, frequency analysis of transient signals and shocks is carried out using real time spectrum analyzers or digital event recorders, whose operation is described in sections 5.4.4 and 5.5.10. In the following, two techniques used for analysing single impulses or shocks [6] are described.

7.4.2.1 Pulse transformation into a pulse train.
In the early (analogic) form of this technique, the pulse is recorded on a magnetic tape which is spliced to form an endless loop. The tape recorder is provided with a device for continuous tape drive so that at playback a continuous pulse train is obtained. This is analysed afterwards as a periodic signal, using a heterodyne analyzer. Electronic gating circuitry or use of time windows (as in Section 5.4.4.3) ensure that the output signal from the tape recorder is

7 Instrument set-ups and techniques

zero except for a certain interval around the pulse; otherwise the discontinuity at the splicing would alter the original spectrum.

In this way, a line-spectrum is obtained. The magnitude of each spectral component is a measure of the ordinate of the pulse Fourier spectrum at the corresponding frequency.

Let T_r be the period of repetition of the pulse. If the periodic signal is expressed by the Fourier series

$$x(t) = C_0 + \sum_{n=1}^{\infty} C_n \cos(n\omega_0 t - \theta_n), \qquad (7.9)$$

then the amplitudes of the series components are given by the integral (2.12)

$$C_n = 2|c_n| = \frac{2}{T_r} \int_{-T_r/2}^{+T_r/2} x(t)\, e^{-n\omega_0 t}\, dt, \qquad (7.10)$$

where $\omega_0 = 2\pi f_0 = \dfrac{2\pi}{T_r}$ is the fundamental angular frequency.

Figure 7.17 b shows the frequency spectrum of a series of rectangular pulses with a pulse duration T (Fig. 7.17 a). As the theoretical spectrum has zeroes at frequency intervals of $1/T$, in order to have a good frequency resolution, it is recommended to have 3 to 5 spectral lines between successive minima, i.e. to use a repetition period $T_r = (3\ldots5)T$.

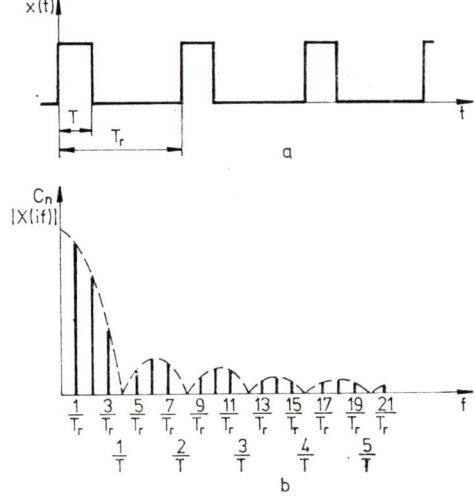

Figure 7.17

7.4 Frequency analysis

On the other side, as the magnitude of harmonic components is usually measured by a time-averaging process, the Fourier spectrum of the single pulse must be defined in terms of spectral density rather than of magnitude.

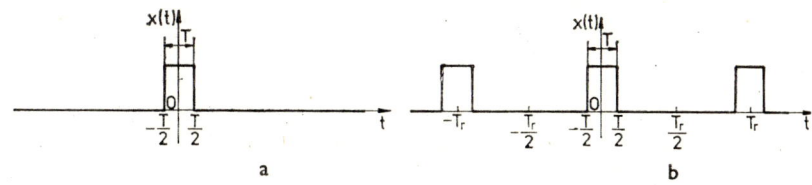

Figure 7.18

The frequency domain description of a single pulse is given by the Fourier integral (2.15). For the pulse shown in Figure 7.18a, one obtains

$$X(i\omega) = \int_{-\infty}^{\infty} x(t)\, e^{-i\omega t}\, dt = \int_{-T/2}^{T/2} x(t)\, e^{-i\omega t}\, dt, \quad \omega = n\omega_0. \quad (7.11)$$

The coefficients in the Fourier series of the periodic signal from Figure 7.18 b have the form (7.10) which becomes

$$C_n = \frac{2}{T_r} \int_{-T/2}^{+T/2} x(t)\, e^{-i\omega t}\, dt \quad (7.12)$$

since within $[-T_r/2, +T_r/2]$ the signal exists only between $-T/2$ and $+T/2$.

Equations (7.11) and (7.12) yield

$$C_n = \frac{2}{T_r} |X(i\omega)|.$$

But in an actual measurement, r.m.s. values for the components of the pulse train are obtained (either on the analyzer display, or on the level recorder paper), so that the magnitude of a spectral line measured at the frequency $f = n\dfrac{\omega_0}{2\pi}$ is directly proportional to

$$C_{rms} = \frac{1}{\sqrt{2}} C_n = \frac{\sqrt{2}}{T_r} |X(i\omega)|.$$

The relationship between the theoretical value of the Fourier spectrum magnitude of a single pulse and the measured r.m.s. value, at $\omega = n\omega_0 = n\dfrac{2\pi}{T_r}$, is

$$|X(i\omega)| = \dfrac{T_r}{\sqrt{2}} C_{rms}. \qquad (7.13)$$

7.4.2.2 *Response of a very narrow bandpass filter.* The Fourier spectrum of a rectangular pulse of duration T is plotted with dashed line in Figure 7.17 b. If the analysis is done using an "ideal" very narrow band filter $\left(B \ll \dfrac{1}{T}\right)$, centred at the frequency f_c, and if it is considered that within the interval $\left[f_c - \dfrac{B}{2}, f_c + \dfrac{B}{2}\right]$ the pulse spectrum is constant, then it can be demonstrated [6] that the maximum value of the filter output signal is

$$y_{max}(t) = 2B |X(if_c)|$$

and the corresponding energy (the squared and integrated filter response) is

$$E_B = \int_0^T y^2(t) dt = 2B |X(if_c)|^2.$$

One can conclude that good approximations of the magnitude of the Fourier spectrum of an impulse, at the frequency f_c, can be obtained : a) by measuring the peak value of the filter output signal and dividing it by twice the filter bandwidth

$$|X(if_c)| = \dfrac{y_{max}(t)}{2B} ; \qquad (7.14)$$

b) measuring the squared and integrated filter response, dividing it by twice the filter bandwidth and square-rooting the result

$$|X(if_c)| = \sqrt{\dfrac{E_B}{2B}}. \qquad (7.15)$$

A block diagram of the instrumentation used to perform these operations is shown in Figure 7.19.

An exhaustive treatment of the techniques used in the frequency analysis of complex signals is given in reference [6].

7.5 *Vibration testing*

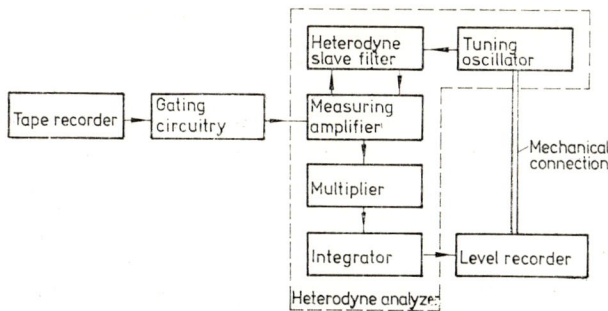

Figure 7.19

7.5 Vibration testing

The purpose of vibration testing is to determine the resistance of equipment to the deleterious effects of natural and induced environments, to find out weak components which could generate failures, malfunctions or improper operation of an equipment during service.

To quote from MIL-STD-810, Method 514 : "The vibration test is conducted to determine that... equipment is constructed to withstand expected dynamic stresses at pronounced vibration susceptible frequencies and that performance degradations or malfunctions will not be produced by the service vibration environment". According to the objectives pursued, one can distinguish : development tests, qualification (design acceptance) tests, quality control tests and trouble-shooting tests. Testing conditions (levels and durations) must be closely related to the anticipated service environmental conditions without duplicating them. The results of (damage caused by) vibration testing are important to be similar to the results of (damage caused by) actual service conditions.

In order to obtain reproducible test results and because the complex waveform of the dynamic loads acting in actual service is difficult to simulate, particular vibration regimes, e.g., sinusoidal, narrowband random, broadband random, impulse, are often used. In this respect, various test specifications exist and are used as : IEEE Standard 344—1975, U.S.NRC Regulatory Guide 1.100, Revision 1 (Aug. 1977), IEC 68—2—6 Test 2F specifications for electronic equipment, BS 2011 British Standard, NF C 20—523 to 529 and NF C 90—163 French Norms, DIN 50100 German Norm, JIS C 0911 and 0912 Japanese Standards, MIL-STD-8106 and MIL-STD-202E Standards (U.S.A.), etc.

7 Instrument set-ups and techniques

7.5.1 Sinusoidal tests

Specifications for vibration tests have been compiled from surveys made on a large number of measurements carried out on many types of vehicles and machines.

Let consider a component which has to be mounted in a vehicle.

First, vibration measurements are made on many similar vehicles, in numerous riding and loading conditions, at various speeds and engine r.p.m., and at different locations. From the analysis of time records or frequency spectra, maximum amplitudes and corresponding frequencies are obtained. The resulting data are plotted as in Figure 7.20 in terms of peak-to-peak displacement versus frequency. Finally, an envelope is drawn enclosing up to 95 % of the data points and which, for convenience in testing, consists of straight line segments. Most often, envelopes have constant acceleration lines at high frequencies and constant displacement (or constant velocity) lines at lower frequencies. These lines give the maximum vibration level which can be obtained during testing at each frequency.

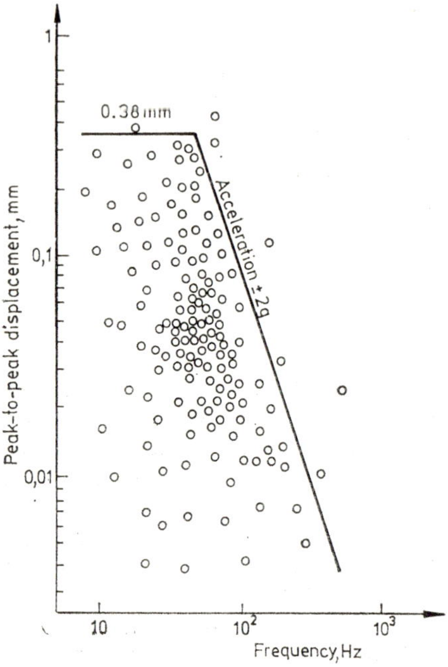

Figure 7.20

With the amplitude vs. frequency relationship once established, frequency must be varied with time, according to some program, based on an imposed value of the component service life, correlated to the number of stress cycles to be applied to it during the test and to the dynamic stress levels.

At the beginning, frequency was manually stepped through the range of interest, with dwells at structural resonances. Newer voltage-controlled oscillators provide either linear or logarithmic sweeping.

As for an actual multi-degree-of-freedom structure, the damping ratios evaluated at resonance do not significantly vary

7.5 Vibration testing

with frequency, it can be seen that the widths of resonance peaks are increasing with frequency. This calls for a higher sweep rate at high frequencies in order not to "overtest" the structure, which is obtained using logarithmic sweeping (Fig. 7.21 a). Linear

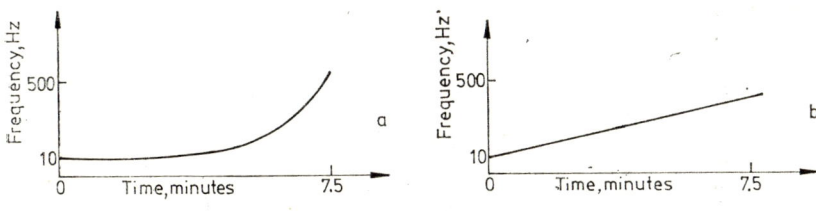

Figure 7.21

sweeping (Fig. 7.21 b) is preferred for slightly damped structures, where a better resolution is required at high frequencies (and at the analysis with tracking filters), and where sweep rates must be slower so that resonances have time to build up [7].

Figure 7.22 shows the essential components of a typical system used for swept-sine testing. The sinusoidal excitation signal is generated by an oscillator whose frequency can be manually controlled or varied by a sweep generator. In older equipment, the sweep generation is performed by an electrical motor which, by intermediary of a gear unit, rotates a variable resistor in the oscillator circuit.

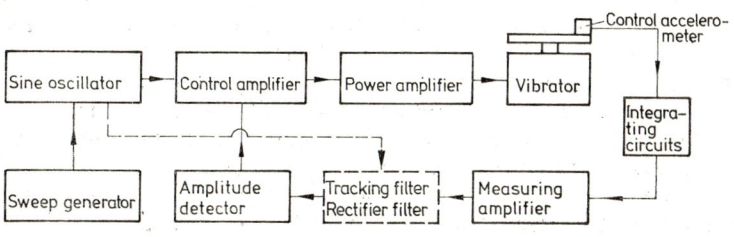

Figure 7.22

Reproducibility of vibration tests requires a stable oscillator, without moving parts and voltage-controlled, so that the sweep is generated by a voltage proportional to the desired frequency. Most sweep generators are capable of operating with either linear or logarithmic sweep, increasing, decreasing or cycling the frequency between given limits. In the simplest circuits, the oscillator output drives, via a power amplifier, the drive coil of an electrodynamic vibrator or the control valve of a hydraulic vibrator.

7 Instrument set-ups and techniques

As has been shown in chapter 6, even when the oscillator output has a constant voltage amplitude, the vibrator table motion has a frequency-dependent amplitude (see Figure 7.23 a — for a nonresonant load and Figure 7.23 b — for a structure with

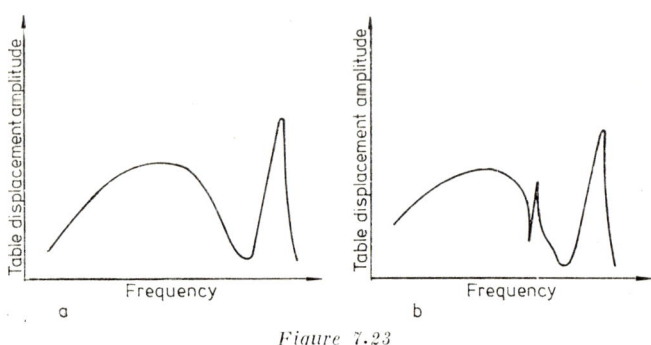

Figure 7.23

a single resonance). It follows that, in order to maintain the vibration level required by test specifications, the level of the oscillator output signal must be changed together with the frequency sweep.

In the case of high sweeping rates and multi-resonant structures, the manual control is too slow and a fast-acting electronic servo-loop is required. It is referred to as an "automatic gain control" but is also called "compression circuit"*.

Usually, the feedback signal is generated by a transducer mounted on the vibrator table (or at any suitable control point of the tested structure) and used to automatically change the gain of a control amplifier connected between oscillator and power amplifier. The compressor circuit also contains a measuring amplifier and an amplitude detector. Since the feedback signal is an a.c. voltage, i.e. a signal variable in time, and the control is performed maintaining constant an average parameter of the vibration, in the compressor circuit the signal is rectified and filtered (smoothed) to extract the average or r.m.s. value, resulting a d.c. control voltage for the compressor amplifier.

The gain control is performed such that when the feedback signal indicates an increasing vibration level on the vibrator table (or test article) the power of the input signal to the vibrator is automatically decreased until the prior level is reached. The power decrease is not instantaneous because averaging is time consuming.

---

* In laboratory language, because this circuit is called upon to "compress" level variations of the vibrator table amplitude of the order of 60 dB down to smaller changes, of the order of 0.5 dB.

7.5 Vibration testing

Selection of the regulation speed, referred to as the "compressor speed", raises some problems. It depends on the estimated values of system damping, on the sweeping rate and on the amount of distortion which can be tolerated. Generally, it is desired that the compressor speed be higher than the speed at which the resonance is built up. In order to have low distortion, low compressor and sweeping speeds have to be chosen. Because the distortion is determined by the ratio between frequency and the compressor speed, it follows that higher regulation speeds (dB/s) can be used at high frequencies.

Increasing the automatic gain control speed, that is lowering the time-constant of the feedback loop, has a destabilizing influence on the loop, which calls for some automatic change of the compressor speed during the sweep, especially when a slave filter is used [8].

The slave filter, whose centre frequency is controlled by the oscillator, is introduced when structural non-linearities determine the distortion of the feedback signal. Older methods used an average rectifier in the compressor circuit and an r.m.s. rectifier in the measuring circuit. If distortion changes during the sweep, large variations in the r.m.s. level appear, while the average level is kept constant. As the distortion increases, the level of the fundamental frequency decreases. This leads to "undertesting" at the fundamental frequency and can be eliminated using a slave filter in the control loop. However, it must be checked that use of the slave filter should not produce unstable operating regimes or over-testing at one or more of the harmonics.

An improvement of the circuit from figure 7.22 can be made by automatic change-over of control from 'constant displacement' to 'constant acceleration', according to the specification given in Figure 7.20. This is accomplished using two measuring amplifiers (or vibration meters). One amplifier, whose gain controls the displacement level, is connected to an accelerometer via a double integrator (or to a displacement pickup); the second amplifier, which controls the acceleration level, is connected directly to an accelerometer. Amplifiers are switched in the control loop by a *level selector*, when the sweeping frequency equals the "crossover frequency". The system can be extended to cover multi-segment control applications.

Vibration testing set-ups are provided with protection circuits which ensure the shut-down when the compressor loses control. This happens owing to either a missing feedback signal (broken cable) or to a too large signal (bad connection or unforeseen resonance approached with too high sweep speed).

7 Instrument set-ups and techniques

Modern set-ups have been developed by using numerically controlled oscillators and more control transducers located in various points of the tested structure. Digital signal processing techniques, implemented in minicomputer-based systems, are successfully applied to the control of vibration tests. In order to

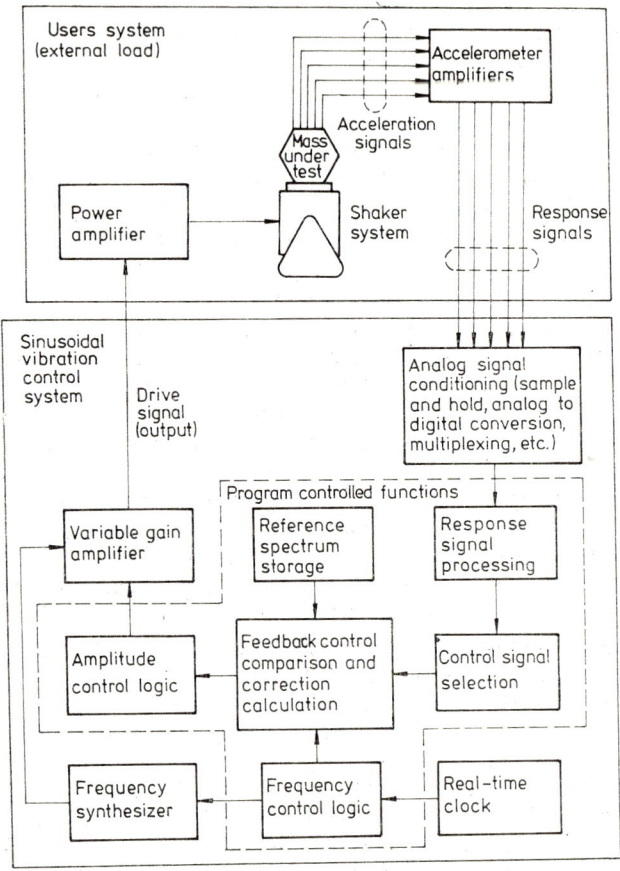

Figure 7.24

achieve high resolution over a wide frequency range, computer-controlled frequency synthesizers are used. Test control is ideally performed by computer, which determines the frequency response characteristics of the system under test and then generates sweeping rates appropriately. A functional block diagram of the fundamental control loop of a multi-channel multi-strategy sinusoidal vibration control system developed by Time/Data is shown in Figure 7.24 [9].

7.5 Vibration testing

7.5.2 Broadband random vibration tests

With the development of jet airplanes, where the energy radiated by jet engines and the high pressures produced in the turbulent boundary layer have a random variation in time, the problem of random testing received more attention [10].

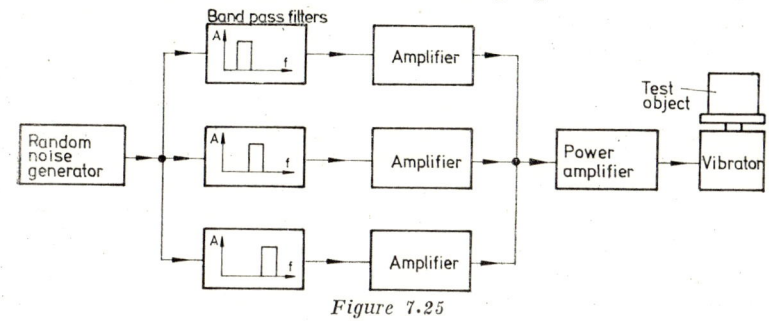

Figure 7.25

Random testing has a number of distinct advantages: a) simultaneous excitation of several resonances of the tested structure (which permits investigation of structures with time variable parameters and also the interaction between different resonances); b) considerable savings in test time with respect to the swept sine analysis (which becomes important for "on-line" analyses); and c) capability to detect structural non-linearities.

The essential parts of an analog random testing system are shown in the block diagram presented in Figure 7.25. The signal provided by a broadband random noise generator is passed through narrow band filters (centred on different frequencies) which divide the continuous spectrum of the signal in convenient bands. Resulting narrow band random signals are fed to amplifiers with different gain settings so that, by summation of their outputs, a broadband signal is synthesized, whose spectrum meets the test specifications. After passing through a power amplifier, the synthesized signal is fed to the drive coil of an electrodynamic vibrator, which excites the test object.

Filter sets and amplifiers are sometimes parts of one instrument called *spectrum equalizer*. Its operation can be easily explained based on Figure 7.26.

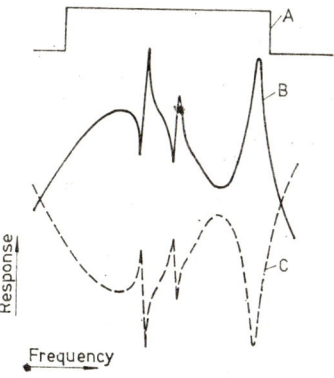

Figure 7.26

7 Instrument set-ups and techniques

If test specifications require generation of a band limited white noise-type random response, say, between 20 and 2000 Hz, the graph of the power spectral density versus frequency is illustrated by line A in Figure 7.26. If the signals from the random voltage source went directly to the power amplifier and vibrator, the power spectral density plot versus frequency would resemble curve B from Figure 7.26, owing to the frequency response characteristic of the vibrator and the resonances of the tested structure.

In order to obtain the response shown by line A, equalizers are used in the circuit to perform the correction shown by curve C from Figure 7.26. This prevents the vibrator motion from being influenced by the test object response, hence the vibrator table will have infinite mechanical impedance. Equalizers are adjusted either manually or automatically by a compressor circuit whose control signal is given by an accelerometer mounted on the vibrator table.

In the case of an instrument or device mounted on a vehicle, where load resonances affect the vehicle dynamic response, specifications of the type shown as line A in Figure 7.26 do not correspond, and a control of the vibrator force rather than vibrator motion is recommended [7].

Modern digital random control systems offer faster test setup, more resolution, wider dynamic range, more flexible frequency coverage, repeatable test excitation and easy conversion to other tasks. A functional block diagram of a typical digital random vibration testing system is presented in Figure 7.27 [11]. The basic system components are shown in Figure 5.72.

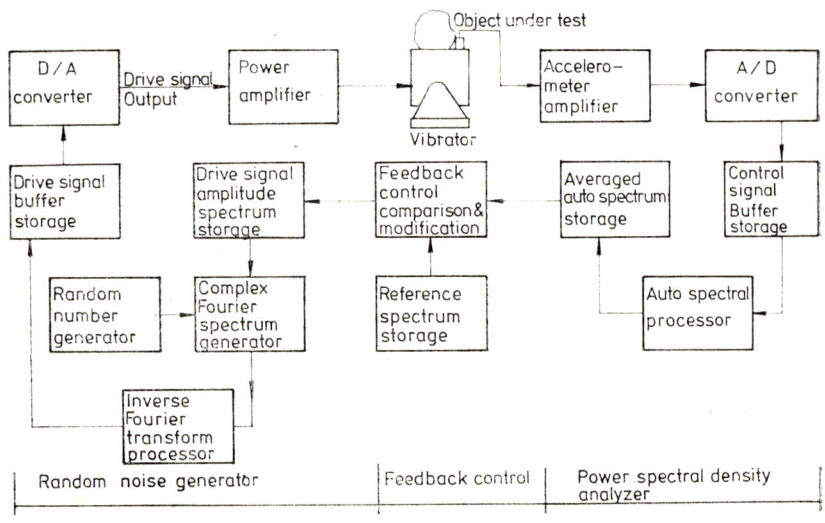

Figure 7.27

7.5 Vibration testing

The procedure is based on a given drive spectrum defined by a certain bandwidth and a number of spectral lines. A complex Fourier spectrum is obtained by adding a random phase angle to each spectral line. Taking the inverse Fourier transform, a frame of a time-domain signal is produced, having the required spectral characteristics. Repeating this operation continuously and connecting serially the resulting frames, a continuous random signal is produced, which is smoothed at the boundary between adjacent frames, to eliminate discontinuities.

Random testing has inherent disadvantages : a) requires more expensive equipment and larger vibrators (due to power limitations when heavy structures have to be excited over a wide frequency range); and b) produces poorer signal-to-noise ratios than swept sine analysis.

7.5.3 Sweep narrowband random vibration tests

The "sweep random" vibration testing, appeared [12] as an 'economic' substitute for the broadband random vibration testing, having the following advantages : a) the same test level can be obtained using smaller power amplifiers and vibrators than for the wideband random testing; b) the statistical character of the test signal is retained ; c) the control of the test level is easy to be done.

Drawbacks consisting of longer testing time and sequential resonance excitation can be partly overcome by using accelerated tests and multiple sweeping.

Generally, specifications intended for broadband random testing are used. In order to have the same number of stress reversals at a given resonance as would result from broadband random excitation, a logarithmic sweep rate is necessary. In order to have the same number of stress reversals in any increment of stress level, compared to the broadband random test, filters are used by means of which the r.m.s. value of the vibration level is increased with the square root of frequency (3 dB/octave). In order to have the same probability distribution of the peaks around the r.m.s. test level, in both cases, graphical methods are used [8].

In recent years, interest has been focused to establish specifications based on measurements performed in the same way as for the sweep sine testing. Various techniques for implementing digital swept sine on broadband random vibration control systems have been presented in reference [13].

7 Instrument set-ups and techniques

7.5.4 Shock tests

The first shock tests were carried out on special hammer or drop-table machines, where simple shock pulses can be obtained and controlled by changing the deformable material placed between the bodies in impact. The main concern was to obtain completely

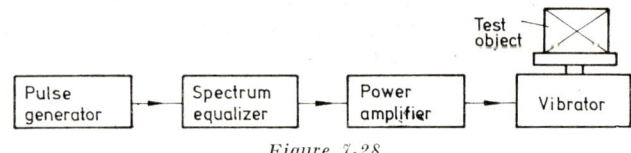

Figure 7.28

reproducible pulse shapes, and more important, reproducible effects.

Later on, shock test specifications were based on shock spectra rather than on test pulse shapes, so that more complex shock waveforms were used and the *shock spectrum synthesis method* was developed. The block diagram of the typical equipment is given in Figure 7.28. The pulse generator produces short duration single pulses fed simultaneously to a set of band filters and attenuators which shape the pulses so as to provide the desired shock spectra. The summed output from the filter bank is fed, via a power amplifier, to the drive coil of an electrodynamic vibrator. Oscillatory transients are recommended instead of single short pulses (of identical shock spectra) due to lower requirements concerning the "input" force magnitude and testing time savings.

7.6 Frequency response measurement

7.6.1 Frequency response functions

Frequency response is one of the most convenient means of describing the dynamic characteristics of a mechanical structure. This property may be simply explained as follows: If a linear time-invariant structure is acted upon by a harmonic force of frequency ω, then the structure will respond by vibrating harmonically at the same frequency, but in general, due to energy losses in the system, the motion will lag the force.

In order to define the relationship between the exciting force (the *input*) and the structure response (the *output*), two parameters are needed: a) the ratio of the response amplitude to that of the force, and b) the phase angle between these two harmonic

7.6 Frequency response measurement

quantities. Both these parameters vary with the frequency and together constitute the *frequency response* of the structure.

There are several frequency response functions, depending on whether the response is a displacement, velocity or an acceleration and whether the cause-effect ratio is taken under the form (output)/(input) or (input)/(output).

Many different symbols and inconsistent denotations can be encountered in the literature. The following definitions are now almost generally accepted and even standardized [14]:

Displacement/Force	= Receptance (Compliance, Admittance),
Velocity/Force	= Mobility,
Acceleration/Force	= Inertance,
Force/Displacement	= Apparent Stiffness,
Force/Velocity	= Mechanical Impedance,
Force/Acceleration	= Apparent Mass.

All these functions contain basically the same information about the structure, their selection being determined by the available equipment.

Usually, motion parameters in the forementioned ratios are measured in the direction and at the point of force application. Otherwise, in order to make distinction, the ratio y_1/F_1 is called a *direct receptance* (driving point receptance), the ratio y_2/F_1 is termed a *transfer receptance*, and x_1/F_1 and φ_1/F_1 are referred to as *cross receptances* (Fig. 7.29).

It is also convenient to express these ratios in complex notation. For example, if excitation $F(t) = F_0 e^{i\omega t}$ produces the response $y(t) = y_0 e^{i(\omega t + \varphi)}$, then the *complex receptance* can be expressed in terms of its vector components

$$\alpha = \frac{y}{F} = \frac{y_0}{F_0} e^{i\varphi} =$$

$$= \frac{y_0}{F_0} \cos\varphi + i \frac{y_0}{F_0} \sin\varphi =$$

$$= \alpha_R(\omega) + i\alpha_I(\omega). \quad (7.16)$$

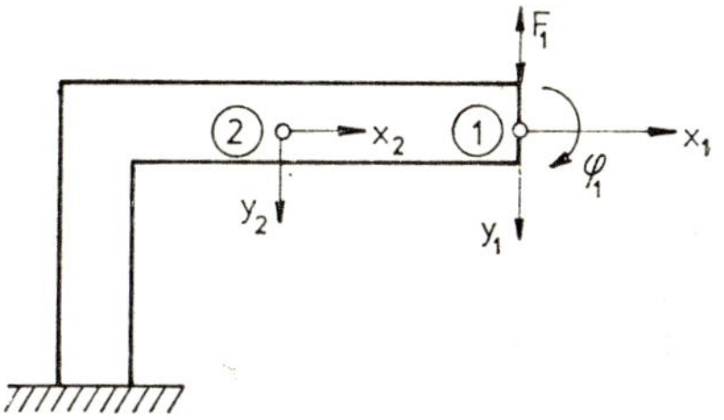

Figure 7.29

Use of similar frequency response functions, defined as ratios of a harmonic couple to the corresponding rotational response, or between forces and rotational response, couples and translational response, is the subject of more recent works, the specific equipment being less developed [15].

7 Instrument set-ups and techniques

7.6.2 Sinusoidal test techniques

7.6.2.1 Single-point excitation. The oldest and the most accurate method for determining the frequency response functions is the so-called "steady state" technique in which point by point measurements are made, at discrete frequencies, using harmonic excitation of constant amplitude.

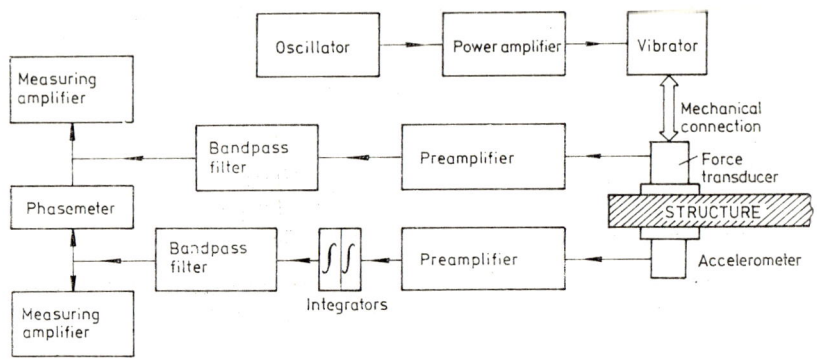

Figure 7.30

The basic components of an analog instrumentation system are shown in the block diagram presented in Figure 7.30. Force transducers and accelerometers are mounted on the vibrating structure, being preferred for their small size and lightweight which does not substantially modify the tested structure.

In some arrangements the force transducer is missing, in others it is embodied in an impedance head. Preamplifiers are used to lower the high output impedance of piezoelectric transducers, extending the linear operating range at lower frequencies. Integrators give signals proportional to the displacement or velocity at the measurement point. Band filters (continuously tunable) ensure the harmonic purity of signals, required for accurate phase measurement using the phasemeter. Measuring amplifiers are used for determining the level of excitation and response signals, optionally expressed as peak values or r.m.s. values. A spectrum analyzer can also be used after preamplifiers.

For each point of a frequency response curve, the following sequence of events is repeated : 1 — selection of the oscillator frequency and corresponding filter tuning ; 2 — adjustment of signals to the desired level (for example, at measurements with harmonic forces of constant amplitude, the output voltage of the power amplifier is adjusted until the required force level is read on the voltmeter of the corresponding measuring amplifier) ;

7.6 Frequency response measurement

3 — measurement of amplitude and phase shift of force and response signals; 4 — computation of the required ratio of the two quantities, taking into account the prior calibration (see Chapter 8); 5 — data presentation.

Measurement of the frequency response functions by discrete values, with stopping at each frequency until steady state is reached, and plotting the corresponding graphs, point by point, is a time consuming procedure so that by the year 1965 equipment was developed for automatically plotting these curves using swept sinusoidal excitation.

Frequency sweep used in conjunction with the wattmeter method (see Section 5.4.7) enables continuous frequency response curves to be plotted while the test is in progress. This reduces the test time and serves to point out significant details of the response diagram which would be overlooked if data points were plotted at discrete frequencies.

A detailed description of this equipment is beyond the purpose of this book so that only some basic features will be presented, illustrated by the block diagram shown in Figure 7.31.

The structure is subjected to sinusoidal excitation by the usual instrumentation chain, consisting of the sine wave oscillator, the power amplifier and the electrodynamic vibrator. The frequency is automatically swept, usually through a range covering up to three decades (1 — 1000 Hz), by a sweep generator. If the sweep rate is sufficiently slow (as determined by structural damping and the time constants of electronic circuitry), the frequency response functions obtained by this technique are identical to those measured by "dwelling" at discrete frequencies.

The instrumentation system from Figure 7.31 permits direct plotting of the modulus and phase of the frequency response functions on an analog $X-Y$ recorder. The recorder's X axis is driven by a d.c. voltage proportional to the excitation frequency. In order to plot the amplitude-frequency curve, the Y axis of the recorder is driven by the detected output of a tracking filter, tuned to the excitation frequency, which filters the noise polluted signal received from the impedance head via the preamplifier (usually a charge amplifier).

The impedance head is attached to the tested structure and to the vibrator drive coil by a threaded stud or a longer rod threaded at both ends. A laterally flexible link is generally recommended to protect the drive coil suspension from side loads. The vibrator is suspended by very soft springs or rubber strings.

In order to trace the phase-frequency curve, the Y axis of the recorder is driven by a d.c. voltage proportional to the phase angle measured with the phasemeter.

7 Instrument set-ups and techniques

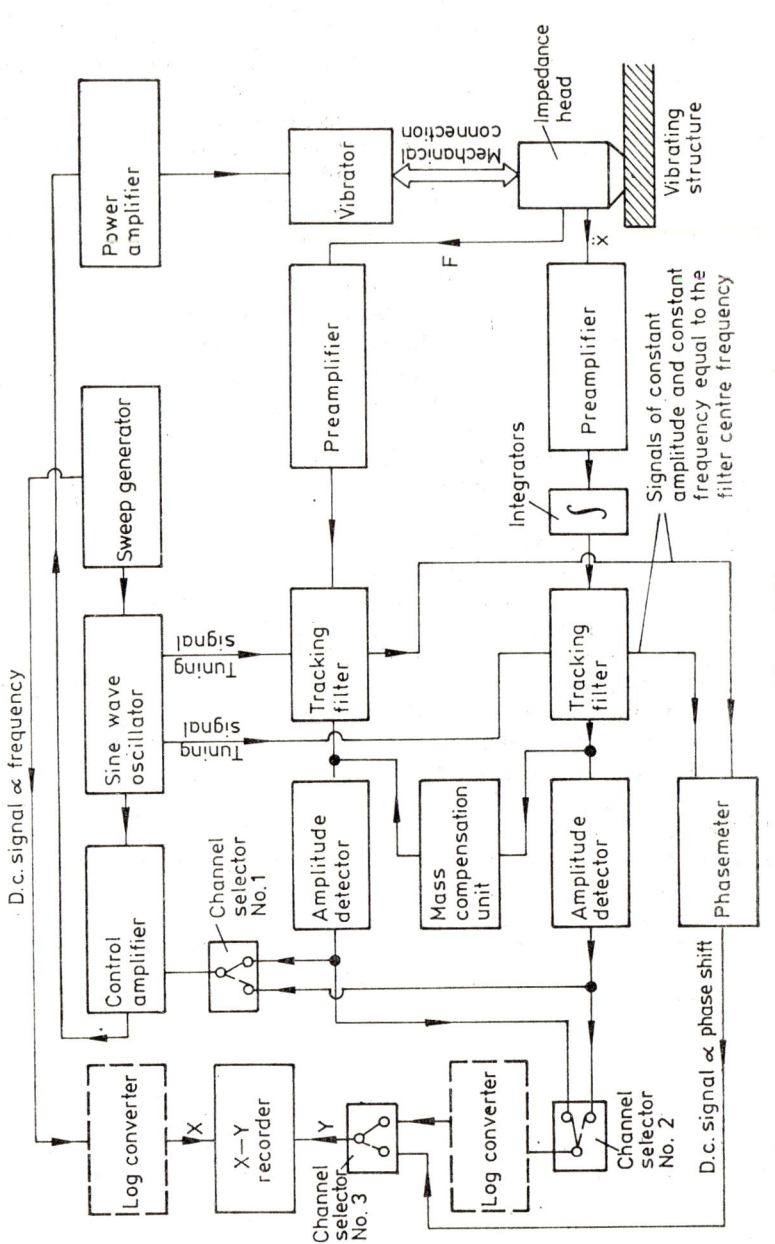

Figure 7.31

244

7.6 Frequency response measurement

The channel selector No. 1 connects the gain control amplifier into the compressor circuit of the excitation, while that of the response compensates the excitation system for changes in impedance caused by the dynamic response of the test structure.

At the measurement of receptance, mobility or inertance, force feedback is employed and the channel selector No. 2 drives the Y axis of the recorder with a signal proportional to the displacement, velocity or acceleration of the vibration. When measuring apparent stiffness, displacement feedback is used. Velocity feedback is used at the measurement of mechanical impedance, and acceleration feedback — when the apparent mass is measured. In the three last cases, channel selector No. 2 commutes the force signal to the recorder Y axis.

Typical plots for the modulus-versus-frequency and phase-versus-frequency characteristics of the driving point mechanical impedance of an elastic structure are shown in Figure 7.32. A ratio of 10,000 :1 exists between values measured at resonance and at antiresonance, which corresponds to a continuous dynamic range of 80 dB that has to be covered by the measuring equipment. These large variations of the frequency response functions (especially for lightly damped structures) have called for plotting the diagrams in log/log format and use of logarithmic converters in the instrumentation system (Fig. 7.31).

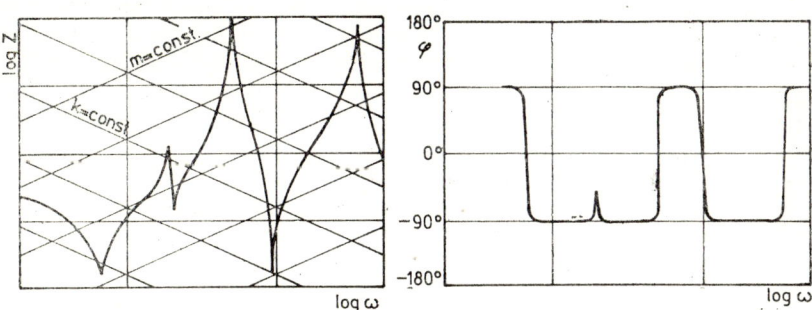

Figure 7.32

An integrator has been (symbolically) introduced in the accelerometer signal path (Fig. 7.31) to indicate the simple or double integration performed when measurement of displacement or velocity is required. Common electronic integrators cannot perform these operations and "synthetic integration" is employed [16].

By far the most difficult problem encountered at the determination of the frequency response is the accurate phase measurement.

At Spectral Dynamics SD 1002 transfer function analyzer, the phase lag is measured between two pure sinusoids of constant amplitude, extracted from the high frequency stage of the heterodyne analyzers, and of fixed frequency, independent of that of the analysed signals (equal to the centre frequency of the narrow band filter) but having the same phase shift.

Another problem is the so-called "mass compensation". The impedance head is attached to the structure by means of a threaded stud or an adapter. The mounting elements together with the "mass below the force transducer" (the mass of the driving platform of the impedance head, located between the sensing element of the force transducer and the tested structure) behave — over the usual operating frequency range — as an additional mass attached to the drive-point. It follows that the force applied to the structure is not that sensed by the force transducer, but the vector difference between this and the force required to drive the interposed mass.

The inertia force corresponding to the additional mass is usually very small and introduces a negligible error in the measured values of the excitation force. It is appreciated that if the apparent mass of the tested structure is 10 times larger than the additional mass of an impedance head, then the error produced in the value of mechanical impedance will be less than 1 dB if mass compensation is not performed [18]. However, in the neighbourhood of resonances, this force can have comparable magnitude with the applied force, especially for lightweight slightly damped structures. Without mass cancelation, there is a shift of resonance frequencies to lower values and a change of the value of damping ratios [19].

Therefore an electrical compensation is recommended. After calibration with a known mass, the mass compensation unit continuously subtracts from the force transducer signal a quantity which is proportional to the drive point acceleration (hence to the inertia force of the additional mass).

The mass compensation circuit is calibrated prior to the vibrator attachment to the test structure. A known mass is mounted to the impedance head driving platform. The gain of the compensation network is adjusted until the pen of the $X-Y$ recorder indicates, on a special graph paper, an apparent mass equal to the actual calibration mass attached to the impedance head. A less accurate method is to vibrate the impedance head without any calibration mass and to adjust the gain until the total indicated force is zero.

The system from Figure 7.31 can be supplemented with a circuit like that shown in Figure 5.63 to get the vector components

7.6 Frequency response measurement

of the frequency response functions. By feeding these signals to the inputs of an $X-Y$ recorder, it is possible to plot polar diagrams (Nyquist plots) of the frequency response which contain information about both the magnitude and phase of the system dynamic response.

Steady-state and quasi-steady-state (frequency sweep) techniques for measuring the frequency response of elastic structures have the same advantages on other techniques as have those presented for similar vibration testing methods: a) high accuracy; b) large power input into the structure under test, which is important at the analysis of large and highly damped structures; c) use of "classical" analog equipment. Moreover, by repeating experiments at different displacement amplitude levels, structural non-linearities can be analysed.

The main disadvantage is the long test time, required for setting up the vibrator (or to move it around at different points), for calibration and check-ups during testing, as well as for maintaining slow sweep rates dictated at low frequencies by lightly damped structures. In the neighbourhood of each resonance, the maximum response measured with frequency sweep is less than the corresponding steady-state maximum, and the resonance frequency is displaced in the direction in which the frequency is changing. Additional errors are due to the averaging time of the wattmeter and to the limited dynamic range of analog instrumentation.

Although sine testing was necessitated by analog instrumentation, it is not limited to the analog domain. Sinusoidally measured frequency response functions can be digitized and processed with an FFT analyzer or can be measured directly. The fastest and the most used method utilizes a type of signal called a "chirp". The chirp is a logarithmically swept sine wave that is periodic in the analyzer measurement window T. The swept sine is generated in the computer and output through the digital-to-analog converter every T seconds. Since the signal is periodic, the leakage does not occur. Another technique using a single swept sine signal is presented in Section 7.6.3.3.

7.6.2.2 *Multi-point excitation.* Multiple-vibrator testing is used for experimentally determining the dynamic characteristics of aerospace structures during the so-called "resonance testing". Though complex and costly, the procedure offers a means of excitation of pure undamped natural modes. There is demonstrated [20] that, in the case of damped systems, a particular sinusoidal excitation can be determined, using synchronous and coherently phased forces, applied using several vibrators, that will produce

7 Instrument set-ups and techniques

a forced mode (mode of distortion) in which the displacements of the drive points have the same phase angle with respect to the input forces. The characteristic phase lag is 90° when the excitation frequency equals an undamped natural frequency of the system. In this case, the forced mode takes the shape of a pure undamped natural mode.

These observations form the basis of the multi-point excitation techniques [21], [22]. From a practical point of view there is no possibility to achieve perfect phase resonance in real structures because it is impossible to get a strict equilibrium between applied and damping forces at each point throughout the structure.

Figure 7.33 shows the block diagram of the basic part of a system capable of exciting the structure at an undamped natural frequency so that the displacements of the excitation points are all in phase with one another and 90° out-of-phase with the forces.

Each vibrator operates within a channel of the system, containing a servo loop. Channel No. 1 controls the frequency of all channels, the other channels control the force amplitude only.

The phasemeter generates a d.c. voltage proportional to the phase angle between the force signal and the acceleration signal. This voltage together with a compensating voltage of equal magnitude but opposite polarity are applied to a summation device, resulting a null output voltage which stops the sweep generator at the frequency corresponding to the phase resonance. Any change of the structure dynamic response due to the application of other forces, which changes the phasing between force and acceleration, generates an error signal which is fed to the sweep generator. The oscillator frequency varies until a phase lag of 90° is reestablished, when the sweep is stopped. Usually, the frequency range is searched by manual sweep until a resonance frequency is found (using the 90° phase lag as resonance criterion) and then the automatic control is switched on, to maintain the phase resonance condition.

Then the force level at vibrator No. 2 is manually adjusted until a local 90° phase lag between force and acceleration is obtained, after which the servo control loop is switched on to control the output of the vibrator, in order to keep constant the 90° phase angle. There is a continuous interaction between the first channel, which controls the excitation frequency, and the second channel, which controls the force level, to maintain the local phase resonance.

The same procedure is repeated on the other channels until the phase resonance condition is achieved at all excitation points thereafter the mode shape is recorded (with a rowing transducer moved at different points of interest of the structure).

7.6 Frequency response measurement

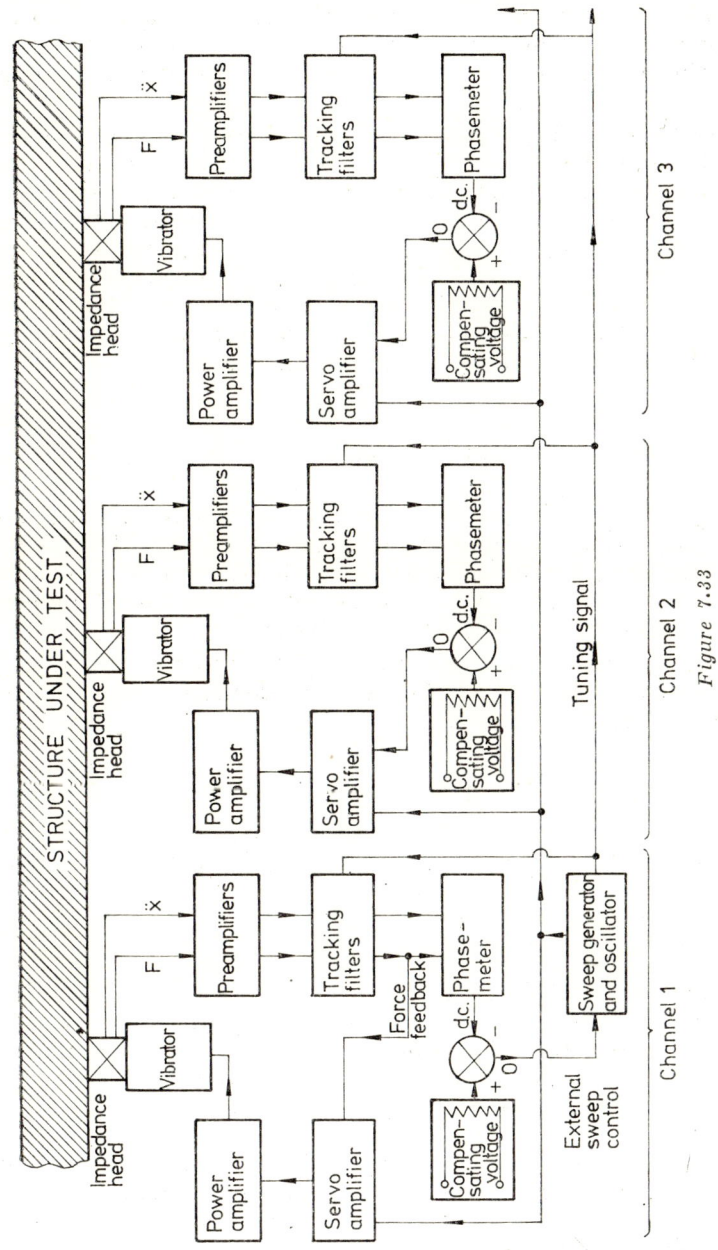

Figure 7.33

7 Instrument set-ups and techniques

The disadvantages of the above tuned-dwell method are the complicated instrumentation required, the inadequate strategy of adjusting excitation forces, the possibility of overlooking modes and the limitation, in some cases, to proportionally damped structures.

GRAMPA [23], MAMA [24], MODAPS [25] and MODALAB [26] are automated test systems developed to improve the force appropriation procedure by using analytically aided or imperfect tuning. Even using these sophisticated systems, difficulties are encountered in maintaining fixed amplitudes and phases of applied forces. These are due to the interaction between vibrators and the tested structure. Limits imposed on the excitation level and inaccessibility of key points produce untunable modes which affect the accuracy of any multi-vibrator method.

7.6.3 Transient test techniques

Owing to the long test time, there is not always possible to use harmonic excitation in frequency response tests. Transient test procedures have been developed for the flight testing of aircraft and missile systems, which were later successfully applied in the automotive and machine tool industries.

In this case, the frequency response function (2.39) may be derived by division of the Fourier transforms of the response and excitation time histories:

$$H(i\omega) = \frac{Y(i\omega)}{X(i\omega)} = \frac{\int_0^\infty y(t) e^{-i\omega t} dt}{\int_0^\infty x(t) e^{-i\omega t} dt} = \frac{\mathscr{F}[y(t)]}{\mathscr{F}[x(t)]}. \qquad (7.17)$$

This can be determined with an FFT analyzer (see Section 5.4.4.4) or using a digital computer after passing the two signals into an analog-to-digital converter. Most recent techniques calculate the function $H(i\omega)$ from the averaged auto- and cross spectra of the digitized signals (see eq. 2.73).

7.6.3.1 *Impact test technique.* The transient test technique based on impulsive force input regained popularity with the development of piezoelectric force transducers and real-time FFT analyzers.

7.6 Frequency response measurement

The structure is impulsed with a hammer. A force transducer is mounted either on the hammer head (Fig. 7.34) or on the impacted portion of the structure. A motion transducer (an accelerometer) is attached to the point where the response is measured.

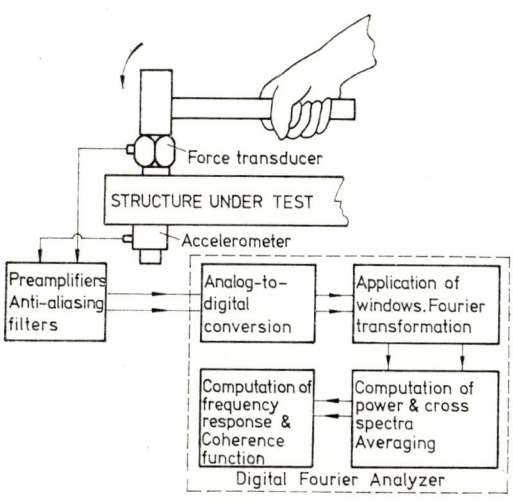

Figure 7.34

A complete set of frequency response functions can be measured using several force input points, acted in turn, and a single response point. After passing through signal conditioning equipment, including appropriate amplifiers and anti-aliasing filters, the force and motion signals are analysed using a dual channel digital processor, preferably an FFT analyzer.

The major drawback of this technique is the uncontrolled shape and magnitude (hence the energy spectrum) of the force pulse which may vary from test to test. As a result, structural non-linearities give various responses even when the same pair of input-output points is used. The measured frequency response functions can be improved by preloading the structure, to take up clearances between parts.

If the pulse has a short duration, the energy spectrum can span a wide frequency range (Fig. 7.35). However, the energy density is not sufficient to excite the whole structure. The only solution to produce more energy is to strike stronger. This may either

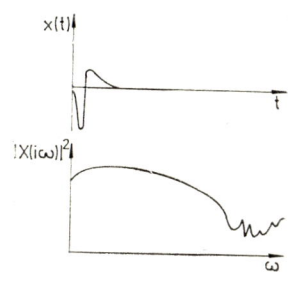

Figure 7.35

7 Instrument set-ups and techniques

damage the structure or give a highly non-linear response. If a resilient material is introduced between hammer and structure, in order to reduce the impact, the pulse duration increases and the frequency range covered by the energy spectrum diminishes (Fig. 7.36).

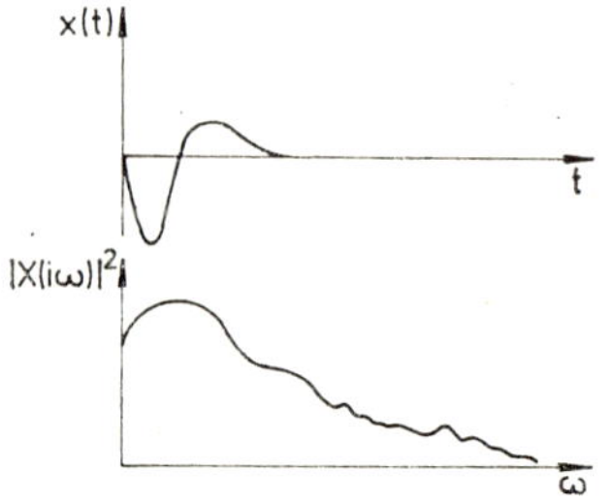

The low amplitude portions of the frequency response function $H(i\omega)$ can be determined with large errors because of the poor signal-to-noise ratio due to the insufficient energy in certain frequency bands. The representation can be improved through the use of ensemble averaging, i.e. recording several functions, summing them and calculating a mean frequency response function.

Figure 7.36

The impact force can be modified by using a softer or a harder hammer head. A hammer with a soft head can be used to concentrate more energy at lower frequencies, whereas a harder head can be used to excite higher frequency modes (with reduced energy density). The method cannot be used to test large, heavily damped structures due to insufficient energy levels obtainable.

On the processing side, the method requires large Fourier transform sizes, because a large number of digital data points are required to represent the signal when adequate frequency resolution has to be obtained for quantifying very narrow band resonance peaks. The problem is alleviated through the use of the "zoom" transform. In this case the Fourier transform is performed over a frequency band with independently selectable lower and upper limits.

Additional problems are related to the use of time windows, especially at lightly damped structures. These tend to vibrate for a long time, so that the transient response may not decay to zero within the sample window. On the other hand, use of a Hanning window could attenuate the first part of the transient, which is the most important. An exponential window has been developed to reduce the truncation errors and the noise effects. However this window increases the apparent damping at resonance, which can be detrimental for structures with closely spaced resonant modes.

The testing program must include preliminary measurements at a number of locations on the structure. Their objectives are the identification of important resonances, the estimation of modal damping and the location of the stationary transducer. Monitoring the force signal helps in rejecting poor measurements

7.6 Frequency response measurement

like overloading or multiple impacts. Monitoring the coherence function is also helpful in observing the quality of the frequency response functions. Curve fitting algorithms may be used at the analysis of mode shapes and to extract the modal coefficients. Use of the peak quadrature component of the frequency response is sufficient in most cases to compute the modal shapes.

The possibility of using short duration impulse excitation has been studied. Single pulses of simple geometric shape (rectangular, symmetrical trapezoidal and triangular, half sine) have been produced by a signal generator and applied to the structure under test by means of an electrodynamic vibrator. The modulus spectra of simple pulses are presented in Table 2.4. A signal having such a Fourier spectrum is inadequate for the excitation of multi--degree-of-freedom systems. Some resonances may not be excited because of the zeroes in the modulus spectrum while resonances outside the range of interest may be excited because the spectrum is (theoretically) infinite in extent.

7.6.3.2 Step relaxation technique.

Another "traditional" form of transient testing consists in the application of a large initial deflection to the structure, followed by quick release and analysis of the resulting free (decay of) vibration. The excitation by release applies especially to constrained (built-in or grounded) structures. Winches, cranes or hydraulic rams can be used to apply the force.

In the modern variant of the method, the force transducer is first mounted to the structure. An inextensible cable is attached to the force transducer and used to preload the structure to a suitable force level. When the cable is cut, the structure "relaxes". A negative force step occurs and the transient response of the structure as well as the transient force input are recorded. The function $H(i\omega)$ is then calculated as the ratio of the Fourier transforms of the two signals using a digital Fourier analyzer.

The force signal has generally higher energy input at low frequencies (below 2 Hz). It is adequate for the excitation of large buildings, towers, columns, bridges and other structures with very low fundamental frequency. When large initial tensioning can be achieved, more energy can be put into the structure than with the previous technique, but the test duration is longer. Repeated tests are recommended, with the excitation applied at various locations, in order not to excite only certain modes. Often more than one mode is excited.

When a multiple-vibrator harmonic excitation is available, it is possible to obtain a transient response in essentially one mode of vibration. This is done by switching off the excitation after a resonance condition has been reached. Interruption of the harmonic excitation can be performed at any convenient level of

7 Instrument set-ups and techniques

the displacement amplitude. In this way, sufficient energy can be put into the structure under test without striking it, which could be beneficial in the case of brittle materials.

7.6.3.3 Rapid frequency sweep excitation technique.

In order to eliminate some of the drawbacks of the previously presented methods, a special type of forcing function was sought, having an essentially uniform frequency spectrum over a limited band of frequencies. Best results have been obtained using swept sine waves of constant amplitude and frequency varying linearly with time (Fig. 7.37), i.e. using a frequency modulated sine wave.

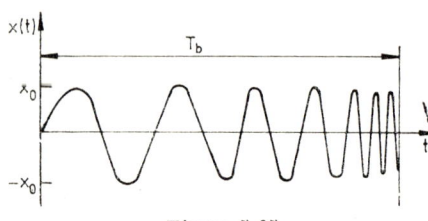

Figure 7.37

It has been shown [27] that a signal of equation

$$x(t) = x_0 \sin \int_0^t \omega(t)\, dt \qquad 0 < t < T_b \qquad (7.18)$$

has a uniform energy spectrum whose level is directly proportional to the sweep duration in the case of a linear variation of frequency with time.

Systematic research work has been carried out at I.S.V.R. Southampton [28] and a special device has been built for generating rapid frequency sweeps of the form

$$x(t) = x_0 \sin(at^2 + bt) \qquad (7.19)$$

where

$$a = \frac{\pi(f_i - f_f)}{T_b},$$
$$b = 2\pi f_i, \qquad (7.20)$$

and T_b is the time duration over which the signal frequency varies continuously from an initial value f_i to a final value f_f.

The power spectrum of a swept sine wave (Fig. 7.38) exhibits the following properties: a) the mean value of the spectrum

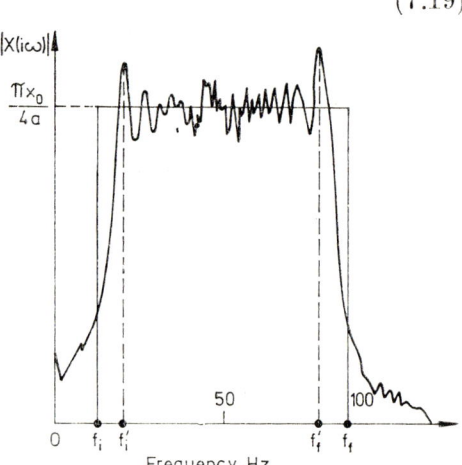

Figure 7.38

is $\frac{\pi x_0}{4a}$; b) two peaks occur at frequencies $f'_i = f_i + 1.2\sqrt{\frac{a}{2\pi}}$ and $f'_f = f_f - 1.2\sqrt{\frac{a}{2\pi}}$, with heights approximately 1.4 times the mean spectrum level; c) the amplitude of the ripple superimposed on the mean spectrum level is proportional to $\frac{1}{\sqrt{T_b}}$; d) the cutoff rate of the spectrum at f_i and f_f is high. The modulus spectrum from Figure 7.38 has been obtained from a sweep between 20 and 100 Hz in 1 second.

The possibility of controlling the highest frequency present in the response ensures the avoidance of aliasing. Good results can be obtained using a manually tuned sine oscillator. The use of a hand-held pressure contact exciter in contrast to the conventional coupled excitation system has been investigated and promising results have been obtained with a standard vibrator. However, results from tests on a very light structure were poor.

In the past, a similar method was used to determine the natural frequencies of large structures. A rotating-mass mechanical vibrator was used. It had been driven up to the maximum speed, then the electric motor was switched off and the structural vibrations were measured during the vibrator rundown. A decreasing frequency sweep can be thus obtained, sufficiently slow as to excite all resonances. Only the response is measured. The input force is calculated approximately based on the readings of a tachometer and the value of the unbalance (eccentric mass times the radius).

7.6.4 Random excitation techniques

A schematic block diagram of the instrumentation used for measuring the structural response to random excitation and the order for computing desired sample functions are shown in Figure 7.39. Using a signal generator, only *pure random* signals can be generated. Other types of broadband random excitation are also used. These are the *pseudo random* and the *periodic random* excitations. They are generated by the processor of a digital spectrum analyzer and output to the structure via a digital-to-analog converter.

The main problem of pure random excitation is the non-periodicity of the measured input and output signals in the time window of the Fourier analyzer. In order to reduce "leakage", the signal is multiplied by a weighting function which still distorts the spectrum of the original signal. But with pure random signals

7 Instrument set-ups and techniques

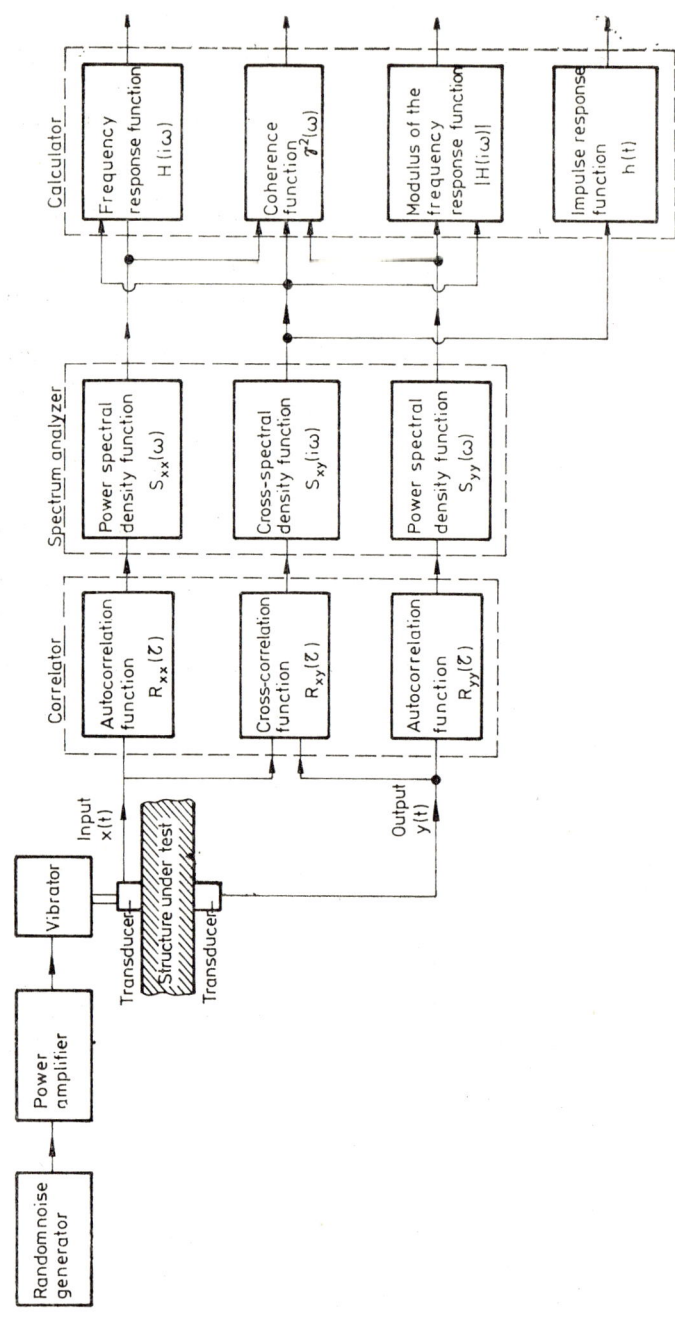

Figure 7.39

7.6 Frequency response measurement

each sampled record of duration T is different from every other sample. Ensemble averaging successive records of frequency domain data removes nonlinear effects, noise and distortion.

Pure random excitation is relatively fast, the force levels being easily and accurately controlled. It has good peak to r.m.s. values and gives the best linear approximation of a non-linear system. A "zoom" transform must be used to improve the frequency resolution when testing lightly damped systems.

Pseudo-random excitation is commonly used to avoid the leakage effects of a non-periodic signal. The signal is created in the frequency domain, having uniform amplitude and random phase angle (zero-variance random noise). It is then Fourier transformed into the time domain and repeatedly output to the vibrator through the digital-to-analog converter every T seconds (T is the analyzer measurement record length), being thus periodic in the measurement window. It is relatively easy to modify the stimulus spectrum to compensate for the vibrator characteristics.

Pseudo-random excitation gives good results for linear systemr in low noise environments. It does not work well in the presence of non-linearities, distortion and periodicities due to rattling of loose parts, which cannot be removed by ensemble averaging.

Periodic random excitation combines the best features os pure random and pseudo-random, but without their disadvantages. First, a pseudo-random signal is output to the vibrator. After a steady-state condition is reached, the first measurement is taken, calculating power and cross-spectra. Then, a different uncorrelated pseudo-random signal is synthesized and output, exciting the structure in a new steady-state condition and a second measurement is made. The procedure is continued, averaging together the power spectra of many different records, so that non-linearities and distortion components are removed from the transfer function estimate. Thus, leakage is eliminated by periodicity and distortions are removed by ensemble averaging [29].

7.6.4.1 *Measurement of the frequency response function.* With a linear system (Fig. 7.40 a), for which the time domain input $x(t)$ and output $y(t)$ are known, one can calculate:

a) the modulus of the frequency response function

$$|H(i\omega)| = \sqrt{\frac{S_{yy}(\omega)}{S_{xx}(\omega)}} \qquad (7.21)$$

where $S_{xx}(\omega)$ and $S_{yy}(\omega)$ are the power spectral density functions of the input and output, respectively;

b) the complex valued frequency response function

$$H(i\omega) = \frac{S_{yx}(i\omega)}{S_{xx}(\omega)} \quad \text{or} \quad H(i\omega) = \frac{S_{yy}(\omega)}{S_{xy}(i\omega)} \tag{7.22}$$

where $S_{yx}(i\omega)$ is the cross spectral density function between the input and the output.

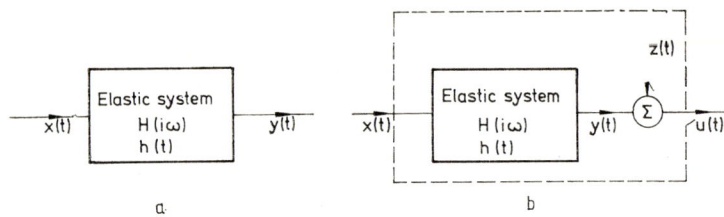

Figures 7.40

The latter form is most commonly used in practice. If averaging is applied to measured data, the first equation (7.22) becomes

$$H(i\omega) = \frac{\overline{S_{yx}(i\omega)}}{\overline{S_{xx}(\omega)}} \tag{7.22 a}$$

where $\overline{S_{yx}(i\omega)}$ denotes the ensemble average of the cross spectra and $\overline{S_{xx}(\omega)}$ represents the ensemble average of the input auto spectra.

Equation (7.22 a) can also be applied to the measurement of the frequency response in the presence of noise [30]. This is advantageous at system identification during normal operation. In this case, the output $y(t)$ due to the excitation $x(t)$ and the noise $z(t)$ due to normal operation are summed together (Fig. 7.40 b).

Let $u(t)$ be the output accessible to measurement. If $S_{yy}(\omega)$ and $S_{zz}(\omega)$ are the auto spectra of the signals $y(t)$ and $z(t)$ respectively, the auto spectrum of the measured output $S_{uu}(\omega)$ is given by the sum

$$S_{uu}(\omega) = S_{yy}(\omega) + S_{zz}(\omega).$$

The cross spectrum between input and output is

$$S_{ux}(i\omega) = S_{yx}(i\omega) + S_{zx}(i\omega)$$

where $S_{zx}(i\omega)$ is the cross spectrum between the noise and the input, and $S_{yx}(i\omega)$ is the cross spectrum between the input and the desired response.

The noise $z(t)$ is generally incoherent with the measured input signal $x(t)$ and has a zero mean value. As the number of ensemble averages becomes larger, the noise term $\overline{S_{zx}(i\omega)}$ becomes smaller, and $\overline{S_{ux}(i\omega)}$ tends to $\overline{S_{yx}(i\omega)}$.

7.6 Frequency response measurement

For the system from Figure 7.40 b, the frequency response function becomes

$$H(i\omega) = \frac{\overline{S_{yx}(i\omega)}}{\overline{S_{xx}(\omega)}} = \frac{\overline{S_{ux}(i\omega)}}{\overline{S_{xx}(\omega)}} - \frac{\overline{S_{zx}(i\omega)}}{\overline{S_{xx}(\omega)}} \cong \frac{\overline{S_{ux}(i\omega)}}{\overline{S_{xx}(\omega)}}. \quad (7.23)$$

The rate at which $\overline{S_{ux}(i\omega)}$ tends to $\overline{S_{yx}(i\omega)}$ depends on the degree of causality between $u(t)$ and $x(t)$.

To determine quantitatively the noise influence on the measured response $u(t)$, the ensemble average coherence function is calculated as

$$\overline{\gamma^2(\omega)} = \frac{|\overline{S_{ux}(i\omega)}|^2}{\overline{S_{uu}(\omega)} \cdot \overline{S_{xx}(\omega)}}. \quad (7.24)$$

The average coherence function takes values between 0 and 1. It is equal to 1 at the frequencies where the measured response power is totally caused by the measured input power. A coherence value less than 1 indicates that the measured response power is greater than that due to the measured input because of the extraneous noise contribution. When the coherence is zero, the output is totally caused by sources other than the measured input. The larger is the number of ensemble averages, the more accurately is estimated the frequency response function and a better estimate of the noise energy in a measured signal is given by the coherence function.

7.6.4.2 *Measurement of the impulse response function.* The time domain description of the dynamic characteristics of a linear system is given by the impulse response (weighting) function $h(\tau)$. This can be determined using random excitation [31]. It is desirable that the input signal $x(t)$ has an impulse-type autocorrelation function of zero mean and approaching that of "white noise". This permits determination of the system impulse response even in the presence of extraneous noise signals. Pseudo-random binary coded test signals (Fig. 7.41) are well suited for practical applications.

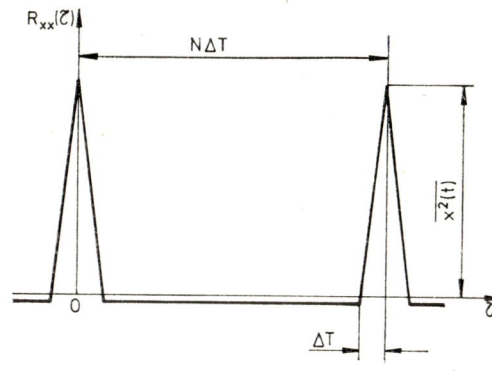

Figure 7.41

7 Instrument set-ups and techniques

In this case, the crosscorrelation function $R_{xy}(\tau)$ of the input $x(t)$ and output $y(t)$ is given by

$$R_{xy}(\tau) = \overline{x^2(t)} \cdot h(\tau),$$

where $\overline{x^2(t)}$ is the mean square of the input pseudo-random signal and $h(\tau)$ is the impulse response function of the system. It follows that the function $h(\tau)$ can be determined based on measured values of $R_{xy}(\tau)$ and given $\overline{x^2(t)}$.

The averaging time T, for the calculation of correlation functions, must equal the period $N \cdot \Delta T$ of the pseudo-random signal; in this case, estimation errors do not occur.

This experimental technique has the following advantages: a) testing can be made during system normal operation, the test signal amplitude being sufficiently small as not to disturb the technological process; b) test results are not contaminated by the noise from the system, which is incoherent with the test signal.

Advantages and disadvantages of some structural frequency response measurement techniques are presented in reference [32]

7.6.5 Experimental modal analysis

Up to now it was shown how a structure can be excited, how the displacements of different points are recorded and how the frequency response functions are calculated and displayed. The problem that arises is to analyze these functions, determining the natural frequencies, the corresponding deformation mode shapes and the characteristic parameters: equivalent masses, stiffnesses and damping factors.

The process of identifying from measured response data the natural frequency, modal mass, damping factor and mode shape of a finite number of predominant modes in an elastic structure is referred to as *Experimental Modal Analysis*. Modal data can be used to completely specify the dynamic behaviour of the structure to arbitrary inputs, by determining the parameters of only a few modes. Modal analysis reduces the number of parameters necessary to predict the vibratory response of an elastic structure, being useful in system identification problems, i.e. in the formulation of a mathematical model of the test structure for the purpose of further analysis.

7.6.5.1 *Single-degree-of-freedom techniques.* For harmonic force excitation, the equation of motion of the system shown in Figure 7.42 is

$$m\ddot{x} + \frac{h}{\omega}\dot{x} + kx = F_0 e^{i\omega t} \qquad (7.25)$$

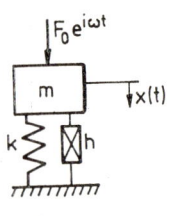

Figure 7.42

7.6 Frequency response measurement

in which m, k and h are respectively the mass, stiffness and coefficient of equivalent hysteretic damping.

Assuming a steady-state solution of the form $x(t) = X^* e^{i\omega t}$, equation (7.25) yields the displacement complex amplitude

$$X^* = \frac{F_0}{k} \frac{1}{1 - \frac{\omega^2}{p^2} + ig} \tag{7.26}$$

in which

$$g = \frac{h}{k} \quad \text{and} \quad p = \sqrt{\frac{k}{m}}. \tag{7.27}$$

This can also be written

$$X^* = X e^{i\varphi} = X_R + i X_I \tag{7.28}$$

where X is the modulus, φ — the phase angle, X_R — the real (in-phase) component, and X_I — the imaginary (in-quadrature) component.

If the plots of X_R versus ω (Fig. 7.43) and X_I versus ω (Fig. 7.44) are traced for $F_0 = \text{const.}$, resonance is located where X_I has a peak and the hysteretic damping factor is calculated from

$$g = \frac{\omega_2^2 - \omega_1^2}{\omega_2^2 + \omega_1^2} \tag{7.29}$$

where ω_1 and ω_2 are the frequencies of the peaks of $X_R(\omega)$.

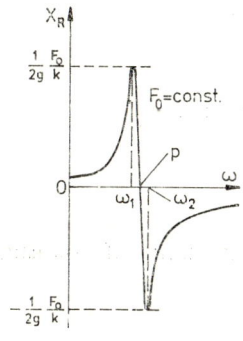

Figure 7.43

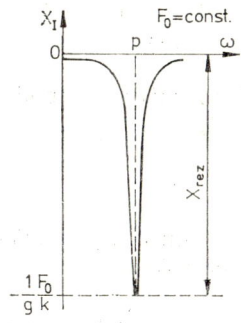

Figure 7.44

The dynamic stiffness can be calculated from the value X_{res} of the amplitude of X_I at resonance

$$k = \frac{1}{g} \frac{F_0}{X_{res}} \tag{7.30}$$

and the mass from the second equation (7.27).

7 Instrument set-ups and techniques

If the plot of the modulus X versus ω (Fig. 7.45) is traced for $F_0 = $ const., the resonance frequency p is defined as occurring at the point where the total response reaches a peak value X_{max}.

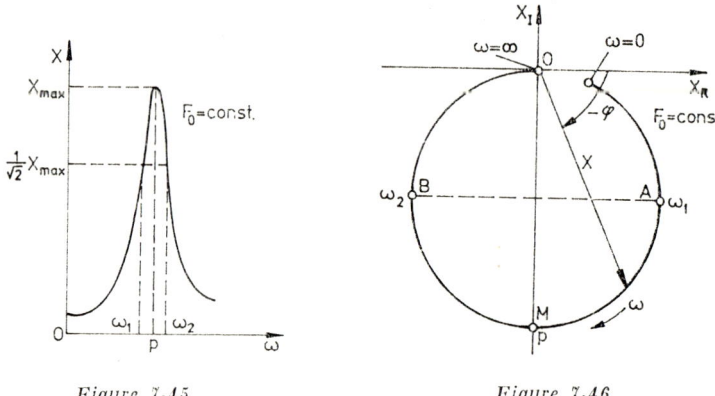

Figure 7.45   Figure 7.46

If ω_1 and ω_2 are the frequencies at which the response amplitude is $\frac{1}{\sqrt{2}} X_{max}$, the damping factor is given by

$$g = \frac{\omega_2^2 - \omega_1^2}{2p^2} \tag{7.31}$$

which, for lightly damped systems, becomes

$$g \cong \frac{\omega_2 - \omega_1}{p}.$$

By eliminating ω between the expressions of X_R and X_I, a circle of equation

$$\left(X_I + \frac{1}{2g}\frac{F_0}{k}\right)^2 + X_R^2 = \left(\frac{1}{2g}\frac{F_0}{k}\right)^2 \tag{7.32}$$

is obtained (Fig. 7.46), which is the locus of the end of the response vector in the complex plane. Resonance occurs at the point M on the imaginary axis, where $X_R = 0$, $\dfrac{dX_I}{d\omega} = 0$, and corresponds to $|X_I|_{max}$ and $|X|_{max}$.

7.6 Frequency response measurement

Because $\dfrac{\mathrm{d}s}{\mathrm{d}(\omega^2/p^2)} = X^2$ is maximum at $\omega = p$, the resonance can be located at the point where $\dfrac{\Delta s}{\Delta \omega}$ is a maximum.

If the system is excited by a harmonic force of constant amplitude and if the displacement response is plotted point by point, at equal frequency increments $\Delta\omega$, the arc length Δs between two successive points is a maximum at resonance (the criterion of Kennedy and Pancu).

The damping factor g is given by equation (7.29) where ω_1 and ω_2 are the frequencies of the half-power points A and B, which are the ends of the diameter AB, perpendicular to OM. The stiffness is then given by equation (7.30) where $X_{\mathrm{res}} = OM$ and the mass is calculated from the second equation (7.27).

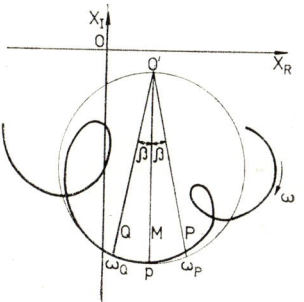

Figure 7.47

If the half-power points cannot be located on the response diagram, the damping factor can be evaluated from the following equation

$$g = \frac{\omega_Q^2 - \omega_P^2}{2p^2} \tan^{-1}\beta \qquad (7.33)$$

where ω_Q and ω_P are the frequencies of two points Q and P in the vicinity of resonance (Fig. 7.47).

When the circles are displaced and rotated due to the contribution of neighbouring modes and to non-proportional damping, the radius and the centre of the circle can be most accurately obtained by using a least squares circular curve fitting technique.

Other techniques for parameter identification of single-degree-of-freedom systems are presented in reference [33].

One-degree-of-freedom analysis techniques are applicable when the overlap between modes is small. A general assumption is that the off-resonance vibration has a (zero or) constant contribution to the total response in the neighbourhood of any resonance.

7.6.5.2 Multi-degree-of-freedom techniques.
In order to represent the frequency response functions in terms of modal characteristics, consider the equation of motion for an N-degree of freedom system with hysteretic damping excited by harmonic forces

$$[M]\{\ddot{q}\} + \frac{1}{\omega}[H]\{\dot{q}\} + [K]\{q\} = \{F\}\mathrm{e}^{i\omega t} \qquad (7.34)$$

7 Instrument set-ups and techniques

where $[M]$, $[H]$ and $[K]$ are the mass, hysteretic damping and stiffness matrices, respectively, $\{q\}$ is the vector of displacements, $\{F\}$ — the excitation vector and ω — the excitation frequency.

In the case of *proportional* damping, the steady-state solution of equation (7.34) is

$$\{q\} = \{Q\}e^{i\omega t} = \sum_{r=1}^{N} \frac{\{\Psi^{(r)}\}^T\{F\}\{\Psi^{(r)}\}}{m_r(p_r^2 + ig_r p_r^2 - \omega^2)} e^{i\omega t} \qquad (7.35)$$

where p_r are undamped natural frequencies, m_r — modal masses, g_r — modal damping factors, $\{\Psi^{(r)}\}$ — real modal vectors, and T indicates transposition.

If a single force $F_l e^{i\omega t}$ is applied at station l, the complex receptance at point j is

$$\bar{\alpha}_{jl} = \frac{Q_j}{F_l} = \sum_{r=1}^{N} \frac{\Psi_j^{(r)} \Psi_l^{(r)}}{m_r} \frac{1}{p_r^2 - \omega^2 + ig_r p_r^2}. \qquad (7.36)$$

In the case of *non-proportional* damping, the complex receptance at point j due to a force applied at point l is

$$\bar{\alpha}_{jl} = \sum_{r=1}^{N} \frac{\Phi_j^{(r)} \Phi_l^{(r)}}{\bar{m}_r} \frac{1}{p_r^2 - \omega^2 + ig_r p_r^2} \qquad (7.37)$$

where $\{\Phi^{(r)}\}$ are complex modal vectors and $\bar{m}_r$ — "complex modal masses".

The frequency response functions (7.36) and (7.37) are analysed in their component modes, characterized by their modal parameters: frequency, damping, equivalent mass and relative displacement amplitude. The "curve fitting" techniques consist of finding a set of composing modes which by recombination restitute or approximate the measured frequency response curve "as well as possible".

Polar diagrams of systems with many degrees of freedom are not circles, but curves with many loops, usually one for each resonance. For lightly or moderately damped systems, each loop is separately analysed using single-degree-of-freedom techniques. Resonance is best located at the point of maximum rate of change of arc length with frequency. The "best circle" is then fitted (and drawn) through the points in the neighbourhood of each resonance. The diameter of such a best fit circle gives the response at resonance in one mode of vibration only — the mode that is used to determine the mode shape. Both complex and real modes may be considered.

7.6 Frequency response measurement

If the modes are uncoupled and well separated in frequency, the diameters of the circles fitted in the Argand plane to the near-resonance points of the displacement response are normal to the force reference direction (Fig. 7.48 a). If the modes are uncoupled, but close in frequency, the diameters have a phase offset relative to the force reference direction due to the variation with frequency of the off-resonance modes. If the modes are

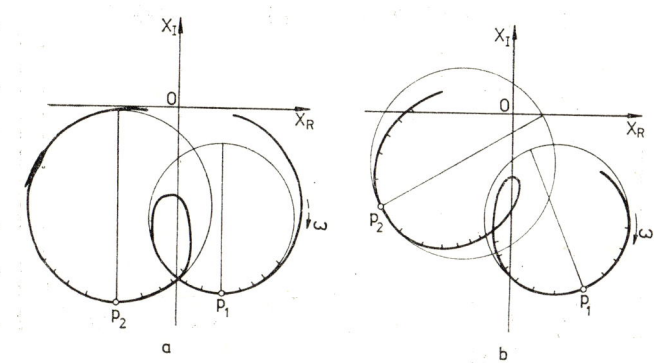

Figure 7.48

coupled, the resonance diameters have additional phase offsets (Fig. 7.48 b). Rotation of the resonance diameter usually indicates the presence of non-proportional damping.

Multi-degree-of-freedom curve fitting is used for parameter estimation. An iterative weighted least square method can be employed in which the quadratic error between measured and analytically derived response functions is minimized [34].

In a given frequency range, the response can be approximately described in terms of the "inertia restraint" of the lower modes of vibration, the sum of the modes of vibration which are resonant in that frequency range and the "residual flexibility" of the higher frequency modes. Equation (7.37) becomes

$$\overline{\alpha}_{jl} = -\frac{1}{\omega^2} Y_{jl} + \sum_{r=N'}^{N''} \frac{\Phi_j^{(r)} \Phi_l^{(r)}}{\overline{m}_r(p_r^2 - \omega^2 + ig_r\, p_r^2)} + Z_{jl}. \quad (7.38)$$

The modal parameters can be extracted by least squares curve fitting equation (7.38) to the frequency response data measured at m excitation frequencies $\omega_1, \ldots, \omega_m$. Normalizing the modal coefficient $\Phi_j^{(r)}\Phi_l^{(r)}$ to 1 and setting up an equation at each excitation frequency, the following set of equations is obtained

$$\left\{\begin{array}{c} \bar{\alpha}_{jl}(\omega_1) \\ \bar{\alpha}_{jl}(\omega_2) \\ \vdots \\ \bar{\alpha}_{jl}(\omega_m) \end{array}\right\} =$$

$$= \begin{bmatrix} \dfrac{1}{\omega_1^2} & \dfrac{1}{p_{N'}^2 - \omega_1^2 + ip_{N'}^2 g_{N'}} & \cdots & \dfrac{1}{p_{N''}^2 - \omega_1^2 + ip_{N''}^2 g_{N''}} & 1 \\ \dfrac{1}{\omega_2^2} & \cdot & \cdot & \cdot \\ \vdots & \vdots & & \vdots \\ \cdot & \cdot & & \cdot \\ \dfrac{1}{\omega_m^2} & \dfrac{1}{p_{N'}^2 - \omega_m^2 + ip_{N'}^2 g_{N'}} & \cdots & \dfrac{1}{p_{N''}^2 - \omega_m^2 + ip_{N''}^2 g_{N''}} & 1 \end{bmatrix} \left\{\begin{array}{c} Y_{jl} \\ \dfrac{1}{m_{N'}} \\ \vdots \\ \dfrac{1}{m_{N''}} \\ Z_{jl} \end{array}\right\}$$

or $\{\alpha\} = [T]\{\beta\}$.

In the case when there are more data points measured than parameters of interest, an error function is introduced

$$\{e\} = \{\alpha\} - [T]\{\beta\}$$

and the error squared is calculated as

$$\{e\}^T\{e\} = \{\{\alpha\} - [T]\{\beta\}\}^T\{\{\alpha\} - [T]\{\beta\}\}. \tag{7.39}$$

A weighting function is introduced into equation (7.39) under the form of a diagonal matrix $[W]$ having the elements proportional to the importance of the data recorded at that frequency. The product

$$\{e\}^T[W]\{e\} = \{\{\alpha\} - [T]\{\beta\}\}^T[W]\{\{\alpha\} - [T]\{\beta\}\}$$

is then minimized by taking its derivative with respect to the vector $\{\beta\}$ and setting the result equal to zero

$$\frac{\partial(\{e\}^T[W]\{e\})}{\partial\{\beta\}} = 0 = 2[T]^T[W]\{[T]\{\beta\} - \{\alpha\}\}.$$

This yields the "weighted" least squares solution

$$\{\beta\} = [[T]^T[W][T]]^{-1}\{[T]^T[W]\{\alpha\}\}, \tag{7.40}$$

wherefrom accurate values of modal masses and residual terms can be obtained starting from initial estimates of the natural frequencies and damping.

All curve-fit methods are based upon specific analytical models, requiring an estimate of the number of composing modes (with

7.6 Frequency response measurement

the risk of eliminating in advance some modes) and starting the iterative process with estimated values of some modal parameters.

Modal analysis programs have been developed by Structural Dynamics Research Corporation, Structural Measurement Systems Inc., Structural/Kinematics, Nicolet etc.

Other analysis techniques of experimental frequency response data are presented in references [35] and [36]. Multiple input estimation of frequency response functions is recently used to increase the consistency of estimated modal parameters [37].

References for Chapter 7

1. Nicolau, E., Beliş, M., *Măsurări electrice şi electronice*, Editura didactică şi pedagogică, Bucureşti, 1972.
2. Bendat, J. S. and Piersol, A. G., *Measurement and Analysis of Random Data*, John Wiley, New York, 1966.
3. Ventsel, M., *Théorie des probabilités* (Transl. from Russian), Mir, Moscow, 1973.
4. Mitchell, L. D., Lynch, G. A., Origins of noise, *Machine Design*, May, 174 (1969).
5. Fieldhouse, K. N., Techniques for identifying sources of noise and vibration, *S/V, Sound and Vibration*, 4, 12, 14—18 (1970).
6. Randall, R. B., *Application of B & K Equipment to Frequency Analysis*, Brüel & Kjaer, Naerum, 1977.
7. Tustin, W., A practical primer on vibration testing, *Evaluation Engineering*, 21—24, 53—54 (Nov.—Dec. 1969).
8. Riis, H., Electronic control of vibration exciters, Brüel & Kjaer Lecture No. *3023*, 1971.
9. Norin, R. S., A multi-channel, multi-strategy sinusoidal vibration control system, *Proc. Inst. Environm. Sci.*, GenRad Reprint, 1974.
10. Morrow, Ch. T., *Shock and Vibration Engineering*, John Wiley & Sons, New York, 1963.
11. Norin, R. S., Pseudo-random and random testing, *Seminar on Understanding Digital Control and Analysis in Vibration Test Systems*, Goddard Space Flight Center, June 17—18, 1975 and Jet Propulsion Lab., July 22—23, 1975.
12. Olesen, M. W., A narrow band vibration test, *Shock and Vibration Bulletin*, **26**, *1* (1957).
13. Stauffer, M. K., Techniques for narrowband random or sine on wideband random vibration testing with a digital control system, *Seminar on Understanding Digital Control and Analysis in Vibration Test Systems*, Goddard Space Flight Center, June 17—18, 1975, and Jet Propulsion Laboratory, July 22—23, 1975.
14. * * * *U.S.A Standard S 2.6—1963* : "Specifying the Mechanical Impedance of Structures".
15. Ewins, D. J., Sainsbury, M. G., Mobility measurements for the vibration analysis of connected structures, *Shock and Vibration Bulletin*, **42**, 105—122 (1972).
16. Keller, A. C., Fundamentals for mechanical impedance analysis, Spectral Dynamics Corp., *Technical Publication* M-2, 6—67.

7 Instrument set-ups and techniques

17. Kerfoot, R. E., Solutions for mechanical impedance measurement problems, Spectral Dynamics Corp., *Technical Publication* M-1, 9—66.
18. * * * *Instruction Manual* — Mass Compensation Unit Type 5565, Brüel & Kjaer Publication 61—795.
19. Radeș, M., Compensarea maselor la măsurarea impedanțelor mecanice, *St. cerc. mec. apl.*, **35**, *4*, 547—568 (1976).
20. Buzdugan, Gh., Fetcu, L., Radeș, M., *Vibrațiile sistemelor mecanice*, Editura Academiei, București, 1975.
21. Hallauer, W. L.Jr. and Stafford, J.F., On the distribution of shaker forces in multiple shaker modal testing. *Shock and Vibration Bulletin*, **48**, Part *1*, 49—63 (1978).
22. Beatrix, Ch., The experimental methods for structural vibration global testing, *La Recherche Aérospatiale*, *109*, 57—64 (1965).
23. Hawkins, F. J., *GRAMPA* — An automatic technique for exciting the principal modes of vibration of complex structures, *TR 65142*, R.A.E. Farnborough, England (1965).
24. Taylor, G. A., Gaukroger, D. R. and Skingle, C. W., *MAMA* — A semiautomatic technique for exciting the principal modes of vibration of complex structures, *TR 67211*, R.A.E. Farnborough, England (Aug. 1967).
25. Knauer, C. D. Jr., Peterson, A. J., and Rendahl, W. B., Space vehicle experimental modal definition using transfer function techniques, *SAE Paper* No. *751069* (1975),
26. Stroud, R. C., Smith, S. and Hamma, G. A., *MODALAB* — A new system for structural dynamic testing, *Shock and Vibration Bulletin*, **46**, *5*, 153—175 (1976).
27. Reed, W. H., Hall, A. W. and Barker, L. E., Analog techniques for measuring the frequency response of linear physical systems excited by frequency sweep inputs, *NASA TN D508* (1960).
28. White, R. G., Evaluation of the dynamic characteristics of structures by transient testing, *Journal of Sound and Vibration*, **15**, *2*, 147—161 (1971).
29. Ramsey, K. A., Effective measurements for structural dynamics testing, *S/V Sound and Vibration*, **9**, *11*, 24—35 (1975) and **10**, *4*, 18—30 (1976).
30. Roth, R. P., Digital Fourier Analysis, *Hewlett Packard Journal*, June 1970.
31. * * * Testing with pseudo-random and random noise, *Hewlett Packard Journal*, Sept. 1967.
32. Broch, J. T., On the measurement of frequency response functions, *Brüel & Kjaer Technical Review*, *4* (1975).
33. Radeș, M., Methods for the analysis of structural frequency response measurement data, *Shock and Vibration Digest*, **8**, *2*, 73—88 (1976).
34. Klosterman, A. and Zimmerman, R., Modal survey activity via frequency response functions, *SAE Paper* No. *751068* (1975).
35. Radeș, M., *Metode dinamice pentru identificarea sistemelor mecanice*, Editura Academiei, București, 1979.
36. Natke, H. G., *Einführung in Theorie und Praxis der Zeitreihen- und Modalanalyse*, Vieweg, Braunschweig/Wiesbaden, 1983.
37. Allemang, R. J. *Investigation of Some Multiple Input/Output Frequency Response Function Experimental Modal Analysis Techniques*, Ph. D. Dissertation, Univ. of Cincinnati, 1980.

8

Calibration of transducers and instrumentation systems

Calibration of instruments used for vibration measurements consists in determining the relationship of the output (electrical or mechanical) to the input (displacement, velocity, acceleration, force, torque). The ratio of these quantities is called *calibration factor* or, more often, *sensitivity* (especially for transducers).

Calibration complexity depends, among other factors, on the intended application, on the conditions under which measurements are performed and on the type of available instruments. Field calibration is generally made at a single frequency and for the overall instrumentation system. Laboratory calibration may be made either separately for each instrument or for the overall instrumentation system.

The following information is generally required : a) the sensitivity over the frequency range of interest ; b) the sensitivity under various environmental conditions (temperature, supply voltage variation, acoustic noise, humidity, electromagnetic fields) ; c) the sensitivity over a given amplitude range ; d) the stability of calibration with time (recalibration at intervals of 1 to 2 years) ; e) the frequency response and linear operating range for different attachment conditions.

8.1 Calibration of vibration pickups

Various techniques of calibrating vibration-measuring pickups are thoroughly described in Chapter 18 of reference [1]. In the following only several methods more frequently used in practice are presented, which can be applied by a large number of laboratories. Each method has a restricted application field owing to mechanical

8 *Calibration*

or electrical limitations. It is recommended to use the calibration method which provides the sensitivity in the range of frequencies and amplitude levels in which the pickup will operate.

8.1.1 Static calibration

There are vibration measuring pickups that can be calibrated statically.

Seismic displacement-sensing pickups with variable-resistance or inductance transducers can be calibrated using any device producing a known displacement of the sensing element, within the linear operating range of the pickup. Both linear and angular displacement pickups can be calibrated in this way, the latter having the calibrating device acting tangentially.

Seismic accelerometers with passive transducers can be calibrated using the method of earth's gravitational field. When the accelerometer sensing axis is horizontal, the measuring bridge is adjusted to be "in balance", so that the output voltage is zero. Orienting the accelerometer with its sensing axis vertical, an $1g$ constant acceleration is applied and a first reading of the bridge imbalance is made. Rotating the accelerometer through 180°, an $1g$ constant acceleration is applied in the opposite direction and a second reading is made on the measuring bridge meter. Making the difference of the two readings and dividing by $2g$, the sensitivity can be determined.

Strain gauge instrumented *force gauges* can be calibrated by loading them with known weights (when the measuring bridge permits static measurements).

Electrodynamic velocity pickups can be calibrated statically based on the reciprocity relationship (4.20)

$$S_v = \left|\frac{e}{v}\right| = \frac{F}{i} = Bl. \qquad (8.1)$$

The pickup sensitivity S_v, V/(m/s), is calculated from the force F, N, produced when a current i, A, passes through the armature coil. The force is measured using a dynamometer.

8.1.2 "Direct" dynamic calibration

Absolute methods of calibrating vibration pickups make use of physical pendula, rotating tables, mechanical or electromagnetic vibrators, etc. Pickups are subjected to a known vibratory motion, which is measured either optically (observing the displacement of

8.1 Calibration of pickups

some moving marks or targets) or acoustically (based on the chatter of a ball, which occurs when an acceleration of $1g$ has been exceeded). The pickup electrical output is measured with suitable auxiliary instrumentation. The optical interferometric method is used for precision calibration (see Section 8.4.1).

This section describes two calibration methods suitable for low frequency (0 — 2000 Hz) and low acceleration (0.1 g — 100 g) vibrations.

8.1.2.1 *Calibration of a piezoelectric accelerometer.* The (open-circuit) voltage sensitivity

$$S_e = \frac{u}{\ddot{x}} \left[\frac{V}{m/s^2}\right] \text{ or } \left[\frac{V}{g}\right] \quad (8.2)$$

can be determined using the following methods :

a) *Mechanical calibrator.* In the simplest form (Fig. 8.1), the mechanical vibration table is supplied directly from the mains and generates a nearly sinusoidal motion of fixed frequency and displacement amplitude. It is used for field calibration.

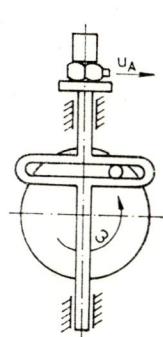

Figure 8.1

As a typical example, the mechanical calibration table MET 1, manufactured by "VEB Schwingungstechnik und Akustik W.I.B.", Dresden, operates at 220 V and for a 50 grams table load has a frequency of 22.7 Hz and an r.m.s. displacement of 1.97 mm ($\dot{x}_{rms} = 0.28$ m/s, $\ddot{x}_{rms} = 40$ m/s^2).

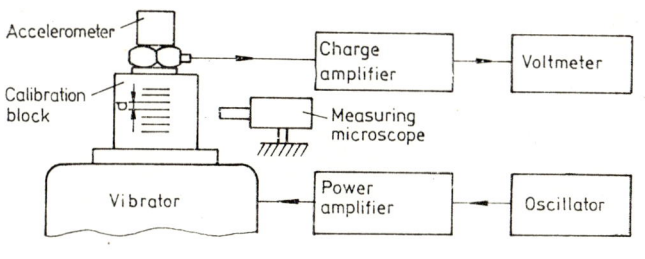

Figure 8.2

b) *Electrodynamic calibrator.* An electrodynamic vibrator energized from a signal generator via a power amplifier (Fig. 8.2) can be used in calibration work. The pickup is mounted on a small metallic block which is attached on the vibrator mounting table.

271

8 Calibration

The excursion of a given point (or "mark") of the block surface can be directly scaled with a measuring microscope equipped with a reticle scale. Both the frequency and amplitude of the block motion can be adjusted over broad ranges so that the frequency response characteristic can also be determined.

If equally spaced lines are engraved on the metallic block, persistence of vision causes each line to appear during vibration as a dark band, of width equal to the peak-to-peak displacement. Adjusting the power amplifier gain, the boundaries of two adjacent lines can be made to just touch one another. If ω is the driving angular frequency and d is the distance between the calibration lines, then the displacement amplitude is $d/2$ and the acceleration amplitude is $\omega^2 d/2$. The corresponding output voltage is read on the voltmeter scale.

Brüel & Kjaer calibrators types 4290 and 4291 are based on this principle when energized from an external oscillator. The calibrator type 4291 has a built-in oscillator which generates an 1 g peak acceleration at 79.6 Hz ($\omega = 500$ rad/s, according to DIN 45666 and ANSI S2.2, 1959).

c) *Calibration with resonant devices.* The pickup to be calibrated is mounted on a resonant structure (e.g., a free-free beam, supported at the nodal points of the fundamental mode of vibration), driven by either a direct mounted vibrator or a non-contacting electromagnet (Fig. 8.3). As for case (b), the excitation is adjusted until the displacement amplitude is sufficiently large so as to be accurately determined using a measuring microscope.

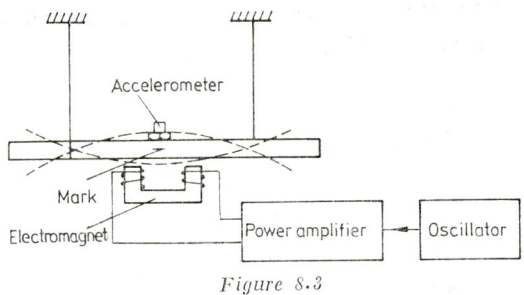

Figure 8.3

Operation at resonance ensures the purity of the sinusoidal motion but is limited to only one frequency. If calibration at several frequencies is required, then additional masses can be mounted to the beam, or different beams can be used. The accelerometer calibrator Brüel & Kjaer type 4292 operates at the resonance frequency of a bronze band whose tension can be externally adjusted and whose motion is calibrated at 1 g peak by a rattling bronze ball.

8.1 Calibration of pickups

With the set-up from Figure 8.4, when the switch is on position 2, the oscillator signal passes through the band filter which is tuned on the signal frequency. When the reference signal and the filtered signal appear 180° phase shifted on the scope

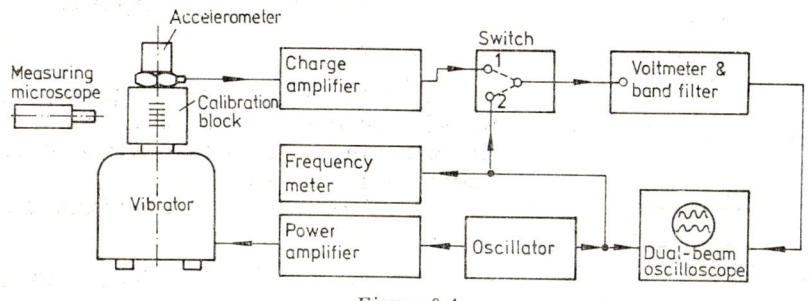

Figure 8.4

screen, the tuning is accomplished. The switch is turned on position 1 and the voltmeter indication is read for a known value of the acceleration.

In this way, the voltage sensitivity is measured for the overall system accelerometer-amplifier-filter-voltmeter. The accelerometer sensitivity can be determined after a simulation calibration of the auxiliary circuits (see Section 8.2).

After determining the voltage sensitivity S_e, using the set-up from Figure 8.4, it is possible to measure the charge sensitivity

$$S_q = \frac{q}{\ddot{x}} \quad \left[\frac{\text{pC}}{\text{m/s}^2}\right] \text{ or } \left[\frac{\text{pC}}{g}\right]. \tag{8.3}$$

If the oscillator generates a sine wave of angular frequency ω and the power amplifier is adjusted so that the vibrator produces a displacement of amplitude x (measured optically), then the accelerometer is vibrated with an acceleration $\omega^2 x$. Its output is $u_e = S_e \omega^2 x$. Because, initially, the voltmeter indicates a different value, the knobs of the charge sensitivity S_q adjustment network of the charge amplifier (e.g., Brüel & Kjaer 2626 or 2635) are turned until the voltmeter readout indicates the voltage u_e. At that moment S_q is the required value.

8.1.2.2 *Calibration of a piezoelectric force transducer.* The arrangement from Figure 8.5 is used for determining the voltage sensitivity

$$S_{eF} = \frac{u_F}{f} \quad \left[\frac{\text{V}}{\text{N}}\right]. \tag{8.4}$$

8 Calibration

Let M be the total mass of the accelerometer and the calibration weight, S_{eA} — the accelerometer voltage sensitivity and u_A — the accelerometer output voltage. The vibrator table has an acceleration $\ddot{x} = u_A/S_{eA}$.

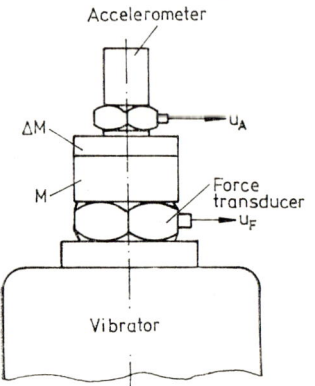

Figure 8.5

The force transducer output is

$$u_F = S_{eF} M \ddot{x}. \qquad (8.5)$$

Keeping constant the frequency ω, an additional mass ΔM is attached to M, and the excitation is adjusted to maintain $\ddot{x}$ constant.

The force transducer output becomes

$$u_F + \Delta u_F = S_{eF}(M + \Delta M)\ddot{x}. \qquad (8.6)$$

Subtracting equation (8.5) from equation (8.6), there is obtained

$$\Delta u_F = S_{eF} \Delta M \ddot{x}.$$

The voltage sensitivity is given by

$$S_{eF} = \frac{\Delta u_F}{\Delta M \ddot{x}} = \frac{\Delta u_F}{u_A} \frac{S_{eA}}{\Delta M}. \qquad (8.7)$$

The total mass m acting on the piezoelectric transducer is

$$m = \frac{f}{\ddot{x}} = \frac{u_F}{S_{eF} \ddot{x}} = \frac{u_F}{u_A} \frac{S_{eA}}{S_{eF}}. \qquad (8.8)$$

In order to afford simple calculations, the phase angle between u_F and u_A should be zero. Otherwise corrections are required.

The total mass m consists of the attached mass m_a and the mass m_c above the sensing element of the force transducer. Equation (8.8) can be written under the form

$$u_F = S_{eF} \ddot{x} m = S_{eF} \ddot{x}(m_c + m_a). \qquad (8.9)$$

From equation (8.9) it follows that if u_F is measured for different values of m_a and data are plotted in a diagram of u_F versus m_a (Fig. 8.6), then the straight line fitted through the data points has a slope $\tan^{-1}(S_{eF}\ddot{x})$ and the abscissa intercept is proportional to m_c.

If the force f measured by the transducer has comparable magnitude with the inertia force $m_c \ddot{x}$, corresponding to the

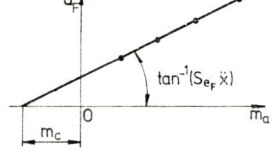

Figure 8.6

8.1 Calibration of pickups

mass of the transducer portion located above the sensing element, then a correction should be applied, the force actually applied to the structure being $(f - m_c\ddot{x})$.

For determining the frequency response characteristic, the ratio u_F/u_A is measured in the range $10 - 5000$ Hz because, according to equation (8.8), $S_{eF} = \dfrac{u_F}{u_A} \dfrac{S_{eA}}{m}$. It is necessary to check that $S_{eA} = $ const. within this frequency range [2].

The charge sensitivity of the force transducer

$$S_{qF} = \frac{q}{f} \qquad \left[\frac{\text{pC}}{\text{N}}\right]$$

can be determined using a set-up similar to that from Figure 8.6 but including a charge amplifier as in Figure 8.4.

A value u_F is chosen and the corresponding acceleration

$$\ddot{x} = \frac{u_F}{mS_{eF}}$$

is calculated from known values of the voltage sensitivity S_{eF} and the total mass m. The power amplifier gain is adjusted until an acceleration $\ddot{x}$ is obtained at the vibrator table, then S_{qF} is varied at the charge sensitivity adjusting knobs of the charge amplifier until a voltage u_F is indicated on the voltmeter readout.

8.1.3 Reciprocity calibration

The technique is applicable to the absolute calibration of vibration standards (reference transducers) used afterwards for performing comparison calibrations on test transducers. It can be used only for the calibration of vibration pickups involving linear bilateral electromechanical transducers, like the velocity electrodynamic transducers and the piezoelectric accelerometers.

The principle of the method is based on the reciprocity relationships (8.1) equating the ratios of force/current and voltage/velocity, where the electrical terms refer to the vibrator drive coil and the mechanical terms refer to the vibrator armature. The drive coil acts as the "reciprocal" transducer whose electrical measurements are used for calibrating the output of the vibration standard attached to the vibrator table. The technique resumes to the determination of the ratio and the product of the sensitivities of two pickups [3].

8 *Calibration*

8.1.3.1 *Reciprocity procedure.* The calibration procedure will be illustrated for an electrodynamic velocity-sensing pickup whose sensitivity will be denoted S_v, Vs/m. It involves two experimental phases and one computational phase.

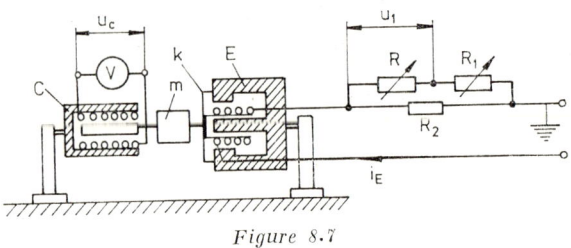

Figure 8.7

a) *Experiment No. 1.* Using the set-up from Figure 8.7, known masses m are rigidly attached between the drive coil of the pickup C to be calibrated and the drive coil of an electrodynamic pickup E used as exciter.

The drive coil of the pickup E is energized by a voltage of angular frequency ω and the transfer admittance between the drive and sensing coils

$$Y = \frac{i_E}{u_C} \tag{8.11}$$

is measured for each mass m. In equation (8.11) i_E is the current in the drive coil and u_C is the voltage generated in the open-circuited sensing coil (or across the terminals of a high impedance voltmeter) of the pickup C.

The resistances from the drive coil circuit are adjusted until $u_1 = u_C$. In this case $|Y| = (R + R_1 + R_2)/R \cdot R_2$, where typical values are $R + R_1 = 1000\ R_2$ and $R_2 = 10\ \Omega$ [4].

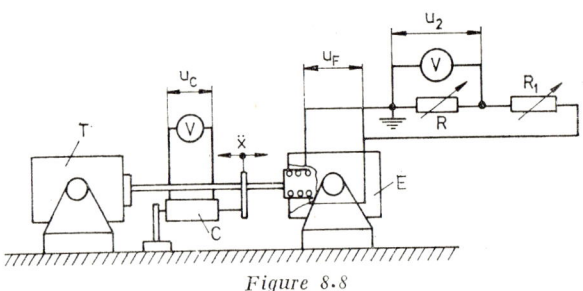

Figure 8.8

b) *Experiment No. 2.* The set-up from Figure 8.8 is used, where the moving coils of pickups C and E are coupled to a vibra-

8.1 Calibration of pickups

tor T which generates a harmonic motion of frequency ω. The ratio

$$K = \frac{u_C}{u_E}$$

of the open-circuited voltage u_C, generated in the sensing coil of the pickup C, and the open-circuited voltage u_E, generated in the drive coil of the pickup E, is measured.

The resistances from the drive coil circuit of pickup C are adjusted until $u_2 = u_C$. In this case $|K| = R/(R + R_1)$ where it is recommended to take $R_1 \cong$ $\cong 10^4 \Omega$ [4].

c) *Computation of sensitivity* S_v. Points corresponding to measurements with different masses m are plotted on a graph (Fig. 8.9) of the mass m, kg, versus the transfer admittance Y, A/V. A straight line is fitted through the data points determining the slope J, kgV/A, and the abscissa intercept Y_0, A/V, which is the value of Y, when $m = 0$.

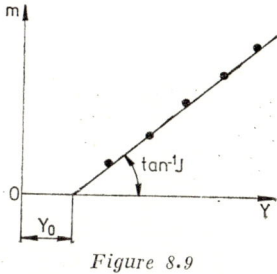

Figure 8.9

The sensitivity of the velocity pickup is given by

$$|S_v| = \sqrt{\omega J K} \qquad \left[\frac{\mathrm{V s}}{\mathrm{m}}\right]. \tag{8.13}$$

If C is a piezoelectric accelerometer, then the sensitivity S_A is given by

$$|S_A| = \sqrt{\frac{JK}{\omega}} \qquad \left[\frac{\mathrm{V s}^2}{\mathrm{m}}\right]. \tag{8.14}$$

8.1.3.2 *Theoretical background*. According to equation (8.1), the force generated by the transducer E in the first experiment is

$$f = B l i_E = S_E i_E, \tag{8.15}$$

where S_E is the sensitivity of the pickup E. This force acts on the system consisting of the moving parts of the two pickups (of mass M) and the attached mass m, supported by the elastic suspension of the pickup E (eventually, of the pickup C), which has a harmonic motion of velocity $\dot{x}$.

The complex mechanical impedance of this vibrating system is

$$Z = \frac{f}{\dot{x}} = Z_0 + i\omega m, \tag{8.16}$$

8 Calibration

where Z_0 is the value of Z when $m = 0$ and $i\omega m$ is the impedance of the mass m.

The voltage across the terminals of the pickup C is

$$u_C = B_C l_C \dot{x} = S_v \dot{x}. \tag{8.17}$$

Substitution of (8.15) and (8.17) into (8.16) yields

$$Z = \frac{f}{\dot{x}} = \frac{S_E i_E}{\frac{u_C}{S_v}} = S_E S_v \frac{i_E}{u_C} = S_E S_v Y. \tag{8.18}$$

According to equations (8.16) and (8.18), when $m \neq 0$,

$$S_E S_v Y = i\omega(m + M) = i\omega m + Z_0$$

and when $m = 0$,

$$S_E S_v Y_0 = i\omega M = Z_0,$$

wherefrom, by subtraction, there is obtained

$$S_E S_v (Y - Y_0) = i\omega m.$$

The product of the two sensitivities is

$$S_E S_v = \frac{i\omega m}{Y - Y_0}. \tag{8.19}$$

During experiment No. 2, the armatures of pickups C and E have velocities of equal magnitude. As $u_E = S_E \dot{x}$, it follows that

$$K = \frac{u_C}{u_E} = \frac{S_v}{S_E}. \tag{8.20}$$

Within the linear operating range of the pickup, this ratio is independent of the velocity $\dot{x}$.

Using equations (8.19) and (8.20), it is now possible to calculate S_v. There is obtained

$$S_v^2 = \frac{i\omega m\, K}{Y - Y_0} \tag{.21}$$

wherefrom, using the notation $J = \dfrac{m}{Y - Y_0}$ and taking the absolute value, relationship (8.13) is easily obtained.

The linear relationship between m and Y, on which the plot shown in Figure 8.9 is based, holds only for systems with negligible damping and at frequencies outside the resonance of the moving coil assembly.

8.1 Calibration of pickups

The sensitivity of an accelerometer is

$$S_A = \frac{u_C}{\ddot{x}} = \frac{u_C}{i\omega \dot{x}} = \frac{S_v}{i\omega} = \sqrt{\frac{JK}{i\omega}}. \qquad (8.22)$$

Equation (8.22) is valid provided that the phase angle of the voltage ratio K (8.12) is 90° and the phase angle of the transfer admittance Y (8.11), where u_C is the output of the piezoelectric accelerometer, is zero. These assumptions can be easily checked using an oscilloscope.

The calibration is performed at 100 Hz or below for most commercial vibrators. The sensitivity is determined with 0.5 percent standard deviation. The vibrator T should generate a sinusoidal motion whose distortion is less than one percent and free of transverse motion components.

The frequency response is measured up to 2000 Hz — at electrodynamic pickups, and up to 10,000 Hz — at piezoelectric transducers used for shock measurements. This is done by a comparison calibration. Several vibrators of different design are used in order to cover broad frequency ranges. Vibration standards are recalibrated on a regular basis, usually once a year. Details on the reciprocity calibration technique are given in references [3, 4, 5].

8.1.4 Optical interferometry calibration

The phenomenon of interference fringe pattern disappearance in an optical interferometer can be used as a precision absolute calibration method. The laser interferometry method is the most widely used today to calibrate standard transducers. The *fringe counting method* will be briefly explained [6].

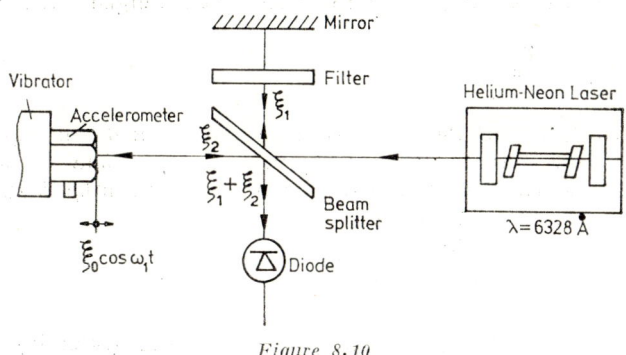

Figure 8.10

As illustrated in Figure 8.10, the laser beam is split into two parts; one is reflected by a fixed mirror, the other by a reflecting

8 Calibration

surface on the mounting plate of the calibrator or on the transducer. The former can be expressed as $\xi_1 = a e^{i\omega_L t}$ and the latter by $\xi_2 = b\, e^{i\omega_L t + \delta}$, where ω_L is the angular frequency of the laser light and δ is the phase shift between the two beams.

The two reflected beams are combined by the beam splitter and the light intensity to the photodiode is $I = (\xi_1 + \xi_2)^2$, both beams being polarized in the same direction.

For time intervals larger than the period of laser light, the intensity can be expressed as

$$I = A + B\cos\delta, \tag{8.23}$$

where A and B are constants. The phase lag can be written

$$\delta = \frac{2\pi}{\lambda}\Delta\varphi = \frac{2\pi}{\lambda} 2(L + \hat{\xi}_0 \cos \omega_1 t),$$

where λ is the laser beam wavelength, $\Delta\varphi$ — the path length difference, $\hat{\xi}_0 \cos \omega_1 t$ the instantaneous displacement of the calibrated transducer and L — the constant path length difference.

The interference between the two beams causes them to shift between maximum and minimum intensity each time the path length difference is changed by half a wavelength. The number of changes per vibration period $T_1 = \dfrac{2\pi}{\omega_1}$ gives the vibration displacement amplitude (from which the acceleration can be calculated).

During a period T_1, the displacement $\hat{\xi}_0 \cos \omega_1 t$ varies from $+\hat{\xi}_0$ to $-\hat{\xi}_0$ and from $-\hat{\xi}_0$ to $+\hat{\xi}_0$, i.e. has an overall variation of $4\hat{\xi}_0$. The variation per cycle of the path length difference is

$$\frac{\overline{\delta}}{T_1} = \frac{4\pi}{\lambda} \cdot 4\hat{\xi}_0. \tag{8.24}$$

According to equation (8.23), a 2π change of δ corresponds to a period of fluctuation of the light intensity on the photodiode. It follows that the frequency of the photodiode output signal is

$$f_{\text{diode}} = \frac{\overline{\delta}}{2\pi T_1} = \frac{8\hat{\xi}_0}{\lambda T_1} = \frac{8\hat{\xi}_0}{\lambda} f_1$$

where $f_1 = 1/T_1$.

The transducer displacement amplitude is given by

$$\hat{\xi}_0 = \frac{f_{\text{diode}}}{f_1} \frac{\lambda}{8}. \tag{8.25}$$

8.1 Calibration of pickups

Set-ups similar to that shown in Figure 8.11 are used for direct reading of the ratio f_diode/f_1 on an electronic counter. By multiplication with the constant $\dfrac{\lambda}{8} = 0.0791$ μm, the displacement amplitude $\hat{\xi}_0$ is determined in μm.

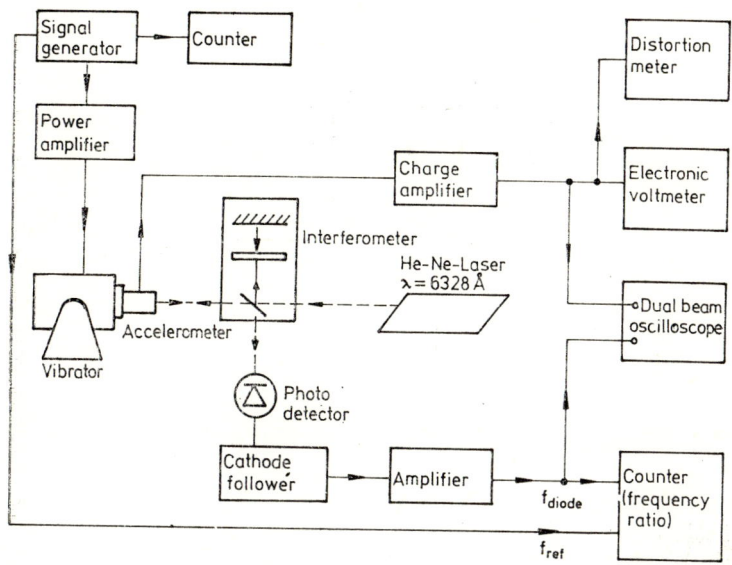

Figure 8.11

The laser interferometry method is used for transducer calibration in the frequency range 20 — 500 Hz.

8.1.5 Comparison calibration

Comparison calibrations are performed using the arrangement from Figure 8.12. The test accelerometer 1, of unknown sensitivity S_1, is mounted as close as possible (preferably back to back) to a reference standard accelerometer 2, of known sensitivity S_2 (determined by prior reciprocity calibration or by the laser interferometry method), on the fixture 3 attached to the mounting table of a vibrator which generates a sinusoidal motion. The output voltages u_1 and u_2 of the two transducers are measured. As $S_1 = u_1/x$ and $S_2 = u_2/x$, the sensitivity S_1 is given by the following equation

$$S_1 = \frac{u_1}{u_2} S_2. \tag{8.26}$$

8 Calibration

Comparison calibration is limited to the frequency and amplitude ranges for which the reference standard transducer 2 has been calibrated. The fixture must minimize the errors due to the relative motion of the two transducers. Additional information can be found in reference [7].

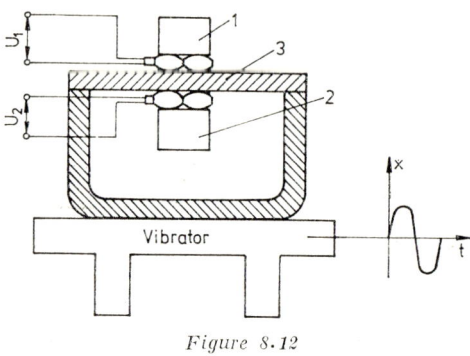

Figure 8.12

8.2 Simulation calibration of auxiliary circuits

It is often desirable to check the electrical portions of an instrumentation system following the transducer (referred to as auxiliary circuits) before, during or after a test. A useful technique, called *simulation calibration*, involves the application of an artificial transducer output signal instead of the transducer signal. Typical applications of this technique are [8]:

a) establishing the optimum calibration factor for a measuring system containing several instruments, for example, the adjustment of the system gain in order to obtain the desired deflection of a recording galvanometer at a specified input level;

b) checking the electrical continuity of transducer and auxiliary circuits (possible shorts);

c) determining the electrical noise of instruments;

d) evaluating the electrical characteristics of instruments under conditions which closely simulate the final measuring situation.

8.2.1 Substitution calibration

The technique of substitution calibration is illustrated in Figure 8.13. The transducer is disconnected from the rest of the instrumentation system. The output of a frequency generator is applied

8.2 Simulation calibration

to the input of the system instead of the transducer signal. The response characteristics of the instruments of the measurement system are determined by varying the frequency and the output voltage of the oscillator.

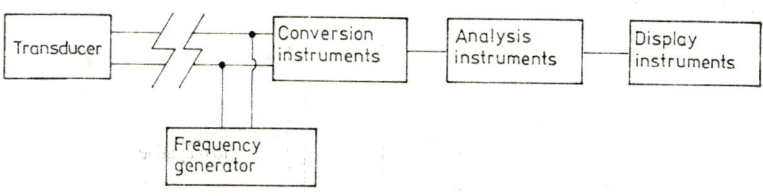

Figure 8.13

The calibration of the overall instrumentation system can be performed if the transducer voltage sensitivity S_e is known. For example, if d is the deflection of the recorder pen or the scope beam produced by a voltage u_g, then an acceleration $\ddot{x} = u_g/S_e$ will correspond to the deflection d when an accelerometer is connected into the circuit.

8.2.2 Insert calibration

The technique involves the insertion of a voltage source in series with an ungrounded transducer (Fig. 8.14). This voltage simulates the output of a self-generating transducer, which acts as a simple impedance (passive element) when there is no mechanical excitation.

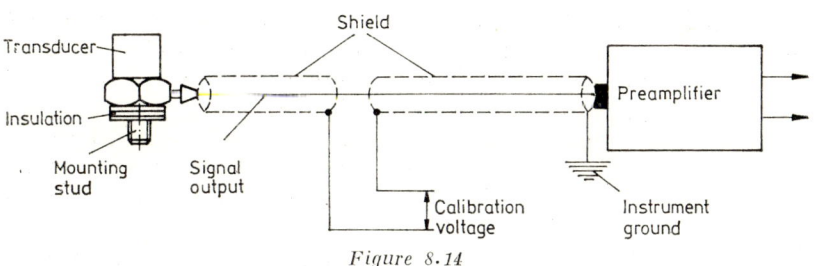

Figure 8.14

In Figure 8.14, the piezoelectric transducer is insulated from the preamplifier electrical ground either by a special insulating mounting stud or by a wax or adhesive layer. The shield of the coaxial cable is broken and a voltage source is connected to the two sections of the shield in series with the transducer.

283

8 Calibration

During calibration simulation, the transducer is usually placed on a table. If the simulation voltage must be applied in the midst of measurements, without dismounting the transducer, steps should be taken to separate the calibration signal from the transducer output.

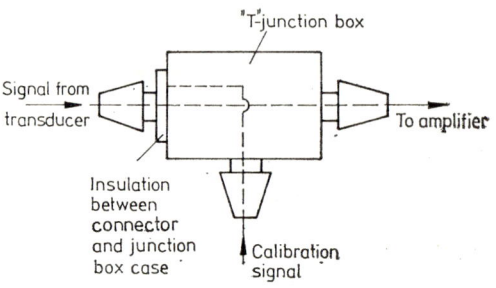

Figure 8.15

Several techniques can be used to realize the insertion [8]:

a) In emergency cases, the shield of the coaxial cable is broken and the voltage source is connected to the two sections of the shield. A portion of the signal cable between transducer and preamplifier may be left unshielded, being a possible source of noise pick-up;

b) Connecting the voltage source through a "T" junction incorporated between transducer and amplifier, and fitted with connectors to match the adjacent cable connectors (Fig. 8.15). With this configuration, the ground circuit is open within the junction box and must be closed externally (hence measurements cannot be made when the calibration input connector is left open). A special plug can be used to short the calibration signal during actual measurement. The "T" junction should be insulated to prevent creation of a ground loop.

c) Use of a "T" junction similar to that presented above but which contains an internal resistor to close the ground circuit and permits measurements with external calibration circuitry connected or disconnected.

d) Introduction of a calibration resistor in the amplifier or even within the transducer.

The equivalent measuring circuit for a piezoelectric transducer with a voltage amplifier is shown in Figure 8.16, where C_{cal} is the shunt capacitance of the calibration voltage source and R_{cal} is its internal resistance (or an actual series resistance connected across its output terminals).

8.2 Simulation calibration

When a voltage amplifier is used, in order that R_{cal} not to affect the circuit performance during measurement, it is necessary that

$$R_{cal} \ll \frac{C_T + C_1 + C_2}{\omega C_2 (C_T + C_1)}$$

at the highest frequency of interest.

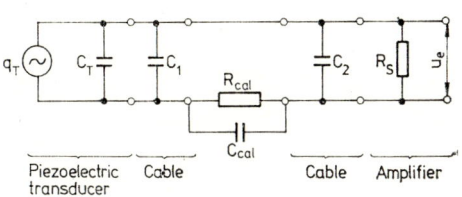

Figure 8.16

When a charge amplifier is used, the above requirement reduces to

$$R_{cal} \ll \frac{1}{\omega(C_T + C_1)}.$$

During calibration insertion, the piezoelectric transducer acts as a passive capacitor and the equivalent circuit reduces to the form shown in Figure 8.17. Therefore, in order to simulate the transducer properly, it is necessary that

$$u_{cal} = \frac{q_T}{C_T + C_1}$$

and the calibration voltage sensitivity should be

$$S_{cal} = \frac{u_{cal}}{\ddot{x}} = \frac{S_q}{C_T + C_1}$$

where S_q is the transducer charge sensitivity.

The insert calibration checks the electrical continuity of the transducer and assesses the effects of transducer cabling.

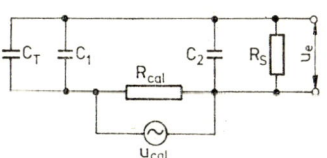

Figure 8.17

8.2.3 Shunt-resistor calibration

With variable-resistance transducers, simulation calibration is performed connecting a shunt resistance R_S across the transducer R_T when no mechanical excitation is applied.

285

8 Calibration

Let R_T be the "active" transducer in a balanced Wheatstone bridge ($u_0 = 0$) having the switch K open (Fig. 8.18). When the switch is closed, the resistance of the arm AB varies and the bridge is unbalanced. The voltmeter indicates u_0 and the resistance change ΔR, which has produced this voltage, is

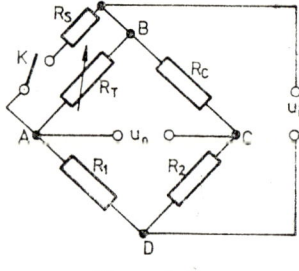

Figure 8.18

$$\Delta R = R_T - \frac{R_T R_S}{R_T + R_S} = \frac{R_T^2}{R_T + R_S}.$$

The bridge sensitivity is

$$S = \frac{u_0}{\Delta R} \quad \left[\frac{V}{\Omega}\right]$$

Other calibration methods, used especially by Hottinger & Baldwin, are described in reference [9]. A critical survey is given in reference [10].

References for Chapter 8

1. Bouche, R. R., *Calibration of Shock and Vibration Measuring Transducers*, SVM-11, Shock and Vibration Information Center, 1979.
2. Bouche, R. R., Instruments and methods for measuring mechanical impedance, Endevco Technical Paper *203* (1961).
3. Trent, H. M., The absolute calibration of electromechanical pickups, *Journal of Applied Mechanics* (Trans. A.S.M.E.), **15**, 49–52 (1948).
4. Levy, S. and Bouche, R. R., Calibration of vibration pickups by the reciprocity method, *Journal of Research National Bureau of Standards*, **57**, 4, 227–243 (Oct. 1956).
5. Bouche, R. R. and Ensor, L. C., Accelerometer calibration with reciprocity vibration standards, Endevco Technical Paper *251* (1970).
6. Hohmann, P. and Martin, R., Kalibrierung eines Schwingungsaufnehmer-Vergleichsnormals, *V.D.I.-Berichte*, *135*, 61–65 (1969).
7. Bouche, R. R., Vibration standards for performing comparison calibrations, Endevco Technical Paper *241* (1967).
8. Rhodes, J. E., Piezoelectric transducer calibration simulation method using series voltage insertion, Endevco Technical Paper *216* (1962).
9. * * * Empfindlichkeit, Massangaben, Kalibrierung, Messbereiche; Betrachtungen über Aufnehmer und Messgeräte, *Hottinger-Baldwin Messtechnische Briefe*, **1**, 1–7 (1969).
10. Licht, T. R. and Zaveri, K., Calibration and Standards. Vibration and Shock Measurements, *Brüel & Kjaer Technical Review*, *4*, 16–27 (1981).

9

Examples of vibration measurements

A detailed treatment of the vibration measurement techniques, applied in various fields of engineering practice, is beyond the frame of this work. Therefore, in the following, only some of the most widely used modern measurement methods will be briefly presented, as well as some techniques used by the authors at the Strength of Materials Laboratory of the Polytechnic Institute of Bucharest, hence connected to their field of interest.

9.1 Identification of vibration sources

During operation, any machine gives rise to characteristic noise and vibrations. The frequency spectrum of the complex signals generated by machines is characteristic of each machine, constituting a unique pattern, referred to as the "machine signature". Analysis of machine mechanical signature facilitates location of vibration sources and monitoring of their evolution in time permits evaluation of their mechanical condition.

The predominant frequencies in the signals produced by machines or their component parts have been studied both analytically and experimentally [1 – 8]. Papers have been published on rotating machinery such as turbines [4], electrical machines [8], pumps [9], fans and blowers [10], reciprocating engines and vehicles [11], as well as on their components — ball bearings [3, 12 – 14], sliding bearings [12], gears [15 – 17].

In order to identify the sources of noise and vibration, the peaks of the measured frequency spectra are correlated with the data pertaining to the possible vibration source components in the machine. This requires knowledge of the operating and geometrical parameters such as rotational speed, number of gear teeth, number of rolling elements of a bearing, number of propeller blades or impeller vanes, number of rotor slots in electric motors

9 Examples of vibration measurement

etc. There are also taken into account the frequency components generated by the unbalance or eccentricity of rotating shafts, by the unbalanced forces and couples produced by reciprocating machines, by the shaft misalignment or journal movement in oil lubricated bearings, as well as by the sidebands generated by gears due to imperfect tooth-meshing [18, 19].

Hence the first step in identifying the source of vibrations is to calculate the frequencies at which peaks may be expected to occur in the measured spectrum during normal operation.

Among the vibration sources of relatively low frequency, which can exist in a machine or product having the rotational speed n, rpm, the following may be mentioned:

Rotary unbalance gives rise to vibrations at the frequency $f_1 = \dfrac{n}{60}$, Hz, having the amplitude proportional to unbalance and largest especially in radial direction.

Misalignment of couplings or bearings as well as the existence of a bent shaft produce vibrations of frequencies f_1 and $2f_1$, with large amplitudes in axial direction at $2f_1$, up to 50% of the radial vibration or more.

Eccentric journals produce vibrations of frequency f_1 and of relatively small amplitudes. Sometimes gears give rise to the same effect. In this case, the largest vibration occurs in line with the gear centres. If the journals or gears belong to a driving motor, vibrations disappear when power is turned off.

Mechanical looseness produce vibrations of frequency $2f_1$.

Faulty drive belts produce vibrations at frequencies $f_1, 2f_1, 3f_1, 4f_1$. The bad belt can be discovered by freezing the image using strobe light.

Electrical motors produce vibrations at the frequency f_1 or at the synchronous frequency (or its double), which suddenly disappear when power is turned off.

Reciprocating machines give rise to vibrations at the frequency f_1 and higher orders of this, which can be reduced by design changes or anti-vibration isolation.

Instability of the journal motion in an oil lubricated bearing produces a subharmonic vibration of frequency $0.5 f_1$ (usually somewhat lower, i.e. $0.42 - 0.48 f_1$) called "oil whip".

Among the vibration sources of relatively high frequency, the following are worth mentioning:

Aerodynamic (or hydrodynamic) forces produce vibrations at the blade-passing frequency, equal to f_1 times the number of blades or vanes. Generally, these are dangerous when causing casing resonance or high level vibrations at some resonant point in the structure.

9.1 Identification of vibration sources

Similar vibrations can be induced by the propeller in the hull or in the rudder of a ship. Vibrations produced by the turbulence of expelled gases of a jet engine on the fuselage have a rather random character, the identification of discrete frequencies being difficult.

Rolling element bearings make noise due to rough spots or identations, irregularities of rings or rolling elements.

Consider a ball bearing having the outer race fixed and the inner race rotating at the speed n, rpm. Denoting by d — diameter of the inner race, D — diameter of the outer race, d_B — diameter of the rolling element, N — number of balls, β — contact angle of the balls to the raceway, $d_m = \dfrac{D+d}{2}$ the pitch diameter, the following discrete frequencies can be calculated:

— the fundamental rotational frequency due to the eccentricity of the inner race

$$f_1 = \frac{n}{60} \quad [\text{Hz}];$$

— the spin frequency of the rotating ball relative to the stationary outer race

$$f_B = \frac{1}{2} f_1 \frac{d_m}{d_B} \left[1 - \left(\frac{d_B}{d_m}\right)^2 \cos^2 2\beta \right]; \qquad (9.1)$$

— the cage frequency relative to the stationary outer race

$$f_c = \frac{1}{2} f_1 \left(1 - \frac{d_B}{d_m} \cos 2\beta \right); \qquad (9.2)$$

— the frequency due to the relative rotation of the cage and the rotating ring

$$f'_c = f_1 - f_c. \qquad (9.3)$$

Using these values one may derive:

— the frequencies due to a rough spot (or identation) of an element which contacts the inner and outer races alternately, once per revolution

$$f_b = 2k f_B; \quad k = 1, 2, 3, \ldots \qquad (9.4)$$

— the frequencies due to an irregularity on the rotating inner raceway

$$f_{\text{int}} = kN f'_c; \quad k = 1, 2, 3, \ldots \qquad (9.5)$$

9 Examples of vibration measurement

— the frequencies due to an irregularity on the stationary raceway

$$f'_{ext} = kNf_c; \qquad k = 1, 2, 3, \ldots \tag{9.6}$$

— the frequency due to ball non-sphericity

$$f'_B = f_c;$$

— the frequencies due to the waviness of the inner raceway

$f'_{int} = kNf'_c \pm f_1$ (according to the linear theory);

$f'_{int} = kf'_c \pm f_c$ (according to the nonlinear theory);

— the frequencies due to the waviness of the outer raceway

$f'_{ext} = 2kf_B \pm f_c$ (according to the linear theory);

$f'_{ext} = (k \pm 1)f_c$ (according to the nonlinear theory).

Gears produce vibrations and noise due to imperfect engagement, manufacturing errors and hereditary errors due to incorrectly machined cutting tools.

If the input shaft, rotating at the speed n, has a pinion with z_1 teeth, and the driven shaft has a gear with z_2 teeth, then the input shaft frequency is $f_1 = \dfrac{n}{60}$ [Hz].

The tooth-meshing frequency is

$$f_a = f_1 z_1. \tag{9.7}$$

The driven shaft has a rotational frequency

$$f_2 = \frac{f_a}{z_2}. \tag{9.8}$$

Besides frequencies f_1, f_2, f_a, the frequency spectrum of a gear shows discrete frequencies which represent sums, differences and products of the fundamental frequencies. For example, due to the runout or eccentricity of the gear shaft, the pitch circles do not mesh correctly and will have an error that varies sinusoidally with the runout. A low frequency (f_1) vibration is superimposed on the relatively high frequency (f_a) vibration. The result is an amplitude-modulation process. For the limit case of 100% modulation, the fundamental component f_1 disappears and two sidebands are generated

$$\begin{aligned} f_{s_1} &= f_a + f_1, \\ f_{i_1} &= f_a - f_1. \end{aligned} \tag{9.9}$$

9.1 Identification of vibration sources

In reality, the modulation process is neither simple nor 100% so that supplementary sidebands are generated

$$f_{s_2} = f_a + 2f_1, \quad f_{s_3} = f_a + 3f_1,$$
$$f_{i_2} = f_a - 2f_1, \quad f_{i_3} = f_a - 3f_1, \ldots \quad (9.10)$$

A simplified example of calculation is taken from [7]. Consider a simple mechanical system consisting of a pair of spur gears and a bearing assembly, as shown in Figure 9.1. The input shaft speed is $n_1 = 1000$ rpm. The pinion has $z_1 = 20$ teeth and the gear has $z_2 = 79$ teeth. A radial-thrust ball bearing type 46305 GOST 831—54 is considered, having the following characteristics: $d = 25$ mm, $D = 62$ mm, $d_B = 11.51$ mm, $N = 10$, $\beta = 26°$.

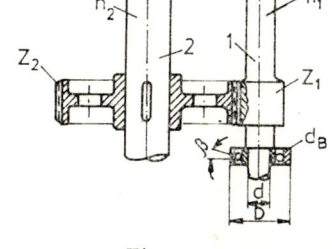

Figure 9.1

Using equations (9.1)—(9.10), the possible frequencies of noise and vibration components are calculated and listed in ascending order as in Table 9.1. Dividing all values by the fundamental frequency f_1, corresponding to the main shaft speed, the "order number" of the various frequency components is calculated.

TABLE 9.1 *Frequency Components of the Vibrations Generated During the Operation of the System from Figure 9.1*

Component	Frequency, Hz	Order number	Component	Frequency, Hz	Order number
f_2	4.22	0.25	f_{i3}	283.3	17
f_c	6.99	0.42	f_{i2}	300.0	18
f_1	16.7	1	f_{i1}	316.7	19
f_B	30.72	1.84	f_a	333.3	20
f_b	61.44	3.68	f_{s1}	350	21
f_{ex}	69.9	4.18	f_{s2}	366.7	22
f_{in}	97.1	5.81	f_{s3}	383.4	23
f_{i4}	266.6	16.0	f_{s4}	400	24

Using such a table, the probable "distance" between the discrete frequency components can be established and the bandwidth of the filter is selected, which at high frequencies has to be at least $\frac{1}{2} f_1$. In order to separate so closely spaced spectral components, narrow band filters should be used, usually constant

9 Examples of vibration measurement

bandwidth filters, as those employed in the modern dynamic analyzers [2].

Figure 9.2 [6] shows the frequency spectrum obtained from a casing-mounted accelerometer on a high speed process air

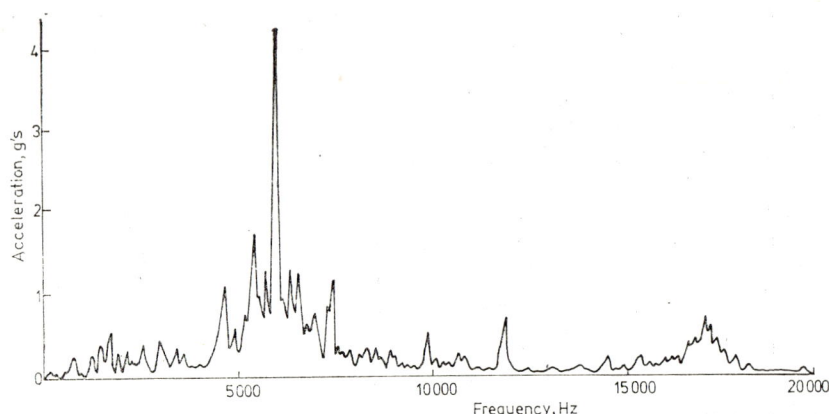

Figure 9.2

compressor, using a real-time spectrum analyzer. On Figure 9.3, which is a magnification of the low-frequency part of the spectrum shown in Figure 9.2, the components corresponding to the operating speeds of the four stages and the input shaft are easy to be identified.

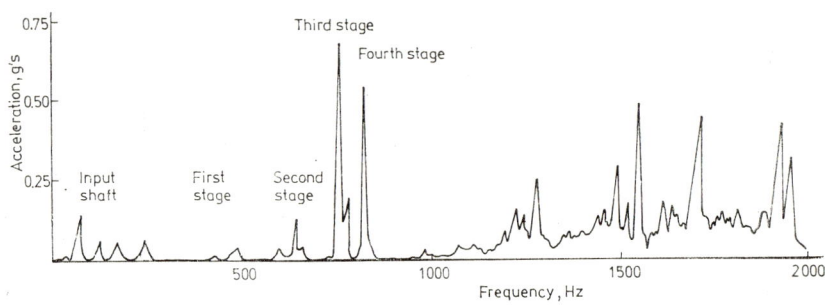

Figure 9.3

A characteristic "signature" for the shaft misalignment in a turbo-compressor is presented in Figure 9.4 [6]. A casing-mounted velocity pickup has been used. The high twice per revolution radial

9.1 Identification of vibration sources

vibration with accompanying characteristic axial vibrations may be noticed. For comparison, the shaft orbits which existed at the same time are displayed, on which the synchronous phase marks are shifted 180° apart.

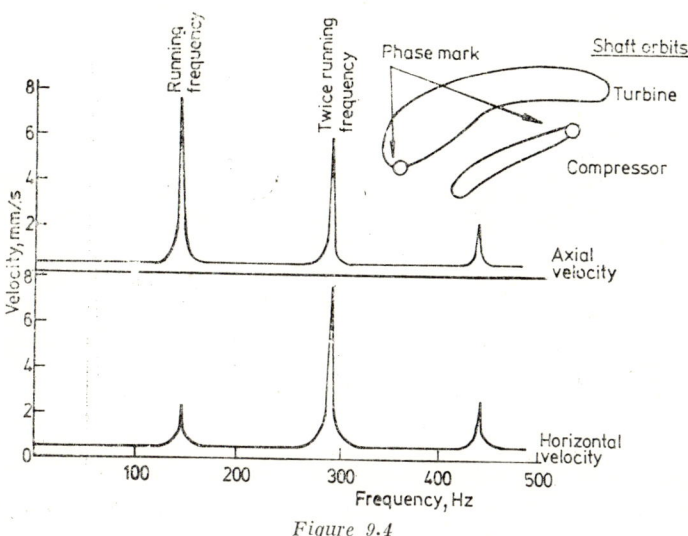

Figure 9.4

Figures 9.5 and 9.6 [6] demonstrate the importance of a suitable selection of pickups used in such measurements. Both figures show spectra taken from identical speed increasing gear-

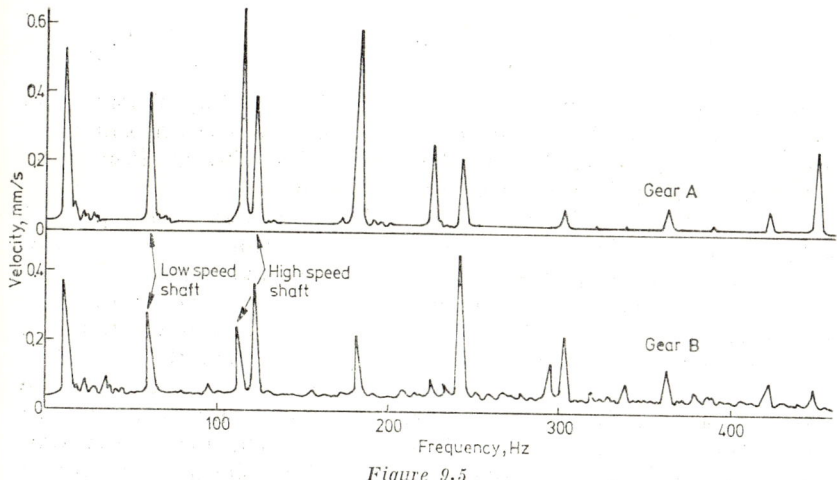

Figure 9.5

9 Examples of vibration measurement

boxes, driving large centrifugal pumps [6]. The low frequency velocity "signatures" from Figure 9.5 are virtually identical, which would lead to the conclusion that both gears are in approximately the same mechanical condition. On the contrary, the high

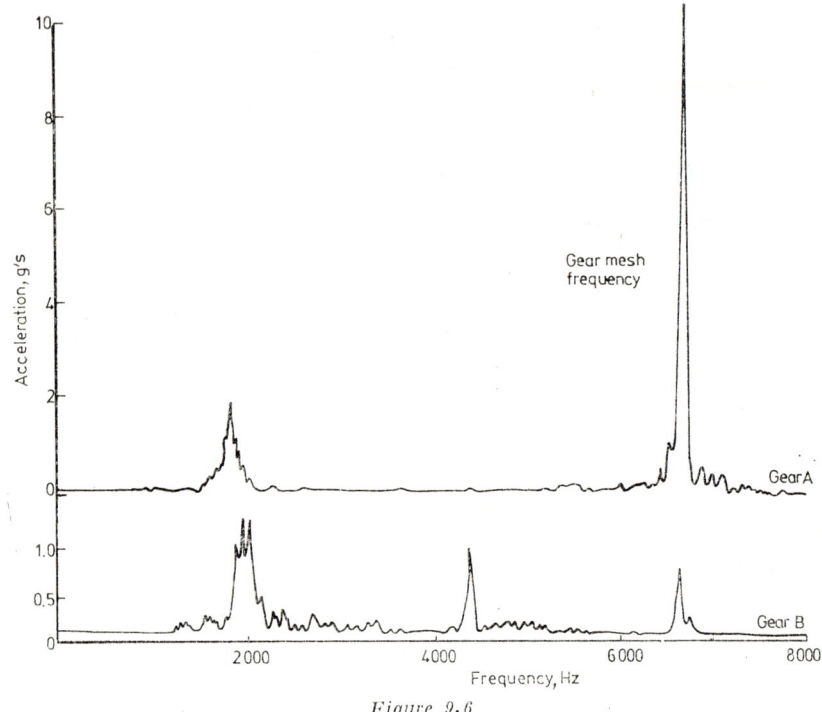

Figure 9.6

frequency acceleration signature from Figure 9.6, obtained at the same time as the velocity signatures, indicate an abnormal gear mesh for gear A. This has been proved at checks after measurements, the gear having a chipped tooth.

Another result which underlines the usefulness of the spectral analysis in identifying high frequency vibration sources is presented in Figure 9.7, reproduced after reference [6].

Increasing shaft vibration on a large axial flow compressor with variable inlet and stator geometry prompted measurements to trace the cause. Use of a displacement pickup and of an oscilloscope revealed an once per revolution unbalance condition, the total signal amplitude being considered acceptable according to the vibration severity charts. On the other hand, use of accelerometers and a frequency analysis resulted in the plot shown in

9.1 Identification of vibration sources

Figure 9.7, where strong frequency components are noted at 45, 47, 94 and 141 times the running speed. Comparison of measurement data with the number of blades have led to the conclusion that the vibration was caused by one or more stator blades which

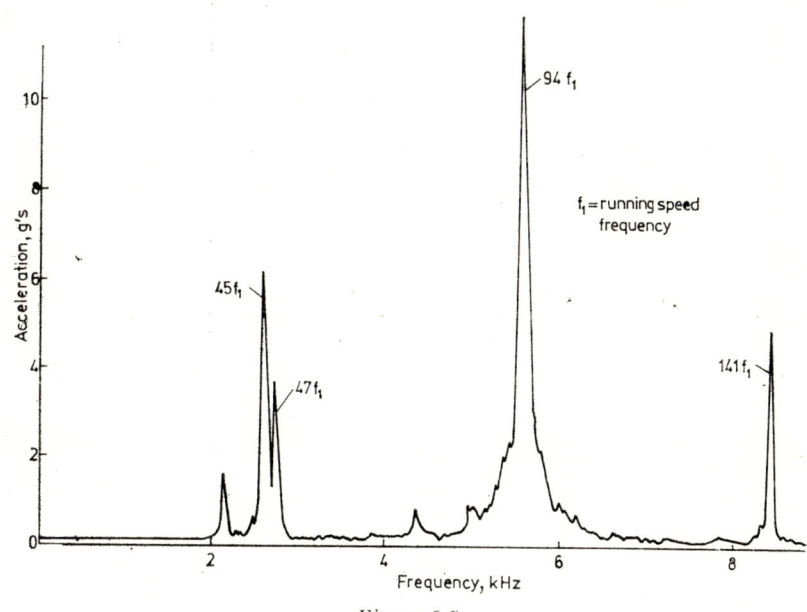

Figure 9.7

had turned across the airflow causing severe cyclic stresses. After the machine was shut down and opened, the diagnosis was confirmed.

Application of the signature analysis methods permits to turn from the simple identification of a malfunction to its prevention.

Monitoring the time changes of frequency spectra, the growth of the amplitude of certain components can be established. Wear, cracks, looseness or incipient local modifications can be detected by correlating their frequencies to the machine operating parameters. A *preventive* maintenance is thus achieved, avoiding sudden failures or emergency shutdowns, extending the time between machinery overhauls and minimizing scheduled maintenance periods.

Figure 9.8 [3] illustrates the mechanical signature of a ball bearing without defects. The mechanical signature of a similar ball bearing showing a ball defect is presented in Figure 9.9. Comparison of the two signatures is an extremely useful tool in

9 Examples of vibration measurement

the production quality control, enabling removal of ball bearings with defects to be made. Introducing a minicomputer into the

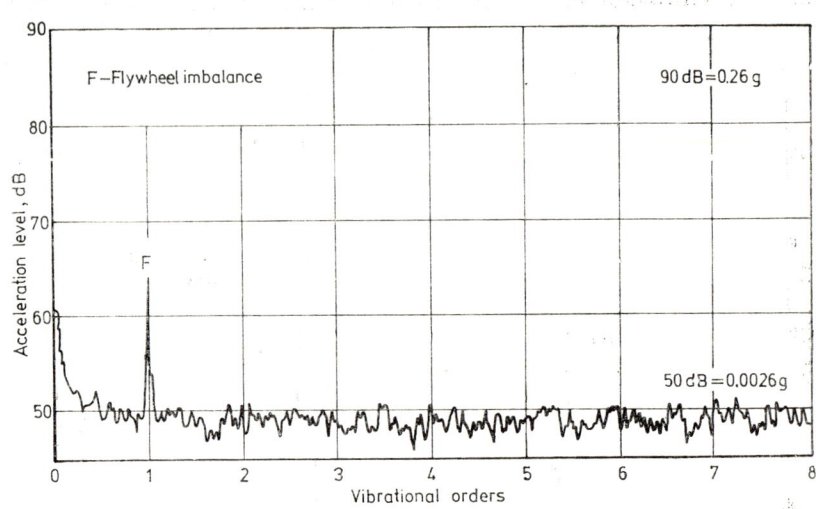

Figure 9.8

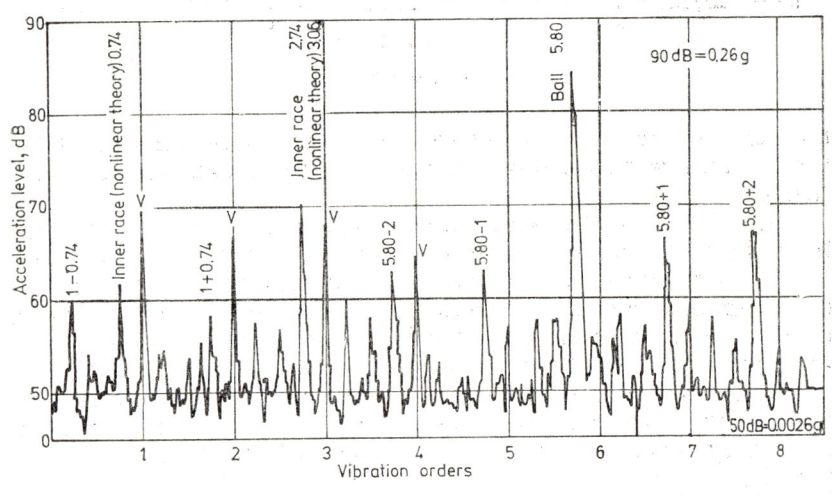

Figure 9.9

instrumentation set-up, the rate and quality of the testing process can be increased.

9.2 Measurements on prototypes

9.2.1 Machine tools

Machine-tool self-excited chatter is the result of dynamic weaknesses in the machine coupled to the cutting process. Possible sources of chatter can be weak structural components (beds, columns, slides, housings, frames, arms), torsional flexibilities in the drive train, and bending weaknesses in drive spindles, bearings, tool holders, jigs or even in the machined workpiece.

The performance of machine tools from the chatter point of view can be improved by increase of the equivalent stiffness of the vibration mode which may generate instability under chatter conditions. This is done after an investigation on test or prototype machines for locating and identifying the weak members causing the machine to chatter. Such an investigation involves a three-phase procedure [21]:

a) First, the machine tool is caused to chatter and a frequency analysis is made while cutting. The chatter should have a relatively constant intensity and should be held no more than 15 seconds (not to damage the machine or to shorten its life). The output signal from a transducer installed near the cutter is either recorded on magnetic tape (for subsequent analysis) or passed through a heterodyne analyzer which is swept through the frequency range of interest (5 — 2000 Hz). The resulting frequency spectrum (Fig. 9.10) exhibits strong peaks at the chatter frequencies.

Measurements are taken for various cutting regimes and positions of machine components (tables, slides, arms), with workpieces from different materials and changing the orientation and geometry of the cutting tool.

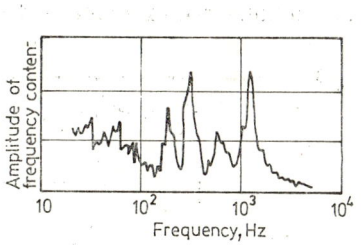

Figure 9.10

b) The second phase of chatter analysis involves frequency response tests performed either under actual machining conditions [22] or with artificial excitation (from an electrodynamic or electrohydraulic vibrator) applied to the machine shut down or with the workpiece (or tool) rotating.

Machine tools with relatively high stiffness are mounted on soft resilient supports, in order to lower as much as possible the "rigid body" resonance frequencies (of the machine on supporting mounts) and to separate them from the structural resonances.

9 Examples of vibration measurement

Simultaneous multi-point excitation is not recommended, owing to the relatively high dynamic stiffness and low accessibility of the important points from the excitation point of view. Accurate isolation of a natural mode of vibration is extremely difficult and the cost of the equipment necessary for multi-point excitation is prohibitive.

Single-point "absolute excitation" of either the workpiece or the cutting tool [23] is employed, using an electrodynamic vibrator suspended by resilient chords from an external frame (or a crane) and mounted successively in the important coordinate directions — vertical, horizontal longitudinal, horizontal transverse, torsional, etc. Orientation in the direction of the resultant cutting force is preferred. Either the relative response between workpiece and cutting tool, or the absolute response of some points on the structure are determined. The type of machine under test determines the points at which these measurements are taken and for which frequency-response plots are traced.

"Relative excitation" between workpiece and tool is recommended when possible. Non-contacting electromagnetic exciters are used in this case, acting on an armature attached to the spindle (or the chuck) which can rotate during the test.

Both rectilinear and torsional vibrations are measured, plotting the frequency response diagrams either as receptance-versus-frequency and phase-versus-frequency plots (Fig. 9.11) or as polar plots.

When harmonic excitation is used, the response is measured at various frequencies — to determine the resonances, and at several levels of the excitation force — to assess the effect of non-linearities. The experimental set-up must ensure a minimum influence of the vibrator and transducers on the dynamic response of the machine tool.

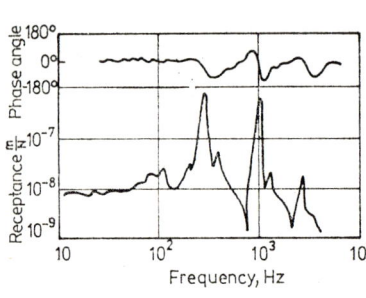

Figure 9.11

Comparing the graphs from Figure 9.10 and 9.11, the frequencies are determined where peaks of the receptance curve, denoting low dynamic stiffness, correspond to peaks of the frequency spectrum of the machine chatter during cutting tests. These are natural frequencies of the modes of vibration in which components of the machine determine low values of the dynamic stiffness.

9.2 Measurement on prototypes

One of these frequencies can also be found on the polar plot of the frequency response, on the loop which corresponds to the mode of vibration producing the maximum value of the negative real (in-phase) component of the receptance. Figure 9.12 shows such a diagram obtained for a milling machine [24] using relative vertical excitation between spindle and table.

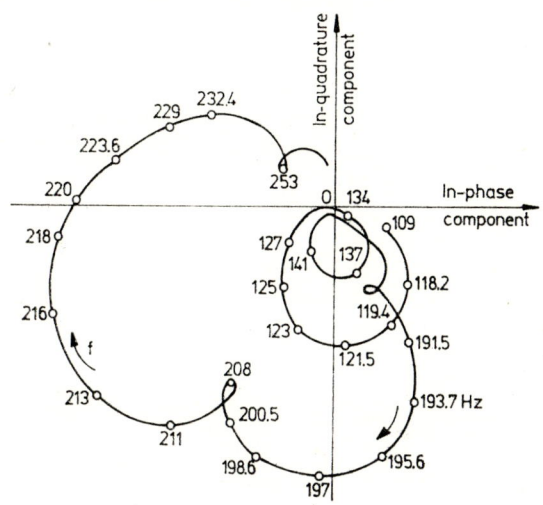

Figure 9.12

c) The third phase of the dynamic analysis involves identification of the weak element causing the chatter. This is done by exciting the structure at each of the resonances found out in the second phase and by determining the mode shape with a rowing transducer.

A vibration pickup is moved from point to point along the main structural elements of the machine and the motion is recorded. Another pickup, held at a fixed point on the machine, provides a signal used as a phase reference (to determine whether the machine motion at various points is in phase or out-of-phase with the force). Connecting the points of equal vibration amplitude, the so-called "modal maps" are obtained. These are particularly useful in the case of welded structures (beds, columns, over-arms) to indicate where stiffening is needed or where dynamic absorbers can be installed to break up resonances.

When the components of the tested structure are predominantly one-dimensional (columns, arms, quills, spindles), a mode shape plot is sketched. By studying the mode plot, the weak elements can be located where sudden slope changes or discontinuities occur.

9 Examples of vibration measurement

Figure 9.13 illustrates the deflected shape (in the vertical plane) at one resonance of the milling machine whose polar plot is given in Figure 9.12. It may be seen that the weak component is at the interface between the table and the vertical column, where the angle differs from 90 degrees. It has been concluded that the mounting screws of the table had a too low stiffness, which leads to the generation of the "unstable" loop with large negative in-phase response on the polar plot from Figure 9.12. Change of the table mounting system eliminated the chatter condition.

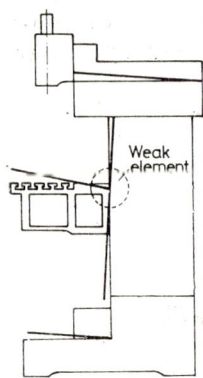

Figure 9.13

9.2.2 Rolling hoisting cranes

When a special type of hoisting crane was put into operation, strain gauge and vibration measurements have been required for their acceptance [25].

Both the crane tracks and the trolley tracks have been manufactured from expanded joists (Fig. 9.14 a). Strain gauges

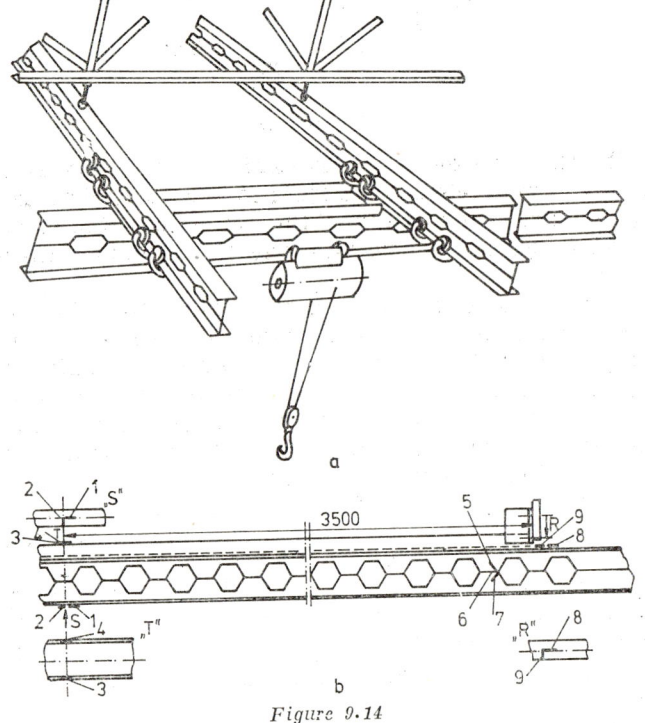

Figure 9.14

9.2 Measurement on prototypes

(numbered from 1 to 9) have been bonded in the middle of the span, on the flanges of the I-section and in some other points, both on web and flanges (Fig. 9.14 b). Stresses occurring during various handlings — load moving upward, horizontally, downward and cable snubbing — have been measured by means of these strain gauges.

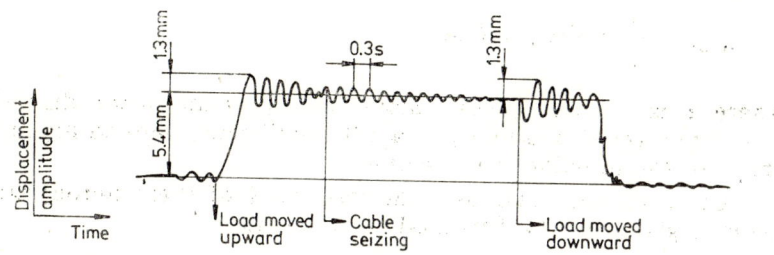

Figure 9.15

The vertical deflections at the measurement points have also been recorded using a Stoppani-type quasi-static vibrograph (see Section 5.2). A typical time record of the crane vertical deflection is presented in Figure 9.15. The load was first lifted and stopped at a certain distance from the soil. The cable was then released until the load fell on the soil. The static and dynamic deflections, the fundamental natural frequency of the crane vertical vibrations and the logarithmic decrement of damping can be measured using such a record.

Figure 9.16 illustrates the similar record of stresses. With the load up, the cable was released, letting the load fall, and then snubbed, stopping the load abruptly at certain distance from the soil. The static stress σ_{st} is marked at the level recorded after the vibrations died out. The maximum dynamic stress σ_d corresponds to the maximum overshot. The ratio of the corresponding segments on the diagram

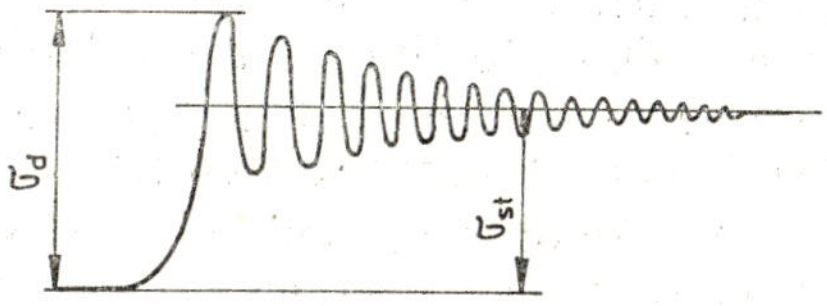

$$\psi = \frac{\sigma_d}{\sigma_{st}}$$

Figure 9.16

is called *dynamic factor* and is usually determined for different manoeuvres. Overloadings of 14.8 percent have been recorded when moving the load upward and of 25.6 percent when moving it downward and suddenly seizing the cable.

9 Examples of vibration measurement

9.2.3 Suspended pipelines

The sections of water or methane gas pipelines, which are passing over rivers, are subject to aerodynamic loads $q(x, t)$ produced by the vortex-shedding phenomenon induced by the wind. The amplitude of these distributed loads is

$$q_{max} = \frac{\rho}{2} D V^2 C_k \quad [\text{N/m}]$$

where ρ is the air specific mass, D — pipeline outer diameter, V — wind velocity and C_k — a lift coefficient, depending on the shape of the pipeline cross-section.

For a circular cylinder, the frequency of these aerodynamic loads is given by the Strouhal relationship

$$f \cong 0.22 \frac{V}{D} \quad [\text{Hz}].$$

After the pipeline structure errection, it is necessary to check its mechanical strength to the wind-induced loads [26]. It is also required to measure the natural frequencies and normal modes of vibration of the pipeline, as well as the corresponding structural damping factors [27] which could be useful design data for other similar pipelines.

First, the free vibrations of the structure are excited using the snapback testing method. The fundamental natural frequency and the logarithmic decrement of damping are thus determined. A static force is applied to this purpose, by means of a notched bar connected to a metallic collar fixed around the pipeline. Increasing gradually the applied force, the pipeline is statically bent until the intermediary bar fails at notch at a predetermined force. The pipeline starts to vibrate freely and after a certain number of cycles its frequency practically equals the system fundamental natural frequency.

Study of the pipeline forced vibrations includes determination of the resonance frequencies and the deflected shapes at resonance, considered to correspond to the shape of normal modes. It also includes plotting of the amplitude-versus-frequency curves near resonance frequencies, in order to calculate the modal damping ratios.

The block diagram of the instrumentation used during the measurements carried out by the authors is presented in Figure 9.17. Excitation of the pipeline was also done using the mechanical vibrator type V1 (see Table 6.1).

9.2 Measurement on prototypes

Selection of the excitation and measurement equipment should be made having in view that suspended pipelines have low (from 1Hz to 30 Hz) and closely spaced natural frequencies. Typical damping ratio values measured on this type of structures re $\zeta = 0.003 - 0.015$.

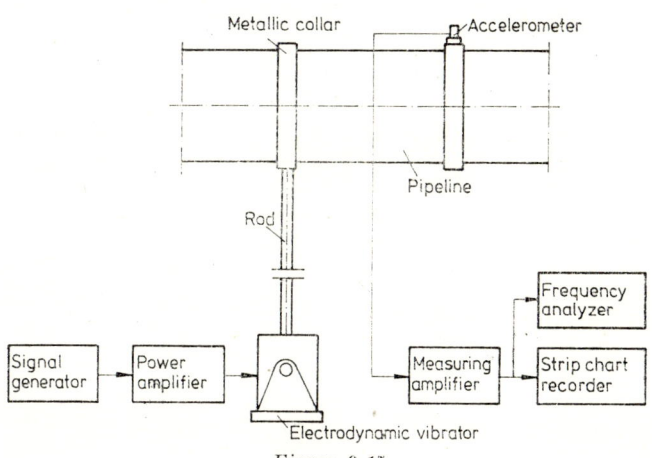

Figure 9.17

Based on the experimental data, one may estimate the amplitude of variable stresses which might occur in the pipeline wall at resonance, when the vortex-shedding frequency equals one of the natural frequencies of the elastic structure.

To this purpose, the most unfavourable case is considered, when the distribution of the aerodynamic loads along the pipeline is identical to the shape of the resonant mode. When the calculated stresses exceed the safe limits, changes in design are recommended, usually by suspending the pipeline on cables, which increases the damping in the system.

9.2.4 Rotating machinery

In the preceding sections, the transducer selection for measurements on rotating machinery has been discussed. A few more problems connected to the analysis of machinery vibrations are examined in the following.

Modern dynamic analyzers (see Section 5.4.3.4), consisting of a narrowband tracking filter based on the heterodyne principle, make it possible to perform a detailed frequency analysis of the complex vibration signals generated by operating machines.

9 Examples of vibration measurement

Adding a *tuner*, a *tracking frequency multiplier* and a *sweep oscillator* (Fig. 9.18), the instrumentation system acts as an *automatic sweep analyzer* capable to produce spectrum plots identical to those generated by real time spectrum analyzers but with lower cost and at a much slower rate. It is well suited for small facilities

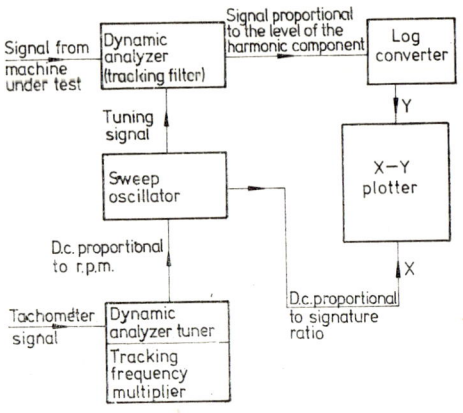

Figure 9.18

or in applications where the frequency band of interest is not very broad.

If the machine runs at constant speed, then the tuner sweeps the frequency of a sinusoidal tuning signal which controls the dynamic analyzer (like Spectral Dynamics SD 101 A). The amplitude-versus-frequency spectrum of the signal generated by a pickup mounted on the machine under test is plotted on the $X-Y$ recorder. The dynamic analyzer operates as a *slave filter*.

If the rotational speed of the machine fluctuates (or is varied from idle to full speed), the tuner converts the signal derived from a tachometer (or a non-contacting pickup) into a sinusoidal signal used to tune the dynamic analyzer which operates as a *tracking filter*.

The *tracking frequency multiplier* makes it possible to multiply the frequency of the reference phenomenon (main shaft speed or firing frequency) by any given ratio. The multiplied frequency tunes the tracking filter to any specific harmonic component of the complex vibration signal. In this way, a selected harmonic order can be plotted as a function of the machine speed. This technique, called *r.p.m. tracking*, can be used to determine *critical speeds*, i.e. speeds at which a specific component reaches its highest vibration level.

9.2 Measurement on prototypes

Figure 9.19 illustrates an r.p.m. tracking profile of the second order component of a vibration signal caused by misalignment [20]. It has been plotted while the main shaft r.p.m. varied from startup to full operating speed. The signature recorded at a given r.p.m. is traced on the same figure.

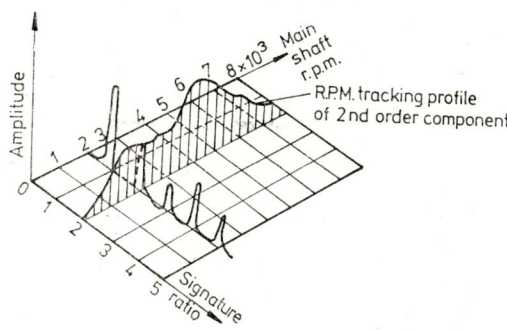

Figure 9.19

As the machine speed varies, so does the machine vibration signature displayed as a frequency spectrum, with frequency across the horizontal axis and amplitude on the vertical axis. Because this variation is determined by both changes in forcing (rotational effects) and excitation of structural resonances, it makes difficult the identification of noise and vibration sources.

A technique called Signature Ratio Analysis was introduced by Spectral Dynamics Corp. in 1966 for the automatic speed normalization of a spectrum display. The horizontal axis is normalized to running speed and presented in multiples (orders) of the main shaft speed rather than frequency. This type of presentation has advantages for variable speed machinery in that harmonics always appear in the same position on the spectrum plot and are thus easier to identify and compare.

The order number equals the ratio between the frequency of a harmonic component of the complex signal and the fundamental frequency corresponding to the main shaft running speed. Order normalized plots are obtained with the aid of the sweep oscillator (Fig. 9.18). The tuner senses the speed variations and develops a correction signal to the sweep oscillator to compensate for those deviations.

With modern real time spectrum analyzers, this is obtained by varying the analyzer sampling rate and the analysis filter bandwidth in accordance with the tachometer signal input. Successively recorded spectra can be thus averaged over a time

9 Examples of vibration measurement

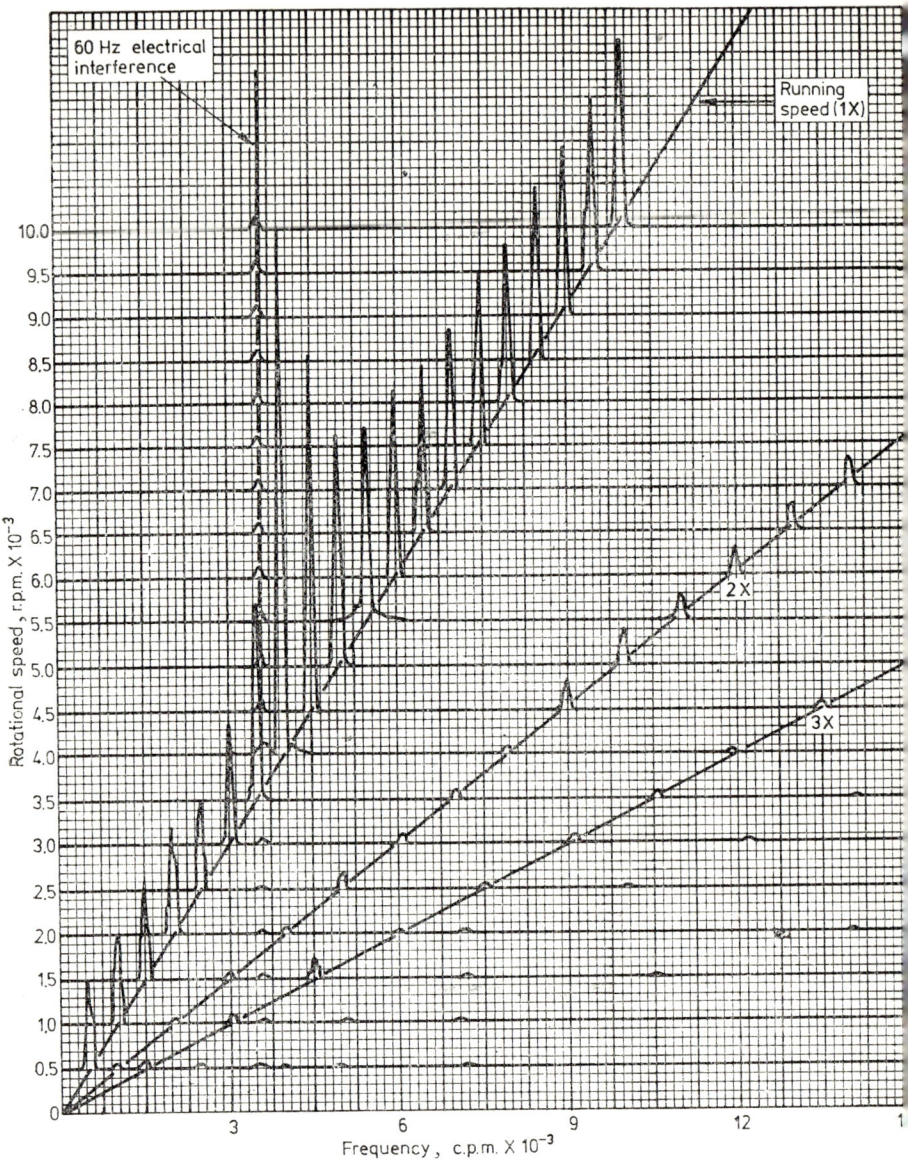

Figure 9.20

9.2 Measurement on prototypes

interval, while the speed is fluctuating, which is necessary for smoothing recorded spectra.

A series of spectrum plots can be combined in a so-called "waterfall" (cascade) presentation, where one appears slightly displaced from the other, as a function of either speed or time (Fig. 9.20). Special instruments have been developed (especially for use with real time analyzers) to form three-dimensional presentations of spectra taken at selected time intervals, including amplitude, frequency and time. Such a display shows changes in vibration frequencies and amplitudes, and very short time increments can be chosen to see the effects of non-stationary phenomena (like subharmonic instabilities). Speed related phenomena will track on a diagonal whereas resonance related responses appear on a vertical line.

Alternatively, *Campbell diagrams* (speed maps) are presented as r.p.m. versus frequency displays. The magnitude of response characteristics is indicated by circles, with the circle calibration factor also given on the display (Fig. 9.21). Structural resonances often appear as being excited at approximately the same frequency over a wide range of speed change [20].

Apart from the foregoing, it is often required to measure the torque on a shaft either as a function of frequency or at a fixed running speed. A widely used technique is based on a set-up involving either strain gauges and slip rings or a contactless signal transmission system (like Philips PR9910 — PR9921). Another technique uses two non-contacting transducers (or two photoelectric tachometer pickups) placed near the shaft at two different sections (Fig. 9.22). The phase angle is measured between two sinusoidal signals in phase with the pulses generated by transducers and a display of phase-versus-frequency is obtained on the $X-Y$ plotter. For small phase angles φ, the torque M_t on the shaft is then calculated from the following equation

$$M_t = \frac{\varphi}{l} G I_p,$$

where l is the distance between measurement locations, G — the shear modulus and $I_p = \dfrac{\pi d^4}{32}$ in which d is the shaft diameter.

During developmental studies, very useful data are obtained by testing machine components, for example — the rotor assembly or a bladed disc, using artificial excitation produced with the aid of an electrodynamic vibrator.

A systematic study of the natural modes of vibration of a bladed disc can be done by the analysis of vibration *nodal patterns* (Chladni figures). These can be obtained either by sprinkling fine

Examples of vibration measurement

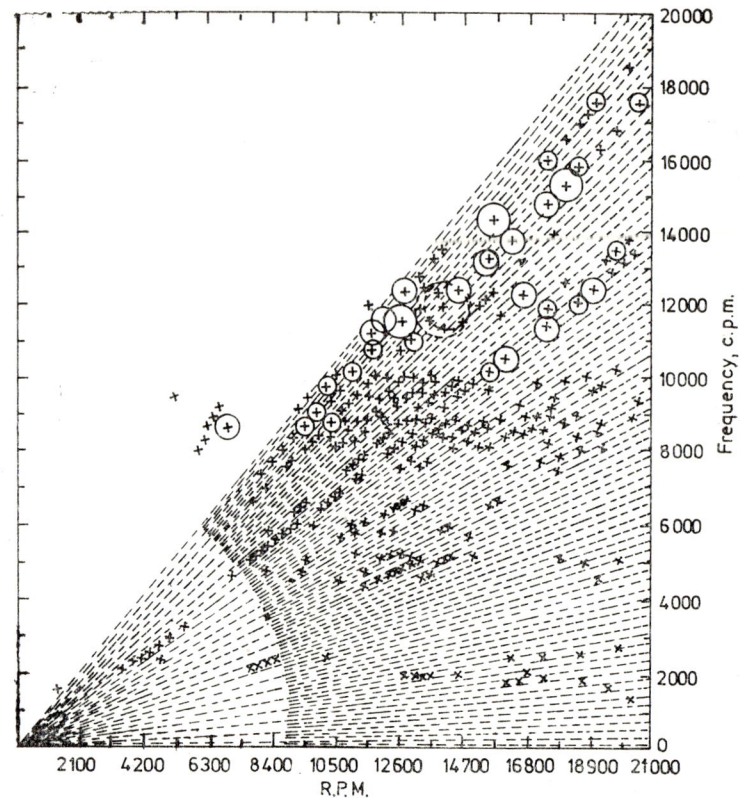

Figure 9.21

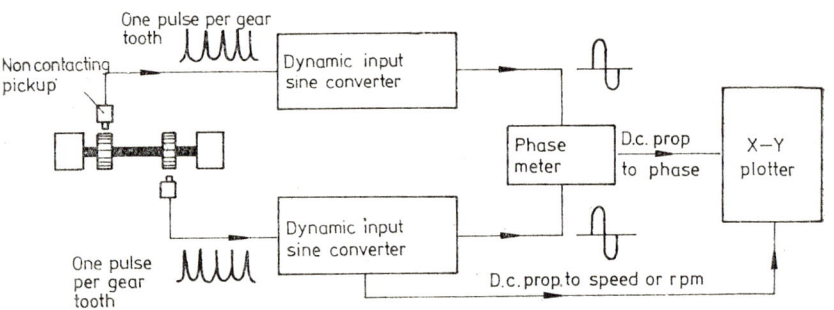

Figure 9.22

9.2 Measurement on prototypes

sand on the horizontal surface of the vibrating body (at resonance the sand stores up along the nodal lines) or, when the sand cannot be used, by holography.

One convenient form for the presentation of natural frequencies of plates and discs is shown in Figure 9.23. Data points

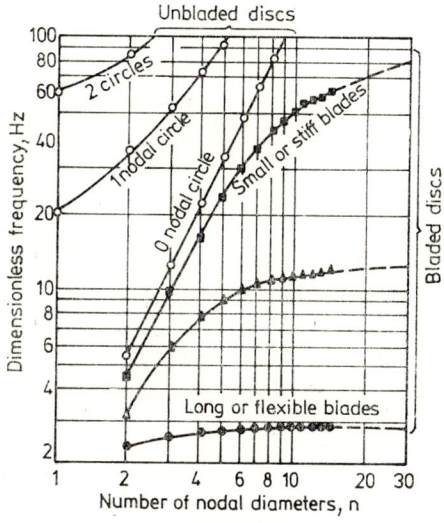

Figure 9.23

are grouped in 'families' of modes having the same number of nodal circles and plotted with the number of nodal diameters across the horizontal axis and the natural frequency on the vertical axis [28].

First, the natural frequency of each mode is plotted against its number of nodal diameters. Then, points corresponding to the same number of nodal circles are connected by solid lines. If one of the natural frequencies has been omitted, then it can be determined as the ordinate of the point located where a mode family line intersects the vertical line corresponding to a given number of nodal diameters.

As it follows from Figure 9.23, different modes of vibration have slightly different natural frequencies and can be excited simultaneously. The lower set of curves, plotted for just one 'family' of modes (with 0 nodal circles), illustrates the effect of blade flexibility on the natural frequencies of the assembly.

The effect of circumferential non-uniformity of the axial pressure in the gas stream of a machine on the bladed disc can

9   *Examples of vibration measurement*

be assessed using the diagram from Figure 9.24. The presence of a $\cos n\theta$ term (where θ defines the circumferential position) in the Fourier series expansion of the static axial pressure distribution will excite the n diameter modes of a bladed disc which rotates through that pressure field. These modes will be excited at an

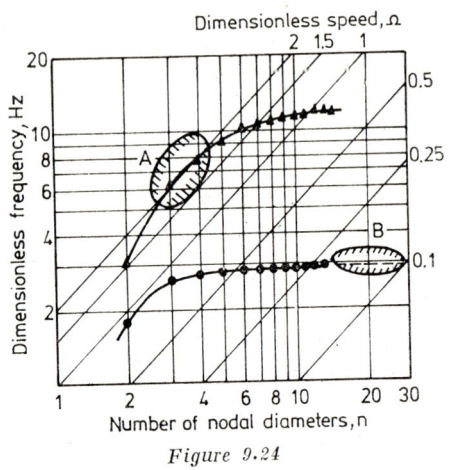

Figure 9.24

angular frequency $n\Omega$ (where Ω is the speed of rotation). The speed at which a given mode is excited to resonance can be deduced from superimposed lines of constant speed on the natural frequency plot (Fig. 9.22) [28].

It can be noticed that for a relatively flexible disc (region A), two different natural modes can be excited at approximately the same speed although their natural frequencies are slightly different. On the contrary, with a relatively stiff disc (or for flexible blades), several modes of vibration have the same natural frequency but each will be excited at a different running speed (region B). If some of these modes are "split" by blade detuning, then a coincidence of resonances will appear, similar to that arising within region A.

9.2.5   Cargo ships

Determination of the hull natural frequencies and mode shapes for the relevant load conditions of a ship is performed in order to check predictions at the design stage using comprehensive computer programs. These enable a main engine to be chosen which will cause a minimum risk for vibration problems due to the hull

9.2 Measurement on prototypes

girder. The influence of wave excitation (springing and whipping) and propeller excitation can also be assessed.

Tests for determining the hull normal modes of vibration have been performed by the authors on cargo ships. The rotating-mass mechanical vibrator VMME-250 (see Table 6.1) has been used to excite flexural vertical and athwartship modes and torsional modes.

Because the forces generated at low frequencies by this vibrator are relatively small, the first two vertical modes of vibration have been excited using the main engine with the ship anchored in calm water.

The vibration generator has been installed on the main deck, at the stern end and on the longitudinal centreline. It has been fastened in a girder with heavy stiffeners above the steering gear in order to prevent excitation of local structures.

The instrumentation system consisted of seismic displacement-sensing pickups (Hottinger B3 and B21), carrier frequency amplifiers (Hottinger KWS T5) and strip chart recorders (Hellige HELCOSCRIPTOR). Another chain of instruments comprising a piezoelectric accelerometer, a vibrometer and a spectrum analyzer (manufactured by Brüel & Kjaer) has been used to obtain amplitude-frequency response curves during frequency sweeps and response amplitudes for engine-generated vibrations.

First, pickups were located at bow and stern of the ship. The displacement amplitude was measured at various vibrator speeds, usually in 10 r.p.m. increments, over the speed range from 90 to 1100 r.p.m. The frequencies corresponding to peak amplitudes of vibration were determined from the frequency response curves (Fig. 9.25). The shape of each mode was obtained by maintaining the vibrator speed constant (equal to the resonance frequency) and by measuring the displacement amplitudes at different frames along the hull. As measurements were taken successively at different points, reference data have been obtained for each record from a stationary pickup installed at the stern end point (for both phase and excitation level reference, the resonance dwell being imperfect).

For athwartship excitation, the position of rotating masses of the vibrator were correspondingly adjusted. To separate athwartship and torsional modes, two horizontally oriented pickups were installed on the main deck and on the double bottom. In-phase signals denote an athwartship mode whilst out-of-phase signals denote a torsional mode. The shapes of the first six vertical modes and the first two athwartship modes together with their resonance frequencies are presented in Table 9.2.

9 Examples of vibration measurement

Underway vibration tests performed by the authors on commercial ships are described in Reference [29].

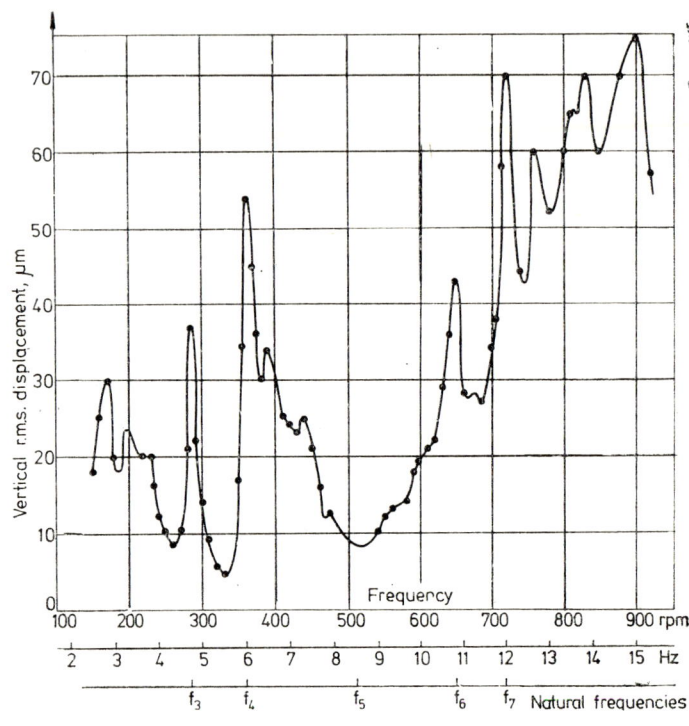

Figure 9.25

TABLE 9.2 *Natural Frequencies of a Cargo Ship*

Lateral vibration	Mode No.	Natural frequency		Mode shape
		Hz	rpm	
Vertical	1	1.53	92	2 nodes
	2	3.8	228	3 nodes
	3	4.66	280	4 nodes
	4	6.16	370	5 nodes
	5	8.50	510	6 nodes
	6	10.75	645	7 nodes
Athwartship	1	3.19	192	2 nodes
	2	6.52	391	3 nodes

9.3 Measurements for production control and acceptance

9.3.1 Machine tools

The "static" quality of a machine tool is presently assessed by the acceptance tests which deal solely with the machine alignment and precision — under test conditions. The "dynamic" quality of a machine tool, i.e. its resistance to vibrations arising as a result of the cutting process, is more difficult to be assessed and no generally recognized dynamic acceptance tests exist.

A method for the evaluation of the stability of radial drilling machines in the case of regenerative chatter has been proposed by Tobias [30] in 1962. He introduced the concept of "coefficient of merit" CoM, which is a quantity proportional to the maximum stable width of cut, h_{m0}. In the case of slot milling, the coefficient of merit is given by

$$CoM = \chi z_c h_{m0},$$

where z_c is the number of cutting edges in continuous cut and χ is a factor dependent solely on the cutting conditions (not affected by the dynamic characteristics of the structure).

On the other hand, the following relationship can be established

$$h_{m0} = \frac{1}{2\chi X_0 z_c},$$

where X_0 is the maximum negative in-phase component of the machine tool harmonic response, plotted as a polar diagram (Fig. 9.26) using artificial excitation.

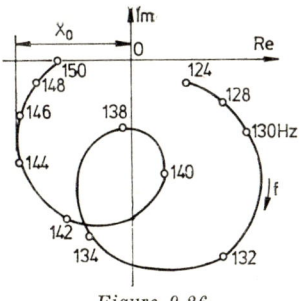 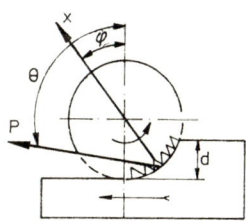

Figure 9.26   Figure 9.27

The technique of determining the coefficient of merit by artificial excitation has been revised by Sadek and Tobias [31] in 1971. They suggested to trace the polar plot of the "operative receptance" α_{xP}, defined as the ratio of the dynamic deflection x normal to the machined surface and the total cutting force P (Fig. 9.27). The latter is determined as the sum of all cutting

9 Examples of vibration measurement

force elements due to individual cutter teeth removing metal and is assumed to act in the centre of the arc of contact between tool and workpiece (in the case of slot milling). The length of the arc of the tool-workpiece interface, hence the position of the centre point and the direction of x, varies with the depth of cut d. The direction of the resultant force P depends on the depth of cut d, on the workpiece material, cutting angles, tool sharpness, etc. The directions P and x can thus vary within wide limits.

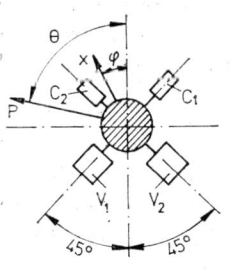
Figure 9.28

The polar plot of the operative response between the directions P and x can be determined by measuring the direct and cross receptances in two mutually perpendicular directions. These can be measured using the arrangement shown in Figure 9.28, where V_1 and V_2 are electrodynamic vibrators, and C_1 and C_2 are vibration pickups.

Using the vibrator V_1 and the pickup C_1, the polar plot of the complex receptance $\alpha_{11} = x_1/F_1$ (Fig. 9.29a) is traced, where x_1 is the displacement in the direction of the pickup C_1, and F_1 is the force generated by the vibrator V_1. The polar plot of the direct receptance $\alpha_{22} = x_2/F_2$ (Fig. 9.29b) is then traced using the vibrator V_2 and the pickup C_2. Finally, the polar plot of the cross receptance $\alpha_{12} = x_1/F_2$ (Fig. 9.29c) is traced using the vibrator V_2 and the pickup C_1. The operative receptance can be plotted

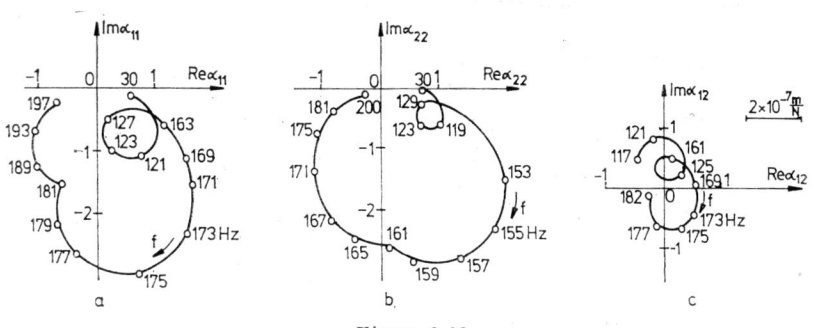
Figure 9.29

point by point, based on the three measured receptance plots, with the aid of the following equation [31]:

$$2\alpha_{xF} = (\alpha_{11} + \alpha_{22}) \cos(\theta - \varphi) + 2\alpha_{12} - (\alpha_{11} - \alpha_{22}) \sin(\theta + \varphi),$$

where θ and φ are angles shown in Figure 9.28.

In this way, different values of the angles θ and φ can be considered, corresponding to different cutting conditions, resulting in diagrams like that shown in Figure 9.30. From this diagram, the segment X_0 is first measured, then the coefficient of merit for the machine is derived as

$$CoM = \frac{1}{2X_0}.$$

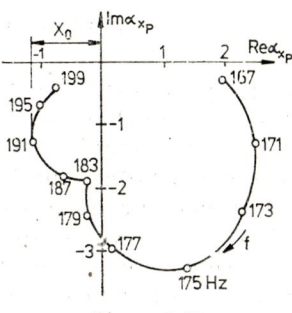

Figure 9.30

Measurements carried out by the authors on a milling machine have resulted in a value $CoM = 2.1 \cdot 10^6 \text{N/m}$ [24].

Comparing the CoM for similar machines, the user will select the machine with the lowest value of this coefficient. This machine is the best of the series, having the highest metal removing capacity before the onset of chatter. The same measurements can be performed by the manufacturer on prototype machines.

Analysing the factors influencing the CoM, four means for improving the dynamic performance of a machine tool can be established [31]:

a) reorientation of the principal axes of vibration with respect to the cutting force direction (angle θ), respectively to the direction of the normal to the machined surface (angle φ); the latter implies change of the cutting tool orientation; b) increase of the equivalent stiffness of the mode which may become unstable under chatter conditions; c) increase of the equivalent damping of this mode of vibration; d) use of passive and/or active vibration absorbers. All these measures lead to a reduction of X_0, hence to an increase of h_{m0}.

It has been shown [32] that the existence of modes with closely spaced natural frequencies and of modal coupling due to non-proportional damping leads to the increase of X_0. Based on measurements, it is possible to determine those positions of tables and slides for which the natural frequencies of the machine tool structure are not too close to each other and the polar plot loops corresponding to "unstable modes of vibration" are displaced in the positive direction of the real axis.

At the same time, the existence of a phase shift within the instrumentation system, between the force and the motion channels, may lead to an apparent increase of X_0 due to the rotation

9 Examples of vibration measurement

of the polar plot in the complex plane. The same type of erroneous data may be obtained when a phasemeter is used for determining the phase angle between unfiltered signals.

9.3.2 Reciprocating compressor piping

During the operation of gas compressor stations, problems occur due to the vibration of pipes and their supports. These become more detrimental as units of greater power are used, capable of conveying important gas flows at high pressures.

Since the excitation forces producing the forced vibrations of the installation equipment originate in the operating process and therefore cannot be removed, a special attention must be paid to problems concerning the vibration of the installation elastic components, even from the design stage. In the following, a short description will be given to a case study concerning a methane gas compressor station [33].

In order to identify the sources of vibration, pipe displacements and dynamic pressures were measured, for a normal operating regime of the compressors, on the inlet (IP) and discharge

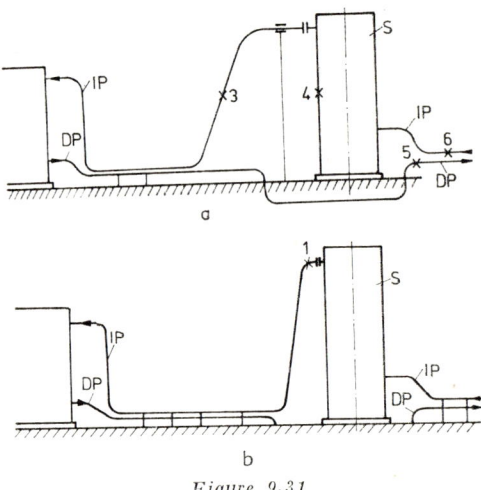

Figure 9.31

(DP) pipes, at the locations shown in Figure 9.31 a. The influence of a series of factors on the vibration level were studied. They are: a) the operating parameters of one compressor (speed, inlet and discharge pressures); b) the operating regime of the station (number of compressors operating simultaneously, running speeds

9.3 Production control and acceptance

of different compressors); c) the shape of inlet and discharge pipes, including the supporting system.

Vibrations have been measured at idle and full load. At point 4 (Fig. 9.31a) on the separator S from the inlet pipeline, the horizontal displacement amplitude increased from 8 μm at idle to 135 μm at full load. A resonance condition was envisaged with the separator vibrating in a rocking mode. The recorded frequency spectra revealed harmonic components being multiples of the compressor rotational frequency.

The resonance method has been used to determine the system natural frequencies. A mechanical exciter type VM1 (see Table 6.1) had been clamped on pipes and the displacements in different points and directions were measured during frequency sweeps.

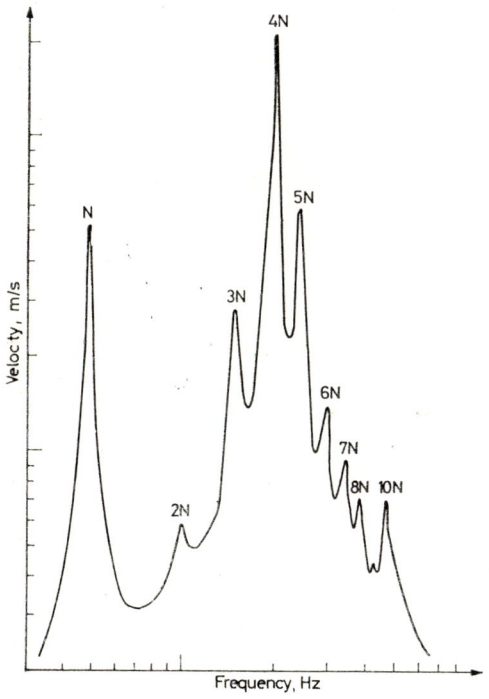

Figure 9.32

It was found out that one of the natural frequencies of the elastic system consisting of the inlet pipe and separator was just within the range of pulsating flow frequencies corresponding to the normal operating speeds. The spectrum from Figure 9.32 is typical for such a resonance condition. In order to avoid the resonance and to diminish the level of vibrations induced by pressure fluctuations, the pipe route was changed as shown in Figure 9.31b. The solution proved to be efficient.

An orifice plate, with an opening about half the pipe inner diameter, placed near point 1 (Fig. 9.31b) restricted the pulsating flow, reducing the amplitudes of the gas pulsations to an unobjectionable level, for a broad range of compressor speeds.

9.4 Measurements during machinery operation

9.4.1 Forge hammers and machine foundations

9.4.1.1 Forge hammers. Measurements on forge hammers involve the determination of natural frequencies and displacement amplitudes.

The fundamental natural frequency of the system consisting of the hammer and its foundation lying either on springs or on a cork layer is determined from the free decay trace of the impact-generated motion produced by a large weight dropped from the crane hook onto the foundation.

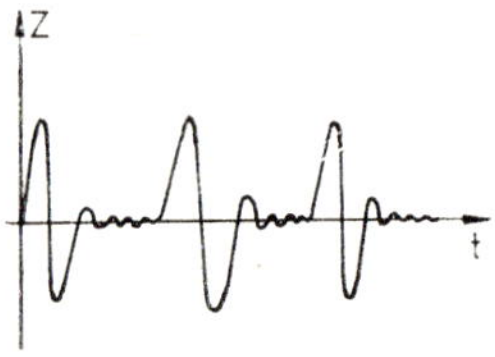

During the hammer normal operation, time records of the type shown in Figure 9.33 are obtained. The peak displacement of the foundation block is compared to the allowable limits given in Chapter 3. Measurement data taken for the anvil can be used in a similar way [34].

Figure 9.33

9.4.1.2 Machine foundations. Measurements on machine foundations are taken for either free vibrations excited by shock or forced vibrations generated by various operating conditions. The frequency response curve can also be measured when the excitation frequency can be changed (variable speed machine, artificial excitation using a vibrator).

Acceptable vibration levels are given in Chapter 3.

The frequency response of foundation blocks which, due to symmetry, have only three degrees of freedom, can be determined using a fairly powerful vibrator and a vibrograph. A typical amplitude-frequency response curve for the vertical vibrations of a model footing is shown in Figure 9.34. Horizontal excitation provides a response curve of the type shown in Figure 9.35, where the two resonances correspond to the two coupled degrees of freedom : horizontal translation and rocking.

The same equipment can be used for testing turbomachinery frame foundations. On Figure 9.36, which illustrates typical

9.4 Machinery monitoring

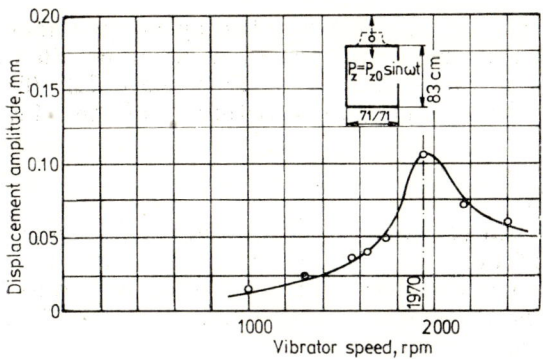

Figure 9.34

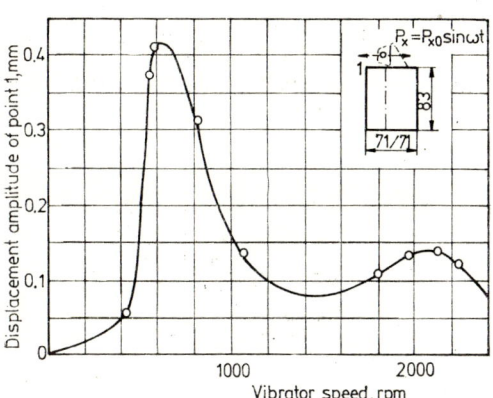

Figure 9.35

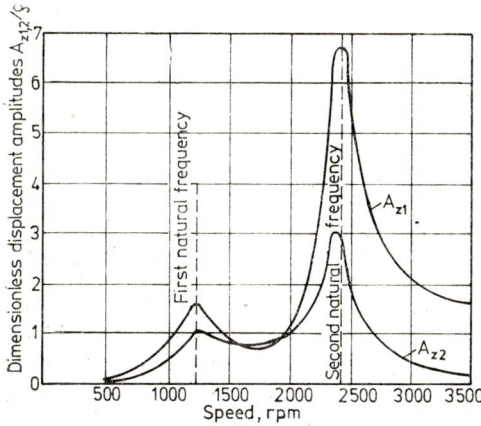

Figure 9.36

9 Examples of vibration measurement

frequency response curves for a turbogenerator foundation, A_{z_1} is the vertical displacement measured in the middle of the frame and A_{z_2} is the vertical displacement of the base slab. The ordinates are dimensionless, ρ being a characteristic of the vibrator

$$\rho = r \frac{m_0}{m_1},$$

where m_0 is the eccentric rotating mass, r — eccentricity, m_1 total mass of machine and top table. The first two resonances correspond to a two-degree-of-freedom model consisting of the upper mass m_1 on elastic columns and the mass of the base slab (considered rigid) resting on soil.

Measurements taken on the frame foundation of a turbo-machine permitted to trace the orbits of different points along the longitudinal girder (Fig. 9.37). The spatial paths described by these points outline the requirement to measure the vibration displacements in three mutually perpendicular directions.

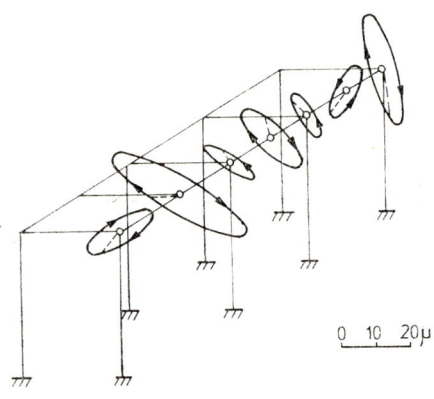

Figure 9.37

9.4.2 Machinery condition monitoring

Vibration and noise levels are useful indicators of machinery condition. During operation, bearings may fail, gear teeth may chip, fasteners may loosen, foundations may settle, blades may be eroded, dirt may build-up on a fan impeller and then break off unevenly, all causing the vibration level to change. Any increase of the vibration level is generally a warning of potential malfunction or failure. Malfunctions can be detected by measuring vibration [5].

9.4 Machinery monitoring

Selection of the proper surveillance and protection system depends on many factors. Noncritical machinery or critical spared machinery in light-duty service can be checked on a regular basis. *Periodic monitoring with portable equipment* can be an efficient procedure when based on a *baseline* measurement program.

For critical machinery, especially new equipment with little or no operating experience, the consequences of unexpected shutdown or failure justify a *permanent monitoring and protection system*. An alarm is activated when the vibration reaches a "warning level" and automatic shutdown sequences are initiated if vibration rises significantly above this level (see Chapter 3). In some cases, permanent monitoring is doubled by periodic detailed analyses to identify long-term trends.

Catlin [35] indicates the following seven steps of a baseline measurement program : 1) selection of machines to be included in the program; 2) selection of measurement points on each machine; 3) selection of the type of measurements (overall level or frequency analysis); 4) selection of measurement interval (from daily to monthly); 5) establishment of a set of standard operating conditions (which must be realized each time measurements are taken) including load, speed, pressure, flow, temperature, etc.; 6) evaluation and check of foundation, piping and associated structures (to be in the final configuration); and 7) establishment of a record-keeping system.

Designing a permanent monitoring system involves [36]: 1) evaluation of the equipment to be monitored (design, construction, service, response to probable malfunctions); 2) identification of potential malfunctions; 3) selection of most responsive operating variables for each potential malfunction; 4) selection of the monitoring method that is most indicative of changing mechanical conditions (shaft displacement, casing acceleration or velocity, rotor axial position); 5) selection of data display and evaluation systems (individual vibration monitors or an automated data-acquisition and trending system with capabilities for automated diagnosis).

A continuous monitoring system on a machine involves permanent monitoring of radial vibration (two eddy probes 90° apart), axial position (one eddy probe) and speed (key phasor plus eddy probe) of the shaft, as well as monitoring of radial and axial vibration of machine casings and piping (piezoelectric accelerometers), with provisions for periodic checks of machine alignment and bearing housing vibration (velocity pickups). Figure 9.38 [37] shows a layout of a complete machine protection system that is common in many petrochemical plants. Case foundation and piping vibrations can also be monitored with

9 Examples of vibration measurement

an additional 'rowing' velocity pickup used for periodic checks. Complete systems for continuous machinery monitoring and protection are manufactured by Bently Nevada (7200 and 9000 Series Monitoring Systems), Dymac, Philips (RMS 700), Shenck (VIBROCONTROL 2000) and Vibro-Meter (VIBRAX).

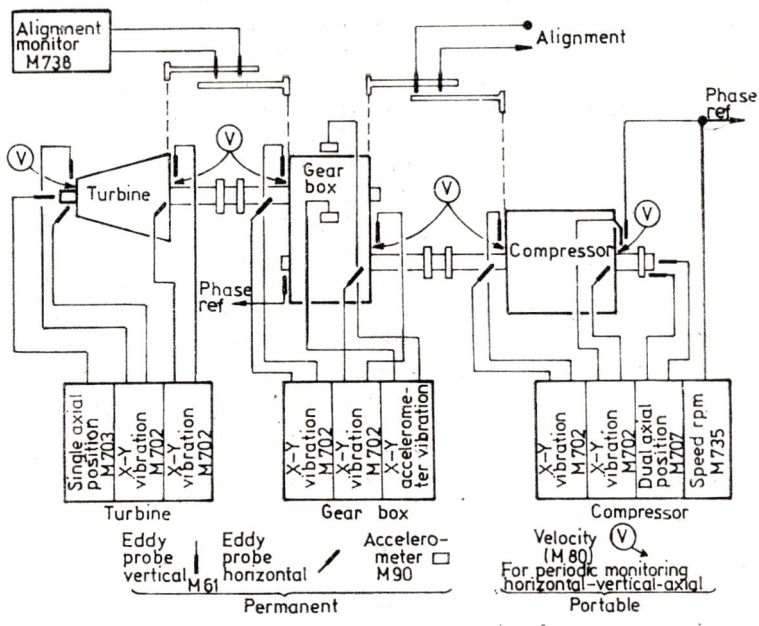

Figure 9.38

In the absence of personal experience, shaft and bearing vibrations can be assessed with the aid of the standards, guidelines and recommendations reproduced in Chapter 3. However, many companies have produced their own guidelines for allowable vibration levels [38] by modifying criteria as V.D.I.2056 or V.D.I.2059 specifications to suit their own particular machinery in the light of their own experience.

In the past, in order to reduce the likelihood of a sudden breakdown, the so-called "planned maintenance" was practiced, in which machines were overhauled at regular and precise intervals, regardless the running condition.

In recent years, this concept has been revised owing to the costs of lost production due to shutdowns of vital machinery in continuously operating process plants. The "on-condition maintenance" has been adopted, where the decision to shut down a machine for maintenance is based on its condition, i.e. on the

9.4 Machinery monitoring

analysis of the variation in time of the most representative parameter for machinery integrity and normal operation.

Figure 9.39 shows a typical machine condition pattern [39]. A decrease of the vibration level during the running-in period can be noticed, followed by a region of relatively constant level.

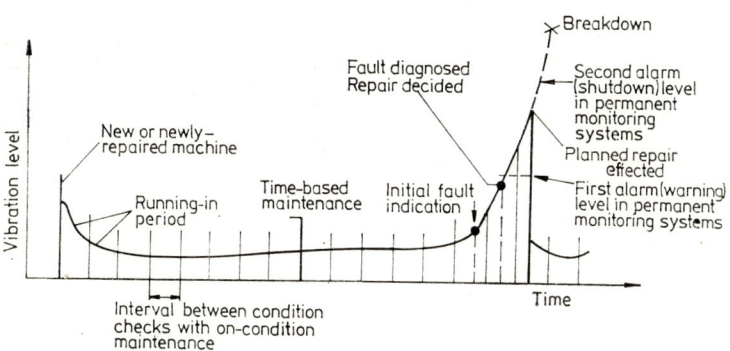

Figure 9.39

Shutdown within this time interval, for a planned maintenance and part replacement, proves to be useless, being scheduled long time before increased vibration level detected by regular condition monitoring indicates the actual need. With "on-condition" maintenance, repairs are carried out only when the machinery condition has deteriorated to a predetermined level. At this stage, a tentative diagnosis is done and there is still time for a more detailed analysis to be carried out in order to completely diagnose the fault, to order the spare parts and to decide the date of repair.

On-condition maintenance is considered to be most suitable where the most important cost factor is the loss in production. Permanent monitoring is recommended where the most important economic factor is the damage to the machine.

The simplest permanent monitoring system is a single channel device designed for maintenance free operation. Figure 9.40a shows the block diagram of the Brüel & Kjaer monitor type 5500. Type 5730 is a version with three separate parallel filter channels (Fig. 9.40b) so that three separate frequency ranges of the input signal can be monitored simultaneously [39]. A multiplexer is used in conjunction with the monitor when a number of frequency bands or a number of input channels are to be permanently monitored.

Conventional monitoring systems have individual monitors for each measurement point. Modern monitoring systems are

9 Examples of vibration measurement

minicomputer based. Individual monitors are replaced by signal conditioning circuits designed solely to prepare the vibration signal for input to the minicomputer.

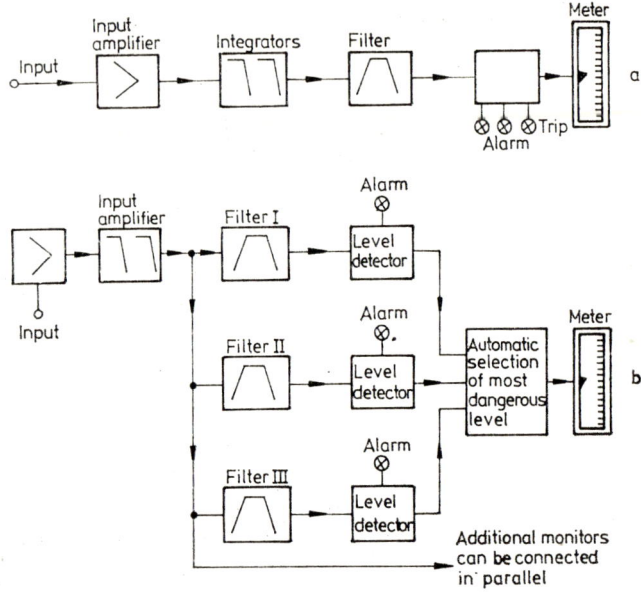

Figure 9.40

The minicomputer provides significantly better machinery protection through : 1) continuous on-line trending; 2) correlation of operating variables; and 3) diagnostics. Data logs can be presented visually on a c.r.t. terminal or printed on either a teletypewriter or a highspeed line printer. Alarm predictions can be obtained by extrapolation of vibration readings versus time into the future. Real time spectrum analysis can also be incorporated in the vibration monitoring system to pinpoint specific faults.

9.5 Measurement of vibrations produced by blasting

Blasting is widely used in construction jobs, quarry operations and open work mining. It is desirable that ground vibration caused by blasting should not be detrimental to persons and neighbouring structures.

9.5 Blasting

Explosive generated ground motions can be characterized by: frequency, displacement, velocity and acceleration of soil particles, as well as by the vibration energy at successive points along the path of the blasting wave.

Reference [40] is a state-of-the-art survey of the published literature regarding the selection of the blast intensity ratings criteria and the maximum allowable values of the pertinent parameters. It is generally recommended to record on magnetic tape the transient signal resulting from blasts and to make a frequency analysis. Digital event recorders can also be used.

It has been shown that at distances beyond 100 m from the centre of a blast, the oscillations with frequencies between 10 and 20 Hz have the lowest attenuation. It can be concluded that for distances above 100 m between the explosion centre and the measuring station, an instrumentation system consisting of seismic pickup, amplifier and strip chart recorder can be used. Measurement data obtained by the authors on different construction sites are presented in Table 9.3.

TABLE 9.3 *Measurement Data of Blast Generated Ground Motions*

Explosive amount, kg	Distance from explosion to measuring station, m	Parameters of the ground motion caused by blasting					
		Displacement, mm		Velocity, cm/s		Acceleration, m/s^2	
		Vertical	Horizontal	Vertical	Horizontal	Vertical	Horizontal
150	150	0.200	*)	1.600	3.120	0.835	1.650
175	162	0.187	—	1.000	3.220	1.710	1.140
118	190	0.075	—	0.300	1.040	0.381	0.505
136	46	—	0.192	—	—	0.450	—
137	56	0.029	0.047	—	—	1.250	—
123	43	0.044	0.232	—	—	0.830	1.380

*) Not measured.

Based on previous experience and measurement data regarding the ground motions produced by blasts, and using extrapolation relationships [42], the charge weight and the distance from the explosion can be determined to avoid structural damage.

9 Examples of vibration measurement

9.6 Measurement of the dynamic characteristics of materials

Most of the existing methods for the determination of the dynamic properties of materials used in vibration isolation are based on relationships derived from the equations of motion of a model of the actual system, in which a sample of the tested material is inserted. The analysis of the dynamic response of such a system, for example — to a harmonic excitation, provides the necessary equations between the dynamic characteristics of the tested specimen and the measurable response parameters. When building up the model, hypotheses are adopted concerning the number of degrees of freedom, the damping mechanisms and the type of force-deflection characteristic, the interaction between the analysed system and the measuring equipment, etc. These are checked by the measurement data and reformulated, if necessary.

Four methods which have been used with good results by the authors for determining the damping and the dynamic stiffness of some anti-vibration mountings are described in the following.

Method I. The tested sample of material is the elastic and dissipating element of the single-degree-of-freedom oscillator. It is bounded between a "rigid" flat base and a rigid mass driven by a harmonic force $F_0 \cos \omega t$ generated by an electrodynamic vibrator through a force transducer (Fig. 9.41).

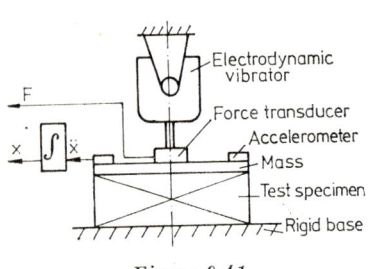

Figure 9.41

Polar diagrams of the mass displacement response are plotted for several values of the exciting force amplitude F_0. The *isochrones* are then traced by connecting data points of constant frequency from different polar plots. If these are straight lines radiating from the origin of coordinates, then the force-deflection and the damping characteristics can be considered linear and the relationships derived for the linear oscillator (see Section 7.6.5.1) can be used for parameter identification [43]. If not, the way in which the isochrones are bent shows whether the system has a non-linear (softening or hardening) stiffness or a non-linear damping (quadratic or dry friction). Formulae derived for the non-linear oscillator [44] should be used in this case. If up to 80 percent of the polar diagram can be fitted to a circle, then a linear damping mechanism can be adopted (supposing good mode isolation). Use of the equivalent hysteretic damping factor leads to simpler

9.6 Dynamic properties of materials

calculations and adoption of a cubic stiffness term in the equation of motion is sufficient in most cases [45].

Figure 9.42 summarizes some of the effects of a non-linear force-deflection characteristic of the form $F_e = k(x + \alpha x^3)$ on the harmonic response of a single-degree-of-freedom system with linear hysteretic damping.

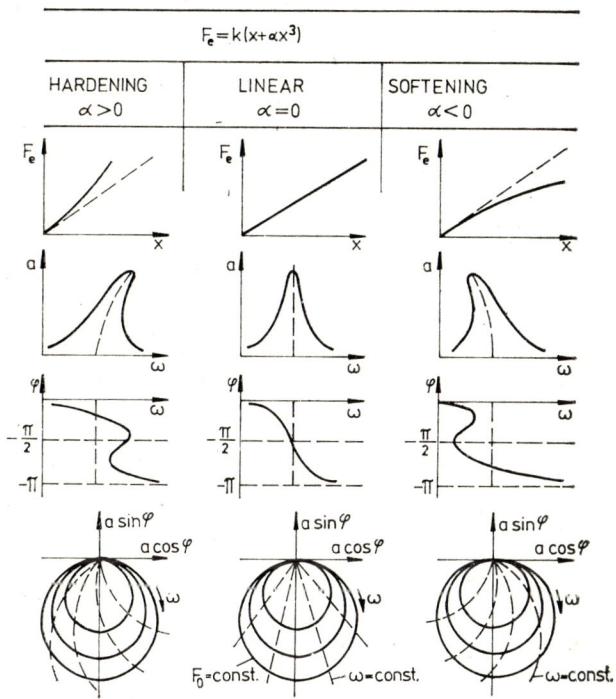

Figure 9.42

When the equipment can produce excitation with forces in quadrature, the "method of the two polar plots" [46] is used. This is a curve fitting method based on the analysis of two polar diagrams plotted using excitation without and with forces in quadrature. The first diagram of the displacement is traced using an exciting force $F_0 \cos \omega t$. The second diagram is plotted by adding to the initial force a component $\lambda F_0 \cos\left(\omega t + \dfrac{\pi}{2}\right)$. The two diagrams cross each other at the point M (Fig. 9.43) defined by the frequency f_r — on the diagram $\lambda = 0$ — and by the frequency f' — on the diagram $\lambda \neq 0$.

The best circle is fitted through the data points of the diagram $\lambda = 0$ in the neighbourhood of point M. The diameter $O'M$ is drawn, then the perpendicular (Δ) to this diameter, which intercepts the circle $\lambda \neq 0$ at the point N, of frequency f_1. The equivalent hysteretic damping factor is given by

$$g = \frac{\lambda(f'^2 - f_r^2)}{(1 + \lambda^2)f_r^2 - f_1^2}.$$

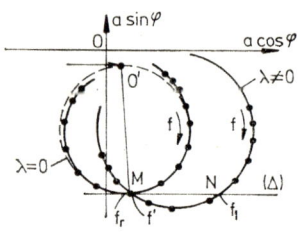

Figure 9.43

The dynamic (in-phase) stiffness is given by

$$k = \frac{1}{g} \frac{F_0}{a_r}$$

where $a_r = O'M$.

In the case of linear systems, $f_1 = f_r$ is the resonance frequency of the system and the damping factor is given by

$$g = \frac{1}{\lambda} \frac{f'^2 - f_r^2}{f_r^2}.$$

Harmonic excitation with forces in quadrature has also been used with the "method of the three circles" to both linear [47] and non-linear [46] systems. Other methods are presented in reference [48].

Method II. Materials whose dynamic properties are required for conditions in compression can be tested using the arrangement

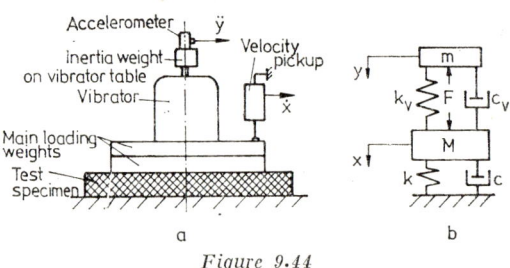

Figure 9.44

shown in Figure 9.44a, based on the use of an electrodynamic exciter as an inertial vibrator [49].

The vibrator body is rigidly attached to the loading weights ("rigid" plates) used to provide the static pressure. The test

9.6 Dynamic properties of materials

specimen (layer or three equal size pads positioned as three points of support) is placed beneath the main loading weights whose motion is sensed by a velocity pickup. The excitation force is created by loading the vibrator table with inertia weights whose acceleration is measured with a piezoelectric accelerometer.

The equivalent two-degree-of-freedom system (Fig. 9.44b) has the following components: m — mass of vibrator table, accelerometer and added masses; M — mass of vibrator body and loading weights; k_v and c_v — stiffness and equivalent viscous damping coefficient of vibrator table suspension; k and c — stiffness and equivalent viscous damping coefficient of test specimen; F — force produced by the vibrator.

The equations of motion for the two masses are

$$M\ddot{x} + c\dot{x} + kx + c_v(\dot{x} - \dot{y}) + k_v(x - y) = F,$$

$$m\ddot{y} + c_v(\dot{y} - \dot{x}) + k_v(y - x) = -F.$$

Adding these equations side to side, the equation of motion of the main system becomes

$$M\ddot{x} + cx + kx = -m\ddot{y},$$

where the right hand side is the inertia force of the moving table plus added masses.

Maintaining the acceleration $\ddot{y}$ constant and measuring the velocity $\dot{x}$ at different frequencies in the resonance region as well as the phase angle between them, the polar plot of the mobility $\dot{x}/(m\ddot{y})$ can be constructed wherefrom the parameters k and c are calculated for the resonance frequency using wellknown relationships.

If the excitation is adjusted so as to maintain the velocity $\dot{x}$ constant and the acceleration $\ddot{y}$ is measured at different frequencies, then the straight line of the mechanical impedance $m\ddot{y}/\dot{x}$ can be plotted in the Argand plane, and the values of k and c can be obtained by graphical methods.

When determining the dynamic properties of non-linear materials it is recommended to plot "inverted resonance curves". They can be obtained keeping the strain amplitude (hence the displacement amplitude x) constant whilst varying the exciting force (hence the acceleration $\ddot{y}$) and plotting the measured acceleration $\ddot{y}$ of the inertia weights against the excitation frequency. The resonance frequency f_r is measured at $\ddot{y}_{min}$, i.e. at the frequency at which a minimum force has to be applied to maintain the constant strain amplitude. The damping ratio is given by

$$\zeta = \frac{f_2 - f_1}{2f_r},$$

9 Examples of vibration measurement

where f_1 and f_2 are the frequencies of the "half power points" located at the curve crossings with the horizontal line of ordinate $\sqrt{2}\ddot{y}_{min}$.

The same calculations are repeated for the other curves, measured at different values of the strain amplitude, and the effects of the displacement amplitude x_{max} on the damping ratio can be determined. The same dependence can be established for the dynamic stiffness (hence for the dynamic modulus of elasticity) based on the relationship

$$k = \frac{m\ddot{y}_{min}}{2\zeta x_{max}}.$$

The method has been successfully used for determining the dynamic properties of soils [50].

Method III. The test specimen is bonded between a mass and a rigid base plate mounted on the table of a mechanical (or

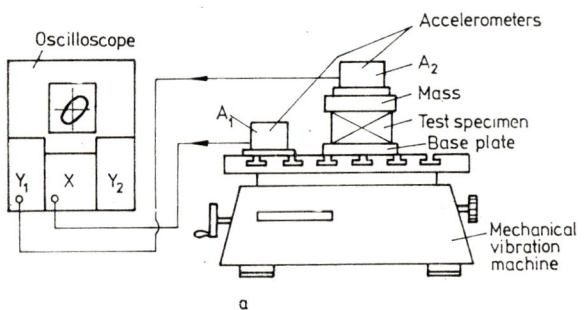

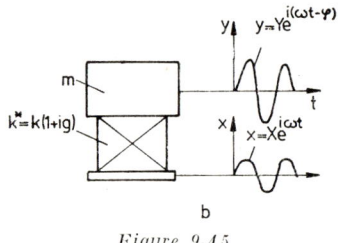

Figure 9.45

electrodynamic) vibration machine (Fig. 9.45 a). Two identical accelerometers are installed, one on the top mass, the other on the table of the vibration machine. The output signals of the

9.6 Dynamic properties of materials

accelerometers are fed to the X and Y plates of an oscilloscope; an ellipse is thus traced on the screen.

If the table of the vibration machine has a harmonic motion $x = Xe^{i\omega t}$ then the supported mass will have a motion $y = Ye^{i(\omega t - \varphi)}$, lagging with an angle φ the input displacement of the vibration table (Fig. 9.45b).

If the test specimen can be modelled by a complex stiffness k^* (2.54), then the dynamic stiffness k and the equivalent hysteretic damping factor g are given by

$$k = m\omega^2 \frac{1 - \frac{X}{Y}\cos\varphi}{1 + \frac{X^2}{Y^2} - 2\frac{X}{Y}\cos\varphi}, \quad g = \frac{\sin\varphi}{\frac{Y}{X} - \cos\varphi},$$

where m is the supported mass (including the accelerometer A_2) and $\frac{X}{Y}$ is the amplitude ratio of the vibration table and supported mass displacements (or accelerations). The quantities $\frac{X}{Y} = \frac{OC}{OB}$ and $\sin\varphi = \frac{OA}{OB}$ can be determined using the elliptical-pattern measuring technique (Fig. 7.8b).

This is a forced non-resonant method, in which measurements are repeated at different values of the excitation frequency, in order to obtain the frequency dependence of k and g.

When the test specimen has the thickness b and the cross section area A, the dynamic modulus of elasticity is given by

$$E_1 = \frac{b}{A} k$$

which holds for small values of b and relatively low excitation frequencies ω.

Method IV. For a complete description of the dynamic behaviour of a visco-elastic material, it is necessary to measure the properties both in compression and in shear, i.e. to determine the moduli $E^* = E_1(1 + ig)$ and $G^* = G_1(1 + ig)$. The shear modulus G^* can be measured by using the apparatus outlined in

9 *Examples of vibration measurement*

Figure 9.46, designed by Grootenhuis [49] for testing elastomers in shear.

Two identical specimens of material, 3, are bonded to a centre drive piece 4, and to two stationary supports 5, clamped by bolts to the base plate 8. The force moving the centre piece is generated by the electrodynamic vibrator 10 and transmitted from the armature 9, through the drive rod 6, to the centre piece 4, which plays the role of an elastically supported mass.

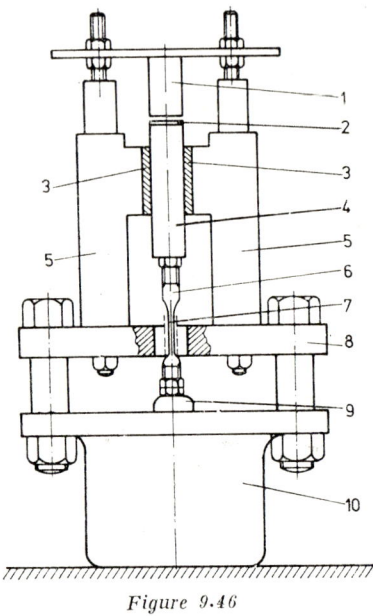

Figure 9.46

Strain gauges 7 are bonded onto the flat surfaces of the rod 6 yielding a force gauge. It can be calibrated statically before bonding the test specimens in place, by turning the apparatus up-side-down and by attaching known weights to the centre piece 4. The displacement of the centre piece 4 is measured with a non-contacting transducer 1. A brass disc 2 on the top of the aluminium centre piece 4 increases the transducer sensitivity.

The three quantities which have to be measured at different frequencies ω are the amplitude F of the force signal measured by the strain gauges, the amplitude X of the displacement measured by the non-contacting transducer and the phase angle φ between these two signals. The dynamic shear modulus is given by

$$G_1 = \frac{b}{A} \frac{F \cos \varphi + M \omega^2 X}{X}$$

and the loss factor by

$$g = \frac{G_2}{G_1} = \frac{F \sin \varphi}{F \cos \varphi + M \omega^2 X}$$

where b is the thickness of the specimen, A — the total area of bonding of both specimens to the centre piece, M — the combined mass of the centre piece 4 and part of the drive rod 6.

For most elastomers, it is considered that $E_1 = 3G_1$ because Poisson's ratio is $\nu \cong 0.5$.

9.7 Soil elastic characteristics

Surveys on the experimental techniques used for measuring the dynamic modulus of elasticity and the loss factor can be found in references [51, 52, 53].

9.7 Measurement of soil elastic characteristics by vibration methods

9.7.1 Spring constants for rigid footings resting on soil

In vibration studies of machine foundations placed directly on the ground, the soil elastic behaviour is usually defined by the spring constants k_z, k_x, k_y, k_{φ_x}, k_{φ_y}, k_ψ. They correspond to the six degrees of freedom of the footing: vertical oscillation, horizontal translation along the principal axes Ox and Oy of the contact surface between foundation and ground, rocking oscillations about these axes and torsional oscillation about a vertical axis passing through the centroid of the contact surface. The spring constants

TABLE 9.4 *Spring Constants for Rigid Base Resting on Elastic Half-Space*

Base shape	Motion	Spring constant	Authors
Circular	Vertical translation (along Oz)	$k_z = \dfrac{4GR}{1-\nu}$	Timoshenko and Goodier
	Horizontal translation (along Ox or Oy)	$k_x = k_y = \dfrac{32(1-\nu)GR}{7-8\nu}$	Bycroft
	Rocking (about Ox or Oy)	$k_{\varphi_x} = k_{\varphi_y} = \dfrac{8GR^3}{3(1-\nu)}$	Borowicka
	Torsion (about Oz)	$k_\psi = \dfrac{16}{3} GR^3$	Reissner and Sagoci
Rectangular	Vertical translation (along Oz)	$k_z = \dfrac{G}{1-\nu} \beta_z \sqrt{Ll}$	Barkan
	Horizontal translation (along Ox)	$k_x = \dfrac{G}{1-\nu} \beta_x \sqrt{Ll}$	Barkan
	Rocking (about Oy)	$k_{\varphi_y} = \dfrac{G}{1-\nu} \beta_{\varphi_y} L^2 \sqrt{Ll}$	Gorbunov-Possadov

9 *Examples of vibration measurement*

are proportionality factors between the soil reactions and the corresponding translational displacements and rotations of the contact surface.

Modelling the supporting medium by the elastic half-space, formulas for the spring constants have been established as functions of the elastic moduli E, G, and Poisson's ratio ν of soil. Table 9.4 gives some of these expressions obtained for footings with rigid circular base of radius R and with rectangular base of length L (along Ox) and width l (along Oy). Coefficients β_z, β_x and β_{φ_y} for rectangular footings are functions of the ratio L/l [54].

Values of shear modulus G and Poisson's ratio ν measured by dynamic methods are needed to evaluate the spring constants for foundations from the formulas of Table 9.4. As reliable values are sometimes difficult to obtain from experimental data, some authors prefer to use some global characteristics of the soil elastic behaviour such as the coefficients of elastic subgrade reaction C_z, C_x, C_φ and C_ψ. These are functions of shape and size of the foundation base.

For large bearing areas, the spring constants for rigid footings are proportional to either the contact area S or the second moment of contact area about the axes Ox, Oy, Oz through the centroid. The proportionality factors are the coefficients of subgrade reactions:

$$k_z = C_z \cdot S\,;\ k_x = k_y = C_x \cdot S\,;$$
$$k_{\varphi_x} = C_\varphi \cdot I_x\,;\ k_{\varphi_y} = C_\varphi \cdot I_y\,;\ k_\psi = C_\psi \cdot I_z. \tag{9.11}$$

Formulas (9.11) also apply for small bearing areas, by using dynamic coefficients of subgrade reactions which are functions of the contact area.

9.7.2 Measurement of elastic constants for soils

As soils have a non-linear behaviour under applied loads, average values of the elastic constants are used in calculations. These are functions of both the mean static contact pressure between the foundation base and soil and the amplitude of dynamic loads. It has been shown that values of dynamic elastic moduli measured by vibration methods are independent of frequency for a relatively broad frequency range [55]. Reliable values of the dynamic elastic moduli can be obtained from measurements carried out after several preliminary loading cycles, in order to eliminate the effect of plastic strains.

9.7 Soil elastic characteristics

Values of the shear modulus G may be evaluated from laboratory tests on soil specimens. Cylindrical specimens of soil are excited in torsional modes of vibration and the excitation frequency is swept locating successive resonances ("resonant-column test"). The dynamic shear modulus is given by

$$G_{\text{dyn}} = 4\rho l^2 \left(\frac{f_i}{i}\right)^2, \qquad (9.12)$$

where ρ is the density of the soil, l — length of specimen, f_i — resonance frequency of torsional vibrations, i — the order of the particular harmonic excited in the specimen.

Similarly, the dynamic Young's modulus of elasticity E_{dyn} can be measured from longitudinal or flexural resonance tests.

The soil specimens must be handled with great care in order not to alter too much the soil natural properties. One must be aware that characteristics determined on small specimens are not representative for the whole supporting medium, and the actual triaxial stress distribution cannot be properly modelled.

Values of dynamic elastic moduli may be obtained using the wave propagation velocities measured *in-situ* in the soil. A disturbance is induced at one point and the wave propagated is picked up by a sensor located some distance away. If the disturbance is a single impulse, the time for the impulse to reach the pickup is measured and the velocities are obtained by dividing the distance by time. If the disturbance is a sinusoidal signal, the time is determined from the phase shift between the excitation and response signals.

The elastic constants may be calculated from the following equations [56]:

$$E_{\text{dyn}} = \rho\, v_s^2\, \frac{3v_c^2 - 4v_s^2}{v_c^2 - v_s^2},$$

$$G_{\text{dyn}} = \rho \cdot v_s^2, \qquad (9.13)$$

$$\nu = \frac{v_c^2 - 2v_s^2}{2(v_c^2 - v_s^2)},$$

where ρ is the mass density of soil, v_c — compression wave velocity and v_s — shear wave velocity.

Another method for *in-situ* measurement of the dynamic properties of soil is based on medium-scale vibrator tests on model foundation blocks. The subgrade coefficients are first obtained from equations relating them to the parameters of the measured vibrations. The same equations are used as for the calculation of actual

9 Examples of vibration measurement

foundation vibration. Equations from Table 9.4 are then used to calculate the dynamic elastic constants E, G and ν. Values of elastic moduli obtained in this way are somewhat lower than those calculated using equations (9.13) and the seismic method [56].

9.7.3 Resonance technique for determining the dynamic coefficients of subgrade reaction

In this approach, a vibrator V is set upon a small foundation block resting on soil (Fig. 9.47) and the frequency of excitation is varied, determining the resonance frequencies. The model from figure 9.47 has a rectangular contact area and two planes of symmetry (xOz and yOz). Point O is the centroid of contact area and G is the combined centre of gravity of the foundation and vibrator. The axes Ox and Oy are located at the contact surface of the foundation and the ground.

Due to the symmetry of the vibrator-foundation system, the vertical translation and torsional (about Oz-axis) vibrations are de-coupled, while the horizontal translation (sliding) and rocking oscillations are coupled.

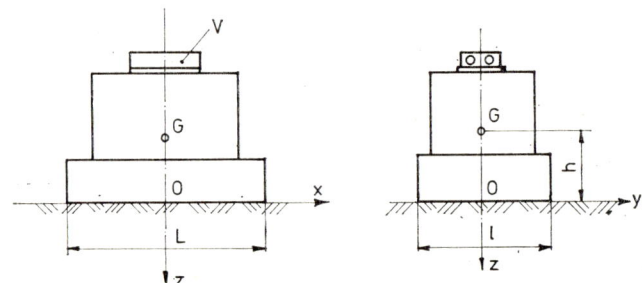

Figure 9.47

Let make the following notations:

$$p_z = \sqrt{\frac{C_z \cdot S}{m}}, \quad p_x = \sqrt{\frac{C_x \cdot S}{m}}, \quad p_\varphi = \sqrt{\frac{C_\varphi \cdot I_y - mgh}{J_y}},$$

$$p_\psi = \sqrt{\frac{C_\psi I_z}{J_z}}, \tag{9.14}$$

where h is the vertical location of the point G above the contact surface, S, I_y, I_z — area, respectively the second moments of con-

9.7 Soil elastic characteristics

tact area about the Oy and Oz axes, m, J_y, J_z — the mass of the foundation and vibrator, respectively their moments of inertia about the Oy and Oz axes, g — acceleration of gravity.

In equations (9.14), p_z is the natural angular frequency of the de-coupled vertical translation and p_ψ is the expression for the de-coupled torsional oscillation. The natural frequencies of the coupled rocking (about Oy-axis) and sliding (along Ox-axis) oscillations are given by [34]:

$$p_{1,2} = \sqrt{\frac{1}{2\gamma}[p_\varphi^2 + p_x^2 \pm \sqrt{(p_\varphi^2 + p_x^2)^2 - 4\gamma p_\varphi^2 p_x^2}]} \qquad (9.15)$$

where $\gamma = J/J_y$, in which J is the moment of inertia of the foundation-vibrator assembly about an axis parallel to Oy and passing through G and J_y is the moment of inertia about the Oy-axis.

If V is a rotating-mass vibrator, the counterrotating masses can be arranged to produce oscillating forces F_ω or moments M_ω of amplitudes (F_0, M_0) proportional to the square of the shafts rotating speed ω:

$$F_\omega = f_0\omega^2 \sin \omega t, \quad F_0 = f_0\omega^2,$$
$$M_\omega = m_0\omega^2 \sin \omega t, \quad M_0 = m_0\omega^2,$$

in which f_0 and m_0 are dynamic unbalances of the rotating weights. Apart from the variation of the rotating speed ω, the size of weights and the eccentricity setting can be adjusted in steps to change the amplitude of the dynamic force or torque.

The unbalance is fixed before each test, then the displacement amplitude a is measured at different points of the model footing by stepping the rotating speed ω through the range of interest.

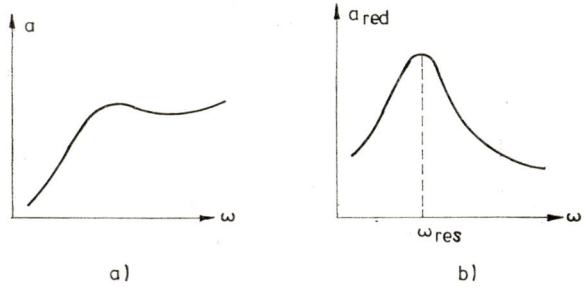

Figure 9.48

For de-coupled oscillations (vertical translation or torsion) a typical amplitude-frequency response curve is shown in Figure 9.48a. For a more precise location of resonance, an easier evalua-

9 Examples of vibration measurement

tion of the frequency corresponding to the peak displacement x_{res} can be done using the response curve from Figure 9.48b, where a 'reduced amplitude'

$$a_{red} = \frac{a}{f_0 \omega^2} \quad \text{or} \quad a_{red} = \frac{a}{m_0 \omega^2}$$

is plotted versus frequency, which is the displacement amplitude corresponding to the unit dynamic force or moment.

After the experimental determination of the natural frequencies p_z and p_ψ (equal to ω_{res}), the coefficients of subgrade reaction C_z and C_ψ are derived from the following equations

$$C_z = \frac{m}{S} p_z^2, \qquad C_\psi = \frac{J_z}{I_z} p_\psi^2.$$

In order to determine the values of C_φ and C_x, coupled rocking and sliding oscillations of the model footing are produced, arranging the rotating masses of the vibrator so as to generate horizontal forces F_ω. If F_ω is along the Ox-axis, then rocking about Oy-axis and sliding along Ox-axis are obtained. The resonance frequencies p_1 and p_2 are determined locating the peaks on the 'reduced amplitude' versus frequency response curve (Fig. 9.49). Frequencies p_x and p_φ are first calculated from equation (9.15)

$$p_{\varphi,x} = \sqrt{\frac{\gamma}{2}\left[p_1^2 + p_2^2 \pm \sqrt{(p_1^2 + p_2^2)^2 - \frac{4}{\gamma} p_1^2 p_2^2}\right]}$$

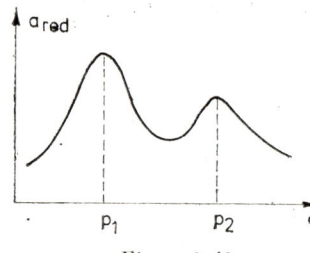

Figure 9.49

then the coefficients of subgrade reaction C_x and C_φ are obtained using equations (9.14)

$$C_x = \frac{m}{S} p_x^2, \quad C_\varphi = \frac{J_y p_\varphi^2 + mgh}{I_y}.$$

At the Strength of Materials Laboratory of the Polytechnic Institute of Bucharest, measurements of the dynamic coefficients of subgrade reactions have been carried out using rotating-mass vibrators and a model footing made from steel plates [57]. Measurements on both medium-size concrete model foundation blocks and real scale foundations have been done using the VMME-250 kN rotating-mass vibrator (see Table 6.1)[58]. Some results published in References [59], [60] are given in Table 9.5.

9.7 Soil elastic characteristics

TABLE 9.5 *Coefficients of Subgrade Reaction for Different Soils*

Soil	Model base area, m²	Mean static contact pressure, $\times 10^5$ N/m²	Coefficient of subgrade reaction			
			C_z	C_x	C_φ	C_ψ
			$\times 10^7$ N/m³			
Sand consolidated by gravel columns	0.5	0.9	18.8			
	1.0	0.48	10.5			
	2.0	0.7	9.3			
	12.25	0.71	3.1—3.8	1.9—2.2	4—5.6	4.6—5.12
	45.5	0.53	4.4	2.1	8.1	
	67.86	0.56	3.5	1.5	6.3	
	197.2	0.54	2.69			
	221	0.48	2.38			
Sand compacted by heavy ram	0.5	0.9	23			
	1.0	0.48	13.1			
	2.0	0.54	8			
Alluvial gravel	0.5	0.67	36.4			
	0.5	0.91	39.1			
	1.0	0.49	20.6	9—10		
	12.0	0.58	11.4	22.15	22.8	
Marl	0.5	0.78	32.7			
	18.1	2.88	88.5			

References for Chapter 9

1. Beranek, L. L. (Ed.), *Noise Reduction*, McGraw-Hill Book Comp., Inc., New York, 1960.
2. Fieldhouse, K. N., Techniques for identifying sources of noise and vibration, *S/V, Sound and Vibration*, *4*, *12*, 14—18 (Dec. 1970).
3. Babkin, A. S. and Anderson, J. J., Mechanical signature analysis of ball bearings by real time spectrum analysis, Nicolet Instruments Ltd., Application Note No. *3* (May 1972).

9 Examples of vibration measurement

4. Bannister, R. and Donato, V., Signature analysis of turbomachinery, S/V, Sound and Vibration, **5**, 9, 14—21 (Sept. 1971).
5. Tustin, W., Measurement and analysis of turbomachinery, Chemical Engineering Progress, **67**, 6, 62—69 (1971).
6. Borhaug, J. E. and Mitchell, J. S., Applications of spectrum analysis to on-stream condition monitoring and malfunction diagnosis of process machinery, Proc. First Turbomachinery Symposium, A & M University, Texas, 1972.
7. Radeș, M., Identificarea surselor de zgomote și vibrații, Comunicările Conferinței "Vibrații în construcția de mașini", Timișoara, 31 Oct. — 2 Nov. 1975, vol. **2**, 355—360 (1975).
8. Miller, T. D., Machine noise analysis and reduction, S/V, Sound and Vibration, **1**, 3, 8—14 (March 1967).
9. Tatge, R. B., Acoustic techniques for machinery diagnostics, Journal of Acoustical Society of America, **44**, 374 (1968).
10. Johnson, N. E., Techniques for reducing fan noise, Machine Design, **34**, 8, 109—115 (1962).
11. Jenkins, S. H. and Walter, J., Identifying engine and vehicle noise sources, Diesel and Gas Turbine Progress (Oct. 1972).
12. Ruffini, A. J., Bearing noise, Machine Design, **35**, 11, 232—235 (1963); **35**, 12, 158—166 (1963).
13. Martin, R. L., Detection of ball bearing malfunctions, Instruments and Control Systems, **43**, 12, 79—82 (1970).
14. Lavoie, F. J., Signature analysis. Product early-warning system, Machine Design, **41**, 2, 149—160 (Jan. 1969).
15. Schlegel, R. G., King, R. J. and Mull, H. R., Gear noise, Machine Design, **34**, 4 (1964).
16. Mitchell, L. D., Lynch, G. A., Origins of noise, Machine Design, **41**, 10, 174—178 (1969).
17. Hugenbruch, E. R., Sounds of bevel and hypoid gears, Design News, **18**, 13, 14—17 (1963).
18. Stein, P. K., Measurement Engineering, **1**, Sec. 17, Stein Engineering Services Inc., Phoenix, Ariz., 1962.
19. Randall, R. B., Cepstrum analysis and gearbox fault diagnosis, Brüel & Kjaer Application Note No. 13—150.
20. Keller, A. C., Real time spectrum analysis of machinery dynamics, Spectral Dynamics Corp. Publication S&V-6 (1975).
21. Barwick, P. and Lemon, J., Experimental method of determining sources of chatter in machine-tool systems, Vibrations Conference, Boston, Mass., March 29—31, 1967.
22. Kwiatkowski, A. W. and Al-Samarai, H. M., Identification of milling machine receptances from random signals during cutting, Annals of C.I.R.P., **16**, 137—144 (1968).

References for chapter 9

23. Olesen, H. P., Measurement of the dynamic properties of materials and structures, Brüel & Kjaer Application Note No. *13—120*.
24. Buzdugan, Gh., Pană, T., Cosac, V., Mihăilescu, E., Radeș, M., Dumitrescu, E., Determinarea răspunsului dinamic al unei mașini de frezat, *A 2-a Conferință națională de mașini unelte*, București, 1976, p. 91—101.
25. Buzdugan, Gh., Segal, H., Blumenfeld, M., Petre, A., Cosac, V., Dumitrescu, E., Mincă, I., Dehnungs- und Vibrationsuntersuchungen der Rollbahnträger, *A 5-a Conferință de Sudură și Încercări de Materiale*, Timișoara, 1965, p. 289—297.
26. Buzdugan, Gh., Sarian, M., Voinea, R., Petre, A., Blumenfeld, M., Pană, T., Barbu, E., Dynamische Stabilität der Hängebrücken zur Überführung von Gasrohrleitungen über Flüsse, *Buletinul Institutului Politehnic* București, **23** (1961).
27. Pană, T., Constantinescu, I., Voicu, C., Mesures de vibrations engendrées dans les structures par vibrateurs, Nauchni dokladi ot seminar "Teoria na mehanizmite i machinite", Varna, Sept. 1976, p. 11—16.
28. Ewins, D. J., Study of resonance coincidence in bladed discs, *Journal of Mechanical Engineering Science*, **12**, *5*, 305—312 (1970).
29. Buzdugan, Gh., Blumenfeld, M., Mihăilescu, E., Radeș, M., Mincă, I., Stoicescu, L. and Ghelmez F., Underway vibration tests on commercial ships, Proc. 4th Conf. "Vibration in Mechanical Engineering", Timișoara, 26—27 Nov. 1982, p. 101—108.
30. Tobias, S. A., Dynamic acceptance tests for machine tools, *International Journal of Machine Tool Design and Research*, **2**, 267—280 (1962).
31. Sadek, M. M. and Tobias, S. A., Comparative dynamic acceptance tests for machine tools applied to horizontal milling machines, *Proc. Inst. Mech. Engrs.*, **185**, 319—337 (1970—1971).
32. Radeș, M., Methods for the analysis of structural frequency response measurement data, *Shock and Vibration Digest*, **11**, *2*, 15—24 (1979)
33. Buzdugan, Gh., Pană, T., Fetcu, L., Mihăilescu, E., Bălan, M., Studiul experimental al vibrațiilor în claviatura de conducte a unei stații de compresoare de gaz metan, *Buletinul Universității din Brașov*, **16**, 129—139 (1974).
34. Buzdugan, Gh., *Izolarea antivibratorie a mașinilor*, Editura Academiei, București, 1980.
35. Catlin J. B., Improved maintenance of machinery through "baseline" vibration measurements, *Journal of Engineering for Industry*, Trans. ASME, 913—918 (1973).
36. Mitchell, J. S., Designing a surveillance system, *Power*, **121**, *3*, 45—50 (May 1977).
37. * * * Application of transducers to rotating machinery monitoring and analysis, Dymac Application Note DA-013 (2—78).
38. Downham, E. and Woods, R., The rationale of monitoring vibration on rotating machinery in continuously operating process plant, *ASME Paper* No. 71-Vibr. *96*.
39. * * * Notes on the use of vibration measurements for machinery condition monitoring, Brüel & Kjaer Application Note *14—227*.

9 Examples of vibration measurement

40. Buzdugan, Gh., Blumenfeld, M., Cosac, V., Radeș, M., Mihăilescu, E., *Relații cantitative privitoare la prevenirea efectelor dăunătoare ale derocărilor prin explozii*, I.N.I.D., București, 1976.
41. White, R. G. and Mannering, M. E. J., Techniques for measuring the vibration transmission characteristics of ground, *Journal of Society of Environmental Engineers*, March 1975.
42. Medvedev, S. V., *Seismika gornych vzryvov*, Nedra, Moskva, 1964.
43. Radeș, M., Analysis of measured structural frequency response data, *Shock and Vibration Digest*, 14, 4, 21—32 (April 1982).
44. Radeș, M., A technique for measuring the dynamic properties of polyurethane foam layers, Second Int. Symp. RILEM "New Developments in Non-Destructive Testing of Non-Metallic Materials", Constanța, 4—7 Sept. 1974, vol. 1, p. 133—140.
45. Radeș, M., On the effects of non-linear stiffness in resonance testing, *Rev. Roum. Sci. Techn.— Méc. Appl.*, 28, 6, 603—614 (1983).
46. Radeș, M., Dynamic testing of non-linear materials using harmonic excitation with forces in quadrature, *Rev. Roum. Sci. Techn.— Méc. Appl.*, 22, 4, 593—606 (1977).
47. Mihăilescu E. and Radeș, M., Dynamic structural measurement using the method of forces in quadrature, Proc. Symp. "Experimental Techniques in Applied Mechanics", București, 1—3 Nov. 1972, p. 21—34.
48. Radeș, M., Parameter estimation of simple non-linear models, Proc. 4th Conf. "Vibration in Mechanical Engineering", Timișoara, 26—27 Nov. 1982, p. 145—152.
49. Grootenhuis, P., Measurement of the dynamic properties of damping materials, Proc. Int. Symp. "Damping of Vibrations of Plates by Means of Layers", Leuven, Sept. 1967.
50. Grootenhuis, P. and Awojobi, A. O., The in-situ measurement of the dynamic properties of soils, Proc. Symp. "Vibration in Civil Engineering", Butterworths, London, 1966, p. 181—187.
51. Payne, A. R. and Scott, J. R., *Engineering Design with Rubber*, Maclaren and Sons Ltd., London, 1960.
52. Buzdugan, Gh., Fetcu, L. and Radeș, M., Physical signification and measurement methods of the dynamic modulus of elasticity, *Proc. 9th Yugoslav Congress on Applied Mechanics*, Split, 3—8 June 1968, p. 349—365.
53. * * * Measurement of the Complex Modulus of Elasticity : A Brief Survey, Brüel & Kjaer Application Note 13—099.
54. Barkan, D. D., *Dynamics of Bases and Foundations*, McGraw Hill Book Co., New York, 1962.
55. Whitman, R. V., Evaluation of Soil Properties for Site Evaluation and Dynamic Analysis of Nuclear Plants, in *Seismic Design for Nuclear Power Plants*, M.I.T. Press, 1970.
56. Richart, F. E. Jr., Hall, J. R. Jr. and Woods, R. D., *Vibrations of Soils and Foundations*, Prentice-Hall Inc., Englewood Cliffs, New Jersey, 1970.

References for chapter 9

57. Buzdugan, Gh., Mincă, I. and Constantinescu, I., Realizarea unei instalații pentru determinarea coeficientului elastic dinamic al solului C_z folosit în proiectarea fundațiilor de mașini, *Construcții*, Nr. 5, 1978.
58. Buzdugan, Gh., Mincă, I., Craifăleanu, D., Constantin, N. and Ionescu B., Vibrator mecanic VMME-250 kN, *Studii și cercetări de mecanică aplicată*, **42**, 2, 1983.
59. Mincă, I., Constantele elastice de cuplare a efectelor corespunzătoare translației orizontale și balansului unui corp rigid rezemat pe mediu elastic, A IV-a Conferință *Vibrații în construcția de mașini*, Timișoara, 26—27 Nov. 1982.
60. Buzdugan, Gh., Mincă I., Determinări experimentale ale caracteristicilor elastice dinamice pentru diferite soluri, Conferința *Vibrații în construcția de mașini*, Timișoara, 1978.

Subject index

A

Accelerometer 4.2.1, 4.2.2.2
Aliasing 5.3.5.1
Amplifiers 5.3.3
Amplitude distribution analyzer 5.4.6
Amplitude modulation 5.3.1.1
Amplitude spectrum 2.2.3.1
Analog meter 5.5.2
Analog-to-digital converter 5.3.5
Analyzers 5.4.3, 5.4.4
Anti-aliasing filter 5.3.5.1
Autocorrelation function 2.2.2
Averaging 5.3.6

B

Bandpass filter 5.4.2
Bandwidth 5.4.2
Blasting 9.5
Bridge circuits 5.3.1.1

C

Calibration of pickups 8.1
Campbell diagram 9.2.4
Capacitive transducer 4.1.1.2
Cathode-ray oscilloscope 5.5.8
Charge amplifier 5.3.3.3
Charge sensitivity 4.2.2.2
Chladni figures 9.2.4
Coherence function 2.2.3.3

Comparison calibration 8.1.5
Computerized analysis systems 5.6
Confidence probability 2.3.1
Constant bandwidth filter 5.4.2.2
Constant % bandwidth filter 5.4.2.1
Correlation function 2.2.2
Correlator 5.4.1
Cross correlation function 2.2.2
Cross spectrum 2.2.3.2

D

Damping ratio 2.2.1
Detection circuits 5.3.1.2
Differential transducer 4.1.1.3
Digital analyzer 5.4.4.4
Digital meter 5.5.3
Digital recorder 5.5.10
Discrete Fourier Transform 2.2.3.4
Displacement measuring pickup 4.2.2.1
Dynamic pressure transducer 4.2.5
Dynamic range 5.1

E

Eddy-current transducer 4.1.1.3, 4.2.2.1
Effects of vibration 3
Effective value 2.2.1
Electrodynamic vibration exciter 6.3
Electrodynamic transducer 4.1.2.2
Electromagnetic vibration exciter 6.2

345

Subject index

Electromagnetic transducer 4.1.2.3
Energy spectral density 2.2.3.2
Ensemble averaging 5.3.6
Estimation errors 2.3
Experimental modal analysis 7.6.5
Exponential averaging 5.3.6

F

FFT analyzer 5.4.4.4
Filter response time 5.4.2.3
Force cell 4.2.4
Fourier Transform analyzer 5.4.4.4
Frequency analysis 2.2.3
Frequency analysis of shocks 7.4.2
Frequency analyzer 5.4.3, 5.4.4
Frequency discriminator circuit 5.3.2
Frequency response function 2.4, 7.6.1, 7.6.4.1
Frequency response measurement 7.6
Frequency sweep excitation 7.6.3.4

G

Geiger vibrograph 5.2.3
Graphic level recorder 5.5.6

H

Hanning weighting 5.4.4.3
Heterodyne analyzer 5.4.3.3
Hydraulic vibration exciter 6.4
Hysteretic damping factor 2.4.1

I

Impact test technique 7.6.3.2
Impedance head 4.2.4
Impedance-transforming amplifier 5.3.3.4

Impulse response function 7.6.4.2
Inductive transducer 4.1.1.3
Initial shock spectrum 2.4.3
Instrumentation system 1.3
Integrator 5.3.4

L

Leakage 5.4.4.4
Level recorder 5.5.6
Linear Fourier spectrum 2.2.3.2
LVD transformer transducer 4.1.1.3

M

Machine condition monitoring 9.4.2
Magnetic oscillograph 5.5.7
Magnetic tape recorder 5.5.9
Magnification factor 2.4.1
Mean square value 2.2.2
Mean square spectral density 2.2.3.3
Measured quantities 1.4
Measuring bridge 5.3.1
Mini-computer based systems 5.4.4.4
Modal analysis 7.6.5
Mounting 4.2.7
Multi-point excitation 7.6.2.2

N

Natural frequencies 2.5
Noise analyzer 5.4.3.1
Non-linear systems 9.6
Non-real time spectrum analyzer 5.4.3
Nyquist frequency 2.2.3.4

O

Optical interferometry calibration 8.1.4
Order analysis 9.2.4

Subject Index

P

Parallel filter analyzer 5.4.4.1
Parseval's theorem 2.2.3.2
Passive transducers 4.1.1
Periodic excitation 2.4.2
Permanent monitoring 9.4.2
Picket fence effect 5.4.4.4
Piezoelectric transducer 4.1.2.1
Potentiometer transducer 4.1.1.1
Power spectral density 2.2.3.3
Probability density function 2.2.4
Pulse 2.4.3

R

R.M.S. detector 5.3.7
R.M.S. value 2.2.1
Random excitation 7.6.4
Random vibration test 7.5.2
Random vibration 1.1
Real-time analyzer 5.4.4
Reciprocating vibration exciter 6.1.1
Reciprocity calibration 8.1.3
Recorders 5.5
Residual shock spectrum 2.4.3
Resolved component indicator 5.4.7
Rotating unbalance vibrator 6.1.2
R.P.M. tracking 9.2.4
Running averaging 5.3.6.1

S

Sampling frequency 5.3.5.1
Seismic instrument 4.2.1
Selective filtering 7.4.1
Self-generating transducer 4.1.2
Shock spectrum analyzer 5.4.5
Shock test 7.5.4
Sidebands 5.3.1.1
Signal waveform 2.2.1, 7.2
Signature analysis 9.2.4
Simulation calibration 8.2
Single-point excitation 7.6.2.1
Sinusoidal test 7.5.1

Slave filter 9.2.4
Spring constants 2.6
Standard error 2.3.2
Step excitation 2.4.3
Step relaxation technique 7.6.3.3
Stoppani vibrograph 5.2.2
Strain gauges 4.1.1.1
Strip chart recorder 5.5.4
Stroboscope 5.5.1
Synchronous filter 5.4.3.4

T

Tape recorder 5.5.9
Tastograph 5.2.1
Time averaging 5.3.6
Time compression analyzer 5.4.4.2
Time window 5.4.4.3
Torque transducer 4.2.4
Tracking filter 9.2.4
Transducers 4.1
Transient excitation 2.4.3
Transient test technique 7.6.3
Transmissibility 2.4.1
Transverse sensitivity 4.2.2.2
Tunable bandpass filter 5.4.3.2

V

Variable resistance transducer 4.1.1.1
Velocity pickup 4.2.3
Vibration intensity 3.1
Vibration standards 3.3
Vibration testing 7.5
Voltage amplifier 5.3.3.1
Voltage sensitivity 4.2.2.2

W

Wave analyzer 5.4.3.3
Weighting 5.4.4.3
Wiener-Khinchin relations 2.2.3.2
Wrap-around error 5.4.4.4

X

X–Y recorder 5.5.5

PRINTED IN ROMANIA